POWER
SEMICONDUCTOR
DRIVES

POWER SEMICONDUCTOR DRIVES

P. V. Rao

Former Professor of Electrical Engineering
JNTU College of Engineering
Anantapur

BSP **BS Publications**
A unit of **BSP Books Pvt., Ltd.**

4-4-309, Giriraj Lane, Sultan Bazar,
Hyderabad - 500 095
Phone : 040 - 23445605, 23445688

Copyright © 2015, by Publisher

Published by :

BSP **BS Publications**
A unit of **BSP Books Pvt., Ltd.**

4-4-309, Giriraj Lane, Sultan Bazar,
Hyderabad - 500 095
Phone : 040 - 23445605, 23445688
e-mail : info@bspbooks.net

ISBN : 978-93-52300-27-3 (HB)

PREFACE

Power Semiconductor Devices play a vital role in electrical power systems. They are widely used in transimission, distribution and control of elecrtical power. They are also used in control of electrical drives used in industries and traction.

This book deals with power semiconductor circuits such as converters, inverters, choppers and cycloconverters used in control of electric drives that use a.c and d.c motors. It covers the topics given in the subject "Power Semicondutor Drives" offered to B.Tech (EEE) degree course students.

It is evolved out of class notes prepared while I was teaching this subject for several years in GNITS, Hyderabad. Many books, written by eminent authors on the subject and listed in bibliography, have been made use of, while preparing the notes and I am indebted to those authors. The subject is explained lucidly so that even an average student can easily understand and develop on his own. Number of problems are worked out giving step by step explanation. Number of objective type questions are given at the end of each chapter answers for which are given separately at the end.

The one semester course is divided into eight chapters. First chapter deals with control of d.c motors with single phase converters. Second chapter deals with control of d.c. motor with three phase converters. Third chapter deals with four quadrant drive with dual converter. Different types of braking d.c motors are also described in this chapter. D.C. chopper drives are given in chapter four. Chapter five deals with speed control of induction motor from the stator with variable a.c. voltage. Chapter six deals with variable voltage and variable frequency control of induction motor from the stator side. Chapter seven deals with speed control of induction motor from rotor side. Chapter eight deals with speed control of synchronous motors.

-Author

ACKNOWLEDGEMENT

I thank the management of G. Narayanamma Institute of Technology and Science for women, Hyderabad for the facilities provided for writing this book. I also thank the staff of EEE department for the encouragement and help rendered by them. My thanks are also due to M/S B.S. Publishers for bringing out the book.

-Author

CONTENTS

CHAPTER - 1

Control of DC Motors by Single Phase Converters

Chapter - 2

Control of DC Motors by Three Phase Converters

Chapter - 3

Four Quadrant Operation of DC Drives

CHAPTER - 4

Control of DC by Motor Choppers

CHAPTER - 5

Control of Induction Motor

CHAPTER - 6

Control of Induction Motor Through Stator Frequency

CHAPTER - 7

Speed Control of Three-phase Induction Motors from Rotor Side

CHAPTER - 8

Control of Synchronuous Motors

1

Control of DC Motors by Single Phase Converters

Control of DC motors by single phase converters : Introduction to thyristor controlled drives, single phase semi and fully controlled converters connected to d c separately excited and dc series motors – continuous current operation – output voltage and current waveforms – speed torque expressions – speed torque characteristics – problems on converter fed dc motors.

1.1 Introduction

Electrical motors are used for driving different types of loads. Power semiconductor drives deal with the use of power semiconductor circuits in controlling the motors used for driving mechanical loads. In the earlier semesters students studied the subjects (i) Power Electronics (ii) Electrical machines and (iii) Control Systems. In this course of Power Semiconductor Drives the students will study how the combination of these three subjects is used for controlling the motors used in driving different types of loads.

1.1.1 Different Types of Motors
    (i)   Separately excited DC motor.
    (ii)   DC Series motors.
    (iii) Induction motors.
    (iv) Synchronous motors.

Types (i) & (ii) require DC supply while (iii) & (iv) require AC Supply.

1.1.2 Types of Loads
    (i)   Constant torque loads like lifts and cranes.
    (ii)   Variable torque loads like fans and pumps.

1.1.3 Power Converters
    (i)   **AC to DC :**
        Single Phase Rectifier and Three Phase Rectifier. They may be controlled or uncontrolled.
    (ii)   **DC to AC :**
        (a)   Voltage Source Inverter.
        (b)   Current Source Inverter.

(iii) **AC to AC :**
    (a) Cycloconverters
    (b) Rectifier - inverter combination
    (c) AC Voltage controller
(iv) **DC to DC :**
    DC Choppers

1.2   Speed Control of Separately Excited DC Motor

The schematic diagram and equivalent circuit of separately excited DC motor are given below in Fig. 1.1.

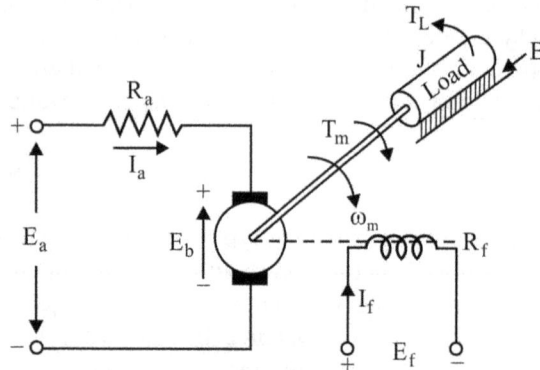

Fig. 1.1 (a) Schematic diagram.

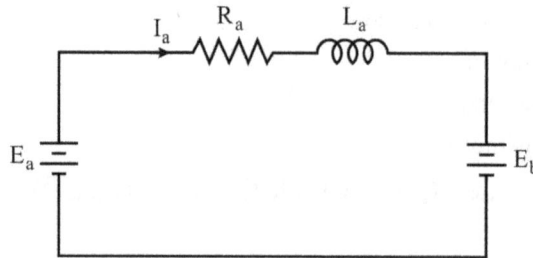

Fig. 1.1 (b) Equivalent circuit.

E_a = Applied voltage in volts.

E_b = Back emf in volts.

I_a = Armature current in Amps.

R_a = Armature resistance in Ohms.

L_a = Armature inductance in Henrys.

T_m = Electromagnetic torque developed by motor in N-m.

ω_m = Speed of motor in rad /sec.

E_f = Applied voltage to field circuit in volts.

I_f = Field current in Amps.

R_f = Resistance of the field winding in Ohms.

L_f = Inductance of the field winding in Henrys.

J = Moment of inertia of rotor & load in Kg-m.

B = Frictional coefficient in N/m/sec.

T_L = Load torque in N-m.

(**Note :** Lower case letters are used for time varying quantities upper case letters are used for average values or RMS values)

Equations describing the dynamic behavior of the motor are given below

$$R_a\, i_a + L_a\, \frac{di_a}{dt} + e_b = e_a \qquad\qquad .....(1.1)$$

$$R_f\, i_f + L_f\, \frac{di_f}{dt} = e_f \qquad\qquad .....(1.2)$$

$$T_m(t) = K_1\, i_a\, \phi_f = K\, i_a\, i_f \qquad\qquad .....(1.3)$$

$$e_b = K_1\, \omega \phi_f = K\omega\, i_f \qquad\qquad .....(1.4)$$

$$J\frac{d\omega}{dt} + B\omega + T_L = T_m \qquad\qquad .....(1.5)$$

Under steady state conditions the above equations will become

$$R_a\, I_a + E_b = E_a \qquad\qquad .....(1.6)$$

$$R_f\, I_f = E_f \qquad\qquad .....(1.7)$$

$$T_m = K\, I_a\, I_f \qquad\qquad .....(1.8)$$

$$E_b = K\omega\, I_f \qquad\qquad .....(1.9)$$

$$B\omega + T_L = T_m \qquad\qquad .....(1.10)$$

The following facts should be noted :

- The back emf 'e_b' is proportional to the product of flux (ϕ) and the speed (ω).
- It is independent of armature current.
- The electromagnetic torque developed by the motor is proportional to the product of armature current 'I_a' and flux 'ϕ'.
- It is independent of speed of motor.

The equations given below are CAUSE and EFFECT equations.

(**Note :** CAUSE is on Right Hand Side and EFFECT on Left Hand Side)

The armature current

$$I_a = \frac{E_a - E_b}{R_a}$$

$$E_b = K\omega\phi$$

$$\phi = K_1 I_f$$

$$T_m = K I_a \phi$$

$$\omega = \frac{T_m - T_L}{B}$$

From the above equations it can be seen that the speed 'ω' of the motor depends on load torque 'T_L', the field current 'I_f', and the applied voltage 'E_a'.

Thus for a given load torque the speed can be controlled by either controlling the applied voltage to armature or the field current 'I_f' or both. For controlling the speed below its rated speed armature voltage is changed. For controlling the speed above its rated value the field current 'I_f' is controlled. The motor can run above its rated speed only at torque less than the rated torque. The drive above rated speed is known as constant Horse Power (HP) drive. The armature current is held at its rated value and the field current is reduced.

The speed versus torque and speed versus power is shown in Fig. 1.2.

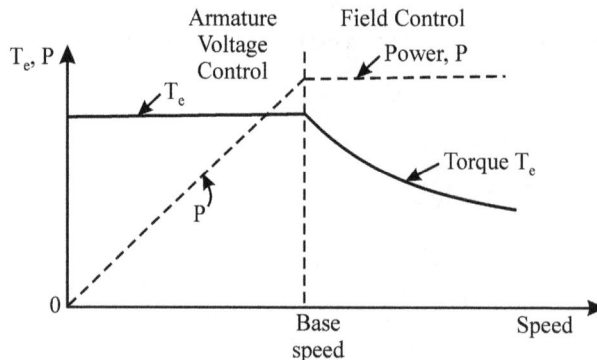

Fig. 1.2 Speed versus torque and speed versus power curves.

***Example* 1.1**

A separately excited DC motor is running at 800 rpm driving a load whose torque is constant. Motor armature current is 500 A. The armature resistance drop and rotational losses are negligible. Magnetic circuit can be assumed to be linear. Calculate the motor speed and armature current if terminal voltage is reduced to 50% and field current is reduced to 80%.

Solution :

$I_a R_a$ drop is negligible

Hence back emf $E_b = V_a$ the applied voltage.

$$\frac{E_{b1}}{E_{b2}} = \frac{N_1 \phi_1}{N_2 \phi_2}$$

$$\frac{N_2}{N_1} = \frac{E_{b2}}{E_{b1}} \cdot \frac{\phi_1}{\phi_2} = \frac{0.5}{0.8}$$

$$N_2 = \frac{800 \times 0.5}{0.8} = 500 \text{ rpm}$$

Torque is proportional to the product of armature current and flux. Hence

$$\frac{T_2}{T_1} = \frac{I_2 \phi_2}{I_1 \phi_1};$$

It is given that $T_2 = T_1$ and $\dfrac{\phi_2}{\phi_1} = 0.8$

Hence

$$\frac{I_2}{I_1} = \frac{T_2 \phi_1}{T_1 \phi_2} = \frac{1}{0.8}$$

$$I_2 = \frac{500}{0.8} = 625 \text{ A}$$

Example 1.2

What will be the answers to the above problem if torque is proportional to square of speed?

Solution :

The speed N_2 will remain same as 500 rpm because back emf E_b is same as applied voltage V_a. Only armature current will change.

$$N_1 = 800 \text{ rpm}; \quad N_2 = 500 \text{ rpm}$$

Since torque is proportional to square of the speed,

$$\frac{T_2}{T_1} = \left(\frac{N_2}{N_1}\right)^2 = \left(\frac{500}{800}\right)^2 = \frac{25}{64}$$

$$\frac{I_2}{I_1} = \frac{T_2}{T_1} \cdot \frac{\phi_1}{\phi_2} = \frac{25}{64} \times \frac{1}{0.8}$$

$$\therefore \qquad I_2 = \frac{500 \times 25}{64 \times 0.8} = 244 \text{ A}$$

Example 1.3

A 220 V, 800 rpm, 80 A separately excited motor has an armature resistance of 0.12 Ω. Motor is driving under rated conditions, a load, whose torque is same at all speeds. Calculate motor speed if the source voltage drops to 200 V.

Solution :

Load torque is same. Therefore the armature current is same at all speeds and equal to 80 A.

Let E_{b1} is the back emf at 800 rpm

E_{b2} is the back emf at new speed N_2

$$\frac{E_{b2}}{E_{b1}} = \frac{N_2}{N_1}\frac{\phi_2}{\phi_1} = \frac{N_2}{N_1} \text{ because flux is constant.}$$

$$E_b = V - I_a R_a$$

$$E_{b2} = 200 - 80 \times 0.12 = 200 - 9.6 = 190.4 \text{ V}$$

$$E_{b1} = 220 - 80 \times 0.12 = 220 - 9.6 = 210.4 \text{ V}$$

∴ $$N_2 = \frac{N_1 E_{b2}}{E_{b1}} = \frac{800 \times 190.4}{210.4} = 723.95 \text{ rpm}$$

Example 1.4

A 220 V, 200 A, 750 rpm separately excited DC motor has an armature resistance of 0.05 Ω. It is driving a load whose torque has an expression $T = (500 - 0.25N)$ N-m where 'N' is the speed in rpm. Speeds below rated are obtained by armature voltage control with full field and speeds above rated are obtained by field control with rated armature voltage.

(i)   Calculate motor terminal voltage and armature current when the speed is 400 rpm.

(ii)   Calculate value of flux as a percent of rated flux when the speed is 1500 rpm.

Solution :

(i) $V_a = 220$ V; $I_a = 200$ A; N = 750 rpm; $R_a = 0.05$ Ω; $T_L = 500 - 0.25$ N

We have to calculate V_a and I_a for N = 400 rpm

At rated speed of 750 rpm $E_b = V_a - I_a R_a$

$$= 220 - 200 \times 0.05 = 210 \text{ V}$$

∴   Back emf at 400 rpm $= \frac{210 \times 400}{750} = 112 \text{ V}$

Load torque at 400 rpm $= 500 - 0.25 \times 400 = 400$ N-m.

Let T_1 be the torque developed at speed of 750 rpm and armature current of 200 A. We know that $T\omega = E_b I_a$

$$\therefore \quad T_1 = \frac{E_b I_a}{\omega} = \frac{210 \times 200 \times 60}{2\pi \times 750} = 534.76 \text{ N-m}$$

To calculate the armature current for torque of 400 N-m we use the equation :

$$\frac{T_2}{T_1} = \frac{I_2}{I_1}; \quad I_2 = \frac{T_2}{T_1} \times I_1 = \frac{400}{534.76} \times 200 = 149.59 \text{ A.}$$

The armature voltage $V_a = E_b + I_a R_a = 112 + 149.6 \times 0.05 = $ **119. 48 V**

When the speed is 1500 rpm, the torque developed $= 500 - 0.25 \times 1500$

$$= 125 \text{ N-m}$$

$$\frac{T_2}{T_1} = \frac{K\phi_2 I_2}{K\phi_1 I_1}; \quad \frac{\phi_2}{\phi_1} = \frac{T_2 I_1}{T_1 I_2} = \frac{125}{534.76} \times \frac{200}{149.6}$$

Speed of 1500 is greater than rated speed of 750 rpm.

$$\therefore \qquad V_a = 220V$$

$$E_{b2} = K\phi_2 \omega_2 = K\phi_2 \frac{1500 \times 2\pi}{60}$$

$$E_{b1} = K\phi_1 \omega_1 = K\phi_1 \frac{750 \times 2\pi}{60} = 210$$

$$\therefore \qquad K\phi_1 = \frac{210 \times 60}{750 \times 2\pi} = 2.673$$

$$\text{Torque constant } K = \frac{2.673}{\phi_1}$$

$$T_2 = K\phi_2 I_{a2}; \quad I_{a2} = \frac{T_2}{K\phi_2} = \frac{125 \times \phi_1}{2.673 \times \phi_2} = 46.76 \times \frac{\phi_1}{\phi_2}$$

$$E_{b2} = V_a - I_{a2} R_a = 220 - 46.76 \times \frac{\phi_1}{\phi_2} \times 0.05 = 220 - 2.338 \frac{\phi_1}{\phi_2}$$

$$E_{b2} = K\phi_2 \frac{1500 \times 2\pi}{60}$$

$$= \frac{2.673}{\phi_1} \phi_2 \frac{1500 \times 2\pi}{60}$$

$$= 419.9 \frac{\phi_2}{\phi_1}$$

$$419.9 \frac{\phi_2}{\phi_1} = 220 - 2.338. \frac{\phi_1}{\phi_2}$$

$$419.9 \left(\frac{\phi_2}{\phi_1}\right)^2 - 220 \frac{\phi_2}{\phi_1} + 2.338 = 0$$

$$\left(\frac{\phi_2}{\phi_1}\right) = \frac{220 \pm \sqrt{220^2 - 4 \times 2.338 \times 419.9}}{2 \times 419.9}$$

$$= \frac{220 \pm 210}{2 \times 419.9} = 0.513$$

% of rated flux = **51.3**

1.3   Separately Excited dc Motor Driven by Controlled Rectifiers

The speed of separately excited dc motor is controlled by the voltage applied to the armature and the field winding. The dc voltage can be changed by changing the firing angle of the controlled rectifier connected to the ac mains supply. Now the operation of dc motor connected to different types of controlled rectifiers will be discussed.

1.3.1 Operation of Separately Excited dc Motor Connected to Single Phase Half-controlled Rectifier

The schematic diagram is shown in Fig. 1.3.

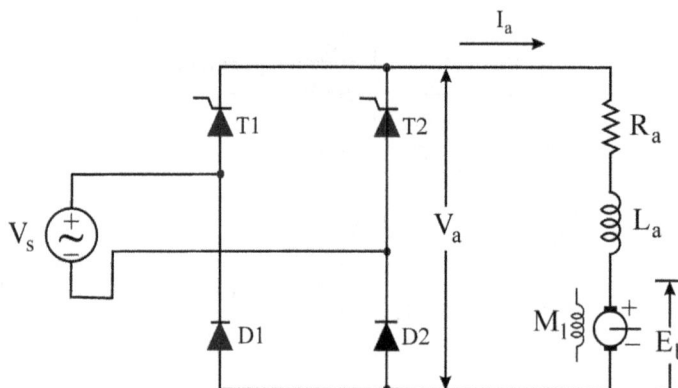

Fig. 1.3 Separately excited dc Motor connected to 1ϕ half controlled rectifier.

The output voltage of the converter is always positive and the current is also positive. The electrical power is drawn from ac mains and delivered to the motor. The motor can operate only in the first quadrant as motor. The equivalent circuit of the system is shown in the Fig. 1.4.

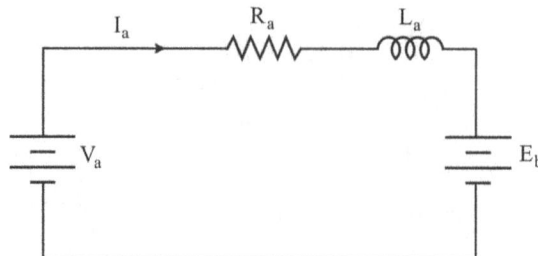

Fig. 1.4 Equivalent circuit.

Thyristor T1 is triggered at firing angle α and thyristor T2 is triggered at angle $\pi + \alpha$.

The current i_a drawn by the armature may be continuous or discontinuous. The motor is said to be operating in discontinuous current mode or continuous current mode based on the nature of current drawn by the armature.

Discontinuous current mode of operation : The waveforms of the voltages and currents for discontinuous mode of operation are shown in Fig. 1.5(a). The assumptions made in drawing the waveforms are :

1. Field current is constant, consequently the induced emf E is constant over one cycle of input voltage. This is true because the speed change is negligible over one cycle of input voltage (i.e., 20 m sec).

2. The supply voltage $V_s = \sqrt{2}\ V \sin \omega t$

3. The firing angle α is such that $\sqrt{2}\ V \sin\alpha > E_b$

4. β is the extinction angle.

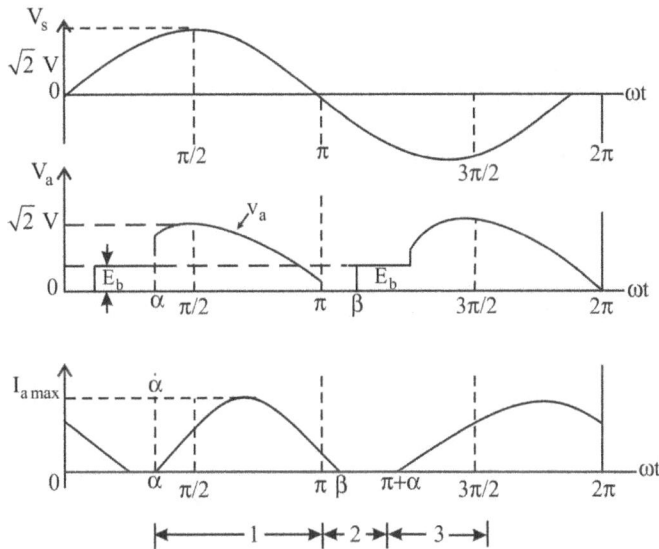

Fig. 1.5 (a) H.C.R waveforms of supply and load voltages and load current (Discontinuos current mode).

Under discontinuous conduction mode there are three distinct intervals over each half cycle. $\alpha < \omega t < (\pi + \alpha)$

1. $\alpha < \omega t < \pi$ is called duty interval.

$$R_a i_a + L_a \frac{di_a}{dt} + e_b = \sqrt{2}\ V \sin\omega t \qquad\qquad .....(1.11)$$

$I_a = 0$ at $\omega t = \alpha$

2. $\pi < \omega t < \beta$ is called free wheeling interval.

$$R_a i_a + L_a \frac{di_a}{dt} + e_b = 0 \qquad\qquad .....(1.12)$$

$I_a = 0$ at $\omega t = \beta$

3. $\beta < \omega t < \pi + \alpha$ is called zero current interval.

$$I_a = 0; \ v_a = E_b \qquad\qquad .....(1.13)$$

The expression for current i_a during duty interval is obtained by solving eqn. (1.11)

$$i_a(\omega t) = \frac{\sqrt{2}V}{Z}\sin(\omega t\text{-}\phi) - \frac{E}{R_a} + K_1\, e^{-\frac{t}{\tau}} \qquad\qquad .....(1.14)$$

where $\qquad Z = \sqrt{R_a^2 + (wL_a)^2}\ ; \ \phi = \tan\text{-}1\dfrac{\omega L_a}{R_a}\ \ ; \ \tau = \dfrac{L_a}{R_a}$

K_1 is obtained by substituting the initial value $I_a(\alpha) = 0$.

$$K_1 = \left[\frac{E}{R_a} - \frac{\sqrt{2}V}{Z}\sin(\alpha - \phi)\right]e^{\frac{\alpha}{\omega\tau}}$$

Substituting this value of K, in eqn. (1.14).

$$I_a(\omega t) = \frac{\sqrt{2}V}{Z}\left[\sin(\omega t - \phi) - \sin(\alpha - \phi)e^{-(\omega t-\alpha)\cot\phi}\right] -$$

$$\frac{E}{R_a}\left[1 - e^{-(\omega t-\alpha)\cot\phi}\right] \qquad\qquad .....(1.15)$$

During free wheeling interval the expression for current is obtained from eqn. (1.12) as

$$i_a(\omega t) = -\frac{E}{R_a}\left[1 - e^{-(\omega t-\alpha)\cot\phi}\right] + K_2\, e^{-\frac{\omega t}{\omega\tau}} \qquad\qquad .....(1.16)$$

i_a at $\omega t = \pi$ is obtained from eqns. (1.15) and (1.16) and the RHS expressions are equated to obtain the value of constant K_2.

$$K_2 = \frac{\sqrt{2}V}{Z}\left[\sin(\pi - \phi) - \sin(\alpha - \phi)e^{-(\pi-\alpha)\cot\phi}\right]e^{\frac{\pi}{\omega\tau}}$$

Substituting the value of K_2 in eqn. (1.16), the expression for current during free wheeling interval $\pi < \omega t < \beta$ is obtained as

$$i_a(\omega t) = \frac{\sqrt{2}\ V}{Z}\left[e^{-(\omega t-\pi)\cot\phi}.\sin\phi - e^{-(\omega t-\alpha)\cot\phi}.\sin(\alpha - \phi)\right]$$

$$\frac{E}{R_a}\left[1 - e^{-(\omega t - \alpha)\cot\varphi}\right] \qquad\qquad .....(1.17)$$

At $\omega t = \beta$ the current $i_a = 0$

Substituting in eqn. (1.17), we get

$$0 = \frac{\sqrt{2}V}{Z}\left[e^{-(\beta-\pi)\cot\varphi}\sin\varphi - e^{-(\beta-\alpha)\cot\varphi}\sin(\alpha - \varphi)\right] - \frac{E}{R_a}\left[1 - e^{-(\beta-\alpha)\cot\varphi}\right]$$

Bringing the terms involving β to the left side and rearranging, we get

$$e^{\beta\cot\varphi} = \frac{\sqrt{2}\,VR_a}{ZE}\left[e^{\pi\cot\varphi}\sin\varphi - e^{\alpha\cot\varphi}\sin(\alpha - \varphi)\right] + e^{\alpha\cot\varphi} \qquad .....(1.18)$$

From eqn. (1.18) the value of β can be obtained by trial and error. For a given value of α, the extinction angle β depends on the induced emf E. When β is equal to $(\pi + \alpha)$, the current just becomes zero and starts increasing i.e., for a particular value of induced emf the current mode changes from discontinuous to continuous. This value of E is called the critical value Ec. For separately excited motor the induced emf is proportional to speed. The speed N_c at which $E = E_c$ is called the critical speed. Thus for a given firing angle if the speed of motor is greater than critical speed, the current is discontinuous. If the speed is less than critical speed, the current is continuous.

Critical speed is calculated from eqn. (1.18) by substituting $(\pi + \alpha)$ for β and $(K\,N_c)$ for E_C

$$e^{(\pi+\alpha)\cot\phi} = \frac{\sqrt{2}\,VR_a}{Z.KN_c}\left[e^{\pi\cot\phi}\sin\phi - e^{\alpha\cot\phi}\sin(\alpha - \phi)\right] + e^{\alpha\cot\iota}$$

$$e^{(\pi+\alpha)\cot\phi} - e^{\alpha\cot\phi} = \frac{\sqrt{2}\,VR_a}{Z.KN_c}\left[e^{\pi\cot\phi}\sin\phi - e^{\alpha\cot\phi}\sin(\alpha - \phi)\right]$$

$$\left[e^{\pi\cot\phi} - 1\right]e^{\alpha\cot\phi} = \frac{\sqrt{2}\,VR_a}{ZKN_c}\left[e^{\pi\cot\phi}\sin\varphi - e^{\alpha\cot\varphi}\sin(\alpha - \varphi)\right]$$

$$\therefore \qquad N_C\left[e^{\pi\cot\phi} - 1\right] = \frac{\sqrt{2}\,VR_a}{KZ}\left[e^{(\pi-\alpha)\cot\varphi}\sin\varphi - \sin(\alpha - \varphi)\right]$$

$$\therefore \qquad N_C = \frac{\sqrt{2}\,VR_a}{KZ}\left[\frac{e^{(\pi-\alpha)\cot\varphi}\sin\varphi - \sin(\alpha - \varphi)}{(e^{\pi\cot\varphi} - 1)}\right] \qquad .....(1.19)$$

When the current is discontinuous, the average voltage V_a applied to the armature is obtained after knowing the value of β.

$$V_a = \frac{1}{\pi}\left[\int_\alpha^\pi \sqrt{2}\,V\sin\omega t\,d(\omega t) + \int_\beta^{\pi+\alpha} E_b\,d(\omega t)\right]$$

$$= \frac{1}{\pi}\left[\sqrt{2}V(1 + \cos\alpha) + E_b(\pi + \alpha - \beta)\right] \qquad .....(1.20)$$

Armature current $\quad I_a = \dfrac{V_a - E_b}{R_a}$ $\qquad\qquad$ .....(1.21)

$E_b = K\omega$ where ω is the speed in rad/sec $= K\dfrac{2\pi N}{60}$ where N is the speed in rpm.

$\qquad$ Torque $T = K\, I_a$

$\therefore \qquad\qquad \dfrac{T}{K} = \dfrac{V_a}{R_a} - \dfrac{2\,\pi\,N}{60}\,\dfrac{K}{R_a}$ $\qquad\qquad$ .....(1.22)

$\qquad\qquad N = \dfrac{60\,R_a}{2\,\pi\,K}\left[\dfrac{V_a}{R_a} - \dfrac{T}{K}\right]$ rpm $\qquad\qquad$ .....(1.23)

V_a is to be calculated using eqn. (1.20).

This equation shows that speed for a given torque depends on firing angle α.

The speed regulation is poor for discontinuous conduction mode. Moreover torque pulsations occur. Hence in practical cases the current is made continuous by connecting some inductance in series with the armature.

1.3.2 Continuous Conduction Mode

The waveforms of voltages and current are shown in Fig. 1.5(b) for **continuous conduction mode** of operation.

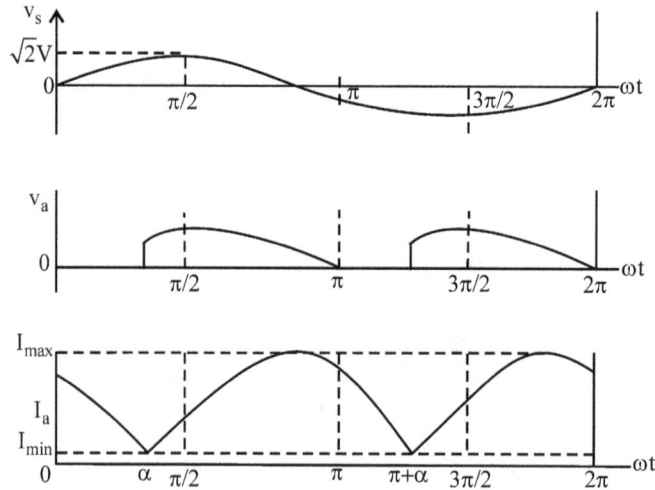

Fig. 1.5 (b) HCR Waveforms of supply and load voltages and load current (continuous current).

The average voltage applied to armature $V_a = \sqrt{2}V(1 + \cos\alpha)$.

Therefore the speed N is given by

$$N = \dfrac{60}{2\,\pi}\left[\dfrac{\sqrt{2}\,V(1 + \cos\,\alpha)}{\pi\,K} - \dfrac{R_a T}{K^2}\right] \qquad\qquad .....(1.24)$$

α can change from 0 to 180^0. For any load torque T, the speed can be changed by changing the firing angle α. If α is reduced the speed will increase.

The torque speed characteristics are shown in Fig. 1.6.

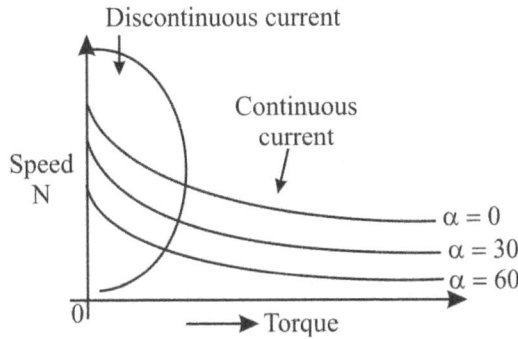

Fig. 1.6 Torque -Speed characteristic.

1.4 Separately Excited dc Motor Connected to Single Phase Fully Controlled Converter

The circuit diagram is shown in Fig.1.7.

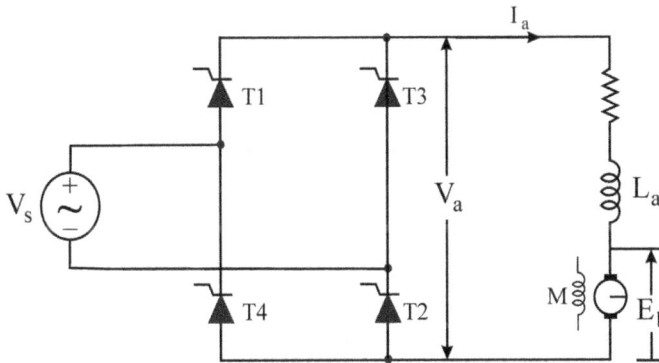

Fig. 1.7 Separately excited dc Motor connected to single phase fully controlled converter.

Thyristors T1 and T2 are triggered at firing angle α and thyristors T3 and T4 are triggered at angle $\pi + \alpha$. The current may be discontinuous or continuous. The waveforms of voltages and currents are shown in Fig. 1.8(a) for discontinuous current mode of operation. Since T1 and T2 are triggered at wt = α the current starts increasing and at wt = β the current becomes zero. T1 and T2 go to off state. β is greater than π but less than $\pi + \alpha$. At wt = $\pi + \alpha$, T3 and T4 are triggered and again the current starts increasing and becomes zero at $\pi + \beta$ and at this instant T3 and T4 go to off state. The operation repeats for every cycle of input voltage. Since the speed of motor will remain almost constant over one cycle of input voltage and field current is constant the induced emf E_b is taken as constant while drawing the waveforms. The waveform repeats in each half cycle. In each half cycle there are two intervals of operation.

There are two modes of operation
(a) discontinuous current mode,
(b) continuous current mode.

(a) *Discontinuous Current Mode of Operation :*

(i) *Derivation of expression for average voltage V_a applied to armature :* The waveforms of supply voltage V_s, armature voltage V_a, and armature current I_a are shown in Fig. 1.8(a). The expression for average voltage applied to armature is derived as follows.

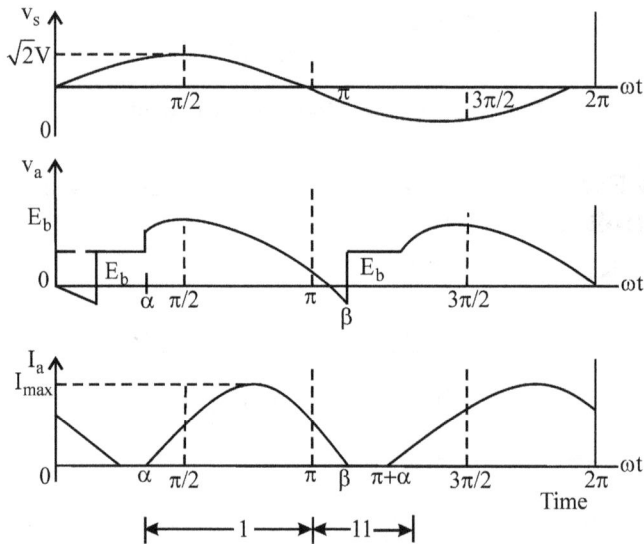

Fig. 1.8 (a) Waveforms of supply and load voltages and load current discontinuous.

(a)   During $\alpha < \omega t < \beta$ the armature is connected to source. Therefore $V_a = V_s$.

$\therefore$ $$V_a = R_a i_a + L_a \frac{di_a}{dt} + E_b = \sqrt{2} \, V\sin \omega t \qquad .....(1.25)$$

$I_a = 0$ at $\omega t = \alpha$ and at $\omega t = \beta$

(b) $\beta < \omega t < \pi + \alpha$ zero current interval

$$V_a = E_b \text{ and } I_a = 0$$

The expression for current during first interval has two components, one due to AC source $\sqrt{2} \, V\sin \omega t$ and other due to $-E_b$. Each has got transient term. The solution is

$$i_a(\omega t) = \frac{\sqrt{2}V}{Z}\sin(\omega t - \phi) - \frac{E_b}{R_a} + k_1 e^{-t/\tau} \qquad \dots (1.26)$$

where $\qquad Z = \sqrt{R_a^2 + (\omega L_a)^2}$; $\phi = \tan^{-1}\dfrac{\omega L_a}{R_a}$; $\tau = \dfrac{L_a}{R_a}$

k_1 is obtained by substituting the initial value $I_a(\alpha) = 0$.

$$k_1 = \left[\frac{E_b}{R_a} - \frac{\sqrt{2}\,V}{Z}\sin(\alpha - \varphi)\right]e^{\frac{\alpha}{\omega\tau}}$$

Substituting this value of k_1, in eqn. (1.26)

$$i_a(\omega t) = \frac{\sqrt{2}V}{Z}\left[\sin(\omega t - \varphi) - \sin(\alpha - \varphi)e^{-(\omega t - \alpha)\cot\varphi}\right] -$$

$$\frac{E_b}{R_a}\left[1 - e^{-(\omega t - \alpha)\cot\varphi}\right] \quad \dots (1.27)$$

$I_a(\omega t) = 0$ at $\omega t = \beta$. Hence

$$0 = \frac{\sqrt{2}V}{Z}\left[\mathrm{Sin}(\beta - \varphi) - \mathrm{Sin}(\alpha - \varphi)e^{-(\beta - \alpha)\cot\varphi}\right] -$$

$$\frac{E_b}{R_a}\left[1 - e^{-(\beta - \alpha)\cot\varphi}\right] \qquad \dots (1.28)$$

β can be evaluated by iterative method. After obtaining the value of β, average value of voltage applied to the armature V_a can be found.

For $\alpha < \omega t < \beta$

$$V_a = \sqrt{2}\,V\sin wt$$

For $\beta < wt < \pi + \alpha$

$$V_a = E_b$$

Average value of armature voltage

$$V_a = \frac{1}{\pi}\left[\int_\alpha^\beta \sqrt{2}V\sin\omega t + \int_\beta^{\pi+\alpha} E_b\, d(\omega t)\right]$$

$$V_a = \frac{1}{\pi}\left[\sqrt{2}V(\cos\alpha - \cos\beta) + E_b(\pi + \alpha - \beta)\right] \qquad \dots (1.29)$$

(ii) ***Derivation of Relation Between Torque and Speed :*** Since the average voltage drop across inductance is zero .

$$V_a = E_b + I_a R_a$$

where V_a and I_a are average values.

Armature current consists of dc component I_a and harmonics. When flux is constant, only DC component produces steady torque. Harmonics produce alternating torque components, the average value of which is zero.

$\therefore$ Motor torque $T = K \, I_a$

$\qquad E_b = K\omega$

$\qquad E_b = V_a - I_a \, R_a$

$$= \frac{\sqrt{2}V}{\pi}(\cos\alpha - \cos\beta) + \frac{\pi + \alpha - \beta}{\pi} \, E_b - I_a \, R_a$$

$$E_b = \frac{\sqrt{2}V}{\pi}(\cos\alpha - \cos\beta) + E_b + \frac{\alpha - \beta}{\pi} \, E_b - I_a \, R_a$$

$$E_b = \left[\frac{\sqrt{2}V}{\pi}(\cos\alpha - \cos\beta) - I_a R_a \right] \frac{\pi}{\beta - \alpha} \qquad \qquad .....(1.30)$$

$\qquad I_a = T/K$

$$\therefore \quad E_b = \frac{\sqrt{2}V(\cos\alpha - \cos\beta)}{\beta - \alpha} - \frac{TR_a \pi}{K(\beta - \alpha)}$$

But $E = K \, \omega_m$

$$\therefore \quad \omega_m = \frac{\sqrt{2}V(\cos\alpha - \cos\beta)}{K(\beta - \alpha)} - \frac{TR_a \pi}{K^2(\beta - \alpha)} \qquad \qquad .....(1.31)$$

Boundary between the continuous and discontinuous conduction is reached when $\beta = (\pi + \alpha)$. Substituting $\beta = \pi + \alpha$ gives the critical speed.

For given value of α, β will be equal to $(\pi + \alpha)$ for a certain value of E.

Substituting $\beta = \pi + \alpha$ in eq. 1.28 and rearranging the terms, we get

$$\frac{\sqrt{2}\,V}{Z} \sin(\pi + \alpha - \phi) - \frac{E_b}{R_a} + \left[\frac{E}{R_a} - \frac{\sqrt{2}V}{Z}\sin(\alpha - \phi) \right] e^{-(\pi + \alpha - \alpha)\cot\phi} = 0$$

$$\frac{E_b}{R_a}\left[e^{-\pi \cot\alpha} - 1 \right] = \frac{\sqrt{2}V}{Z}\sin(\alpha - \phi) + \frac{\sqrt{2}\,V}{Z}\sin(\alpha - \phi)\, e^{-\pi\cot\phi}$$

$$= \frac{\sqrt{2}V}{Z}\sin(\alpha - \phi)\left[1 + e^{-\pi\cot\phi} \right]$$

The critical value of induced emf E_b is E_c.

$$\therefore \quad E_c = \frac{\sqrt{2}\,VR_a}{Z}\sin(\alpha - \phi)\left[\frac{e^{-\pi\cot\phi} + 1}{e^{-\pi\cot\phi} - 1} \right] \qquad \qquad .....(1.32)$$

$\qquad E_c = K\omega_{mc}$

where ω_{mc} is the critical speed in rad/sec.

$$\therefore \qquad \omega_{mc} = \frac{\sqrt{2}VR_a}{KZ}\sin(\alpha - \varphi)\left[\frac{e^{-\pi\cot\varphi} + 1}{e^{-\pi\cot\varphi} - 1}\right] \qquad\qquad .....(1.33)$$

(b) ***Continuous current mode of operation :*** The waveforms of supply voltage V_s, armature voltage V_a, and armature current I_a are shown in Fig. 1.8 (b).

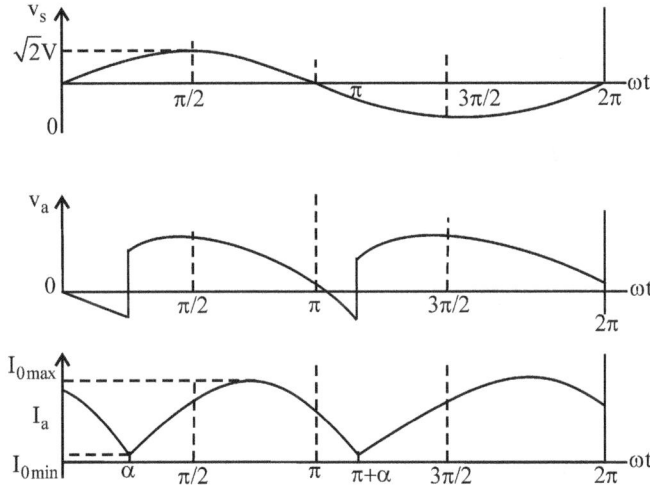

Fig. 1.8 (b) Waveforms of supply and load current (continuous current).

The voltage applied to armature V_a (ωt) during the interval $\alpha < \omega t < \pi + \alpha$ is equal to $\sqrt{2}$ V sin ωt.

Therefore the average voltage V_a applied to armature for continuous conduction is given by

$$V_a = \frac{2\sqrt{2}V}{\pi}\cos\alpha$$

$$\omega_m = \frac{2\sqrt{2}V}{\pi}\frac{\cos\alpha}{K} - \frac{R_a T}{K^2} \qquad\qquad .....(1.34)$$

For controlling the speed of the motor the firing angle is varied from 0 to 90^0 only.

Example 1.5

A 220 V, 960 rpm, 12.8 A separately excited dc motor has armature circuit resistance and inductance of 2 Ω and 150 mH respectively. It is fed from a single phase half-controlled rectifier with an ac source voltage of 230 V, 50 Hz. Calculate

(i)   Motor torque for $\alpha = 60^\circ$ and speed = 600 rpm.

(ii)   Motor speed for $\alpha = 60°$ and torque T = 20 N-m.

Assume continuous current.

Solution :

(i)   Back emf E_{b1} at rated speed of 960 rpm = 220 – 12.8 × 2 = 194.4 V

Back emf E_{b2} at 600 rpm = $\dfrac{194.4 \times 600}{960}$ = 121.5 V

∴ Armature current at speed of 600 rpm

$$I_a = \dfrac{E_a - E_b}{R_a}$$

For $\alpha = 60°$, $E_a = \dfrac{\sqrt{2} \times 230}{\pi}$ (1 + cos 60°) = 155.3 V

∴ $I_a = \dfrac{155.3 - 121.5}{2}$ = 16.9 A

We know that $K\phi\omega = E_b$

E_b at 960 rpm = 194.4 V

∴ $K\phi = \dfrac{194.4 \times 60}{2\pi \times 960}$ = 1.935 Volts/rad/sec

Torque developed T = $K\phi\, I_a$

$$= 1.935 \times 16.9$$

$$= \mathbf{32.70} \text{ N-m}$$

(ii)   T = 20 N-m.

∴ $I_a = \dfrac{20}{1.935}$ = 10.33 A

$\alpha = 60°$ ∴ $E_a = \dfrac{\sqrt{2} \times 230}{\pi}$ (1 + cos 60°)

$$= 155.3 \text{ V}$$

∴ E_{b2} = 155.3 – 10.33 × 2 = 134.7 V

Back emf is proportional to speed.

$$\dfrac{N_2}{N_1} = \dfrac{E_{b2}}{E_{b1}} = \dfrac{134.7}{194.4}$$

∴ $N_2 = \dfrac{134.7}{194.4} \times 960 = \mathbf{665}$ rpm

Example 1.6

A 220 V, 960 rpm, 12.8 A separately excited dc motor has armature resistance and inductance of 2 Ω and 150 mH respectively. It is fed from a single phase half controlled rectifier with an ac source voltage of 230 V, 50 Hz. Calculate motor torque for $\alpha = 60°$ and speed = 600 rpm.

Solution :

To find the voltage at $\alpha = 60°$, we must know whether the current is continuous or not.

For this purpose we have to calculate the critical speed given by the equation

$$\omega_C = \frac{R_a\sqrt{2}\ V}{k\ Z}\left[\frac{Sin\phi\ e^{-\alpha\ cot\phi} - Sin\ (\alpha - \phi)\ e^{-\pi\ cot\alpha}}{1 - e^{-\pi\ cot\phi}}\right]$$

ω_C is the critical speed in rad/sec.
k is the back emf constant.

$$k = \frac{V - I_a R_a}{\omega} = \frac{(220 - 12.8 \times 2)60}{2\pi \times 960} = 1.933$$

$$Z = \sqrt{R_a^2 + (\omega L)^2} = \sqrt{2^2 + (314 \times 150 \times 10^{-3})^2} = 47.16\ \Omega$$

$$\phi = \tan^{-1}\frac{\omega L}{R} = 87.56°$$

$$\alpha = 60°$$

$$V = 230$$

Substituting these values in above equation, we get

$\omega_C = 78.29$ rad/sec.

$$\text{Critical speed} = \frac{78.29 \times 60}{2\pi} = 747.6\ \text{rpm.}$$

Since the motor speed of 600 rpm is less than the critical speed, the current is continuous. Hence voltage applied to armature

$$E_a = \frac{\sqrt{2}V}{\pi}\ (1 + \cos\alpha)$$

$$= \frac{\sqrt{2} \times 230}{\pi}\ (1 + \cos 60) = 155.3\ \text{V}$$

$$E_b \text{ at 600 rpm} = 1.933 \times \frac{600 \times 2\pi}{60} = 121.45\ \text{V}$$

$$I_a = \frac{155.3 - 121.45}{2} = 16.9\ \text{A}$$

Torque developed = $kI_a = 1.933 \times 16.9 = $ **32.67 N-m**

Example 1.7

A single phase half controlled bridge converter drives a 110 V, 3.73 Kw, 1200 rpm separately excited dc motor. The ac supply is 120 V, 50 Hz. The thyristors are fired at delay angle $\alpha = 45°$. The motor current is 15 A and is assumed to be ripple free. The armature resistance is neglected. Assume the converter is loss less. For the condition

(i)   Determine the power drawn by the motor

(ii)   Determine the supply volt amps and supply power factor.

(iii)   Determine the rms value of the thrysistor current and diode current.

Solution :

The circuit diagram is shown below.

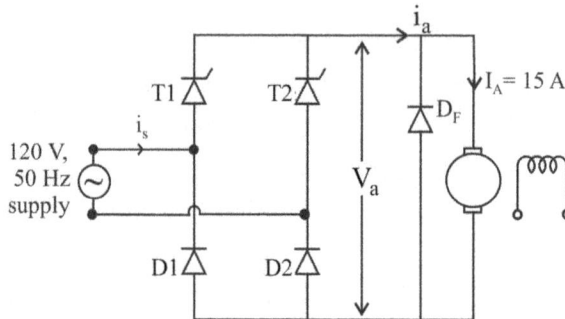

(i) $\alpha = 45°$, $V_S = 120$ V

$\therefore$ Average voltage applied to the motor $= \dfrac{\sqrt{2}V}{\pi} (1 + \cos \alpha)$

$$= \dfrac{\sqrt{2} \times 120}{\pi} (1 + \cos 45°) = 92.21 \text{ V}$$

Power drawn by the motor $= V_a I_a = 92.21 \times 15 = \mathbf{1383 \text{ W}}$

(ii)   The current drawn from the supply is as shown below

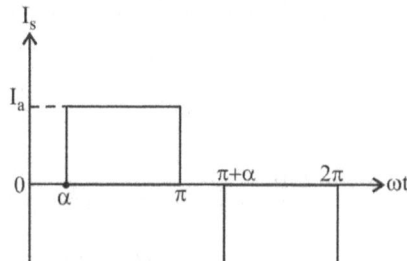

RMS value of the supply current

$$I_s = I_a \sqrt{\dfrac{\pi - \alpha}{\pi}}$$

$$= 15\sqrt{\frac{\pi - \pi/4}{\pi}} = 15\sqrt{\frac{3}{4}}$$

$$= 12.99 \text{ A}$$

Supply volt amps $= V_s I_s = 120 \times 12.99$

$$= 1558 \text{ VA}$$

Active power drawn $= V_a I_a = 1383 \text{ W}$

∴ Supply power factor $= \dfrac{1383}{1558} = 0.887$ lag.

(iii) RMS value of thyristor current :

The wave form of current through thyristor T_1 is as shown below.

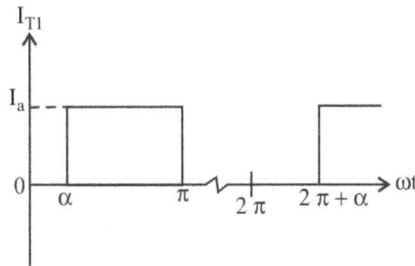

rms value $= \sqrt{\dfrac{1}{2\pi} I_a^2 (\pi - \alpha)}$

$$= I_a \sqrt{\frac{1}{2\pi}\left(\pi - \frac{\pi}{4}\right)} = 15 \sqrt{\frac{3}{8}} = 9.186 \text{ A}$$

rms value of diode current is same as that of thyristor

∴ $I_D = 9.186 \text{ A}$

Example 1.8

A 200 V, 875 rpm, 150 A separately excited dc motor has an armature resistance of 0.06 Ω. It is fed from a single phase fully controlled rectifier with an ac source voltage of 220 V, 50 Hz. Assuming continuous conduction calculate firing angle for rated motor torque and speed of 750 rpm.

Solution :

At rated speed back emf $E_b = V - I_a R_a$

$$= 200 - 150 \times 0.06$$

$$= 191 \text{V}$$

∴ Back emf at 750 rpm $= \dfrac{191 \times 750}{875} = 163.7 \text{ V}$

I_a at rated torque = 150 A

$\therefore \qquad V_a = 163.7 + 150 \times 0.06 = 172.7$ V

For continuous conduction V_a for 1ϕ fully controlled rectifier is given by equation

$$V_a = \frac{2\sqrt{2}V}{\pi} \cos \alpha$$

$$172.7 = \frac{2\sqrt{2} \times 220}{\pi} \cos \alpha$$

$$\cos \alpha = 0.872. \qquad \alpha = 29.3°$$

Example 1.9

The speed of a 10 h.p, 210 V, 1000 rpm separately excited dc motor is controlled by a single phase fully controlled converter. The rated motor armature current is 30 A and the armature resistance $R_a = 0.25$ Ω. The ac supply voltage is 230 V. The motor voltage constant is $K_a\phi = 0.172$ V/rpm. Assume armature current as continuous and ripple free.

For a firing angle of 45° and rated motor armature current determine
- (i) the motor torque
- (ii) the speed of the motor
- (iii) the supply power factor.

Solution :

(i) $K_a\phi = 0.172$ v/rpm $= \dfrac{0.172 \times 60}{2\pi} = 1.64$ V/(rad/sec)

We know that torque $= K_a\phi I_a = 1.64 \times 30 = 49.2$ N-m

(ii)   Voltage applied to armature $V_a = \dfrac{2\sqrt{2}V}{\pi} \cos \alpha$

$$= \frac{2 \times \sqrt{2} \times 230}{\pi} \cos 45° = 146.42 \text{V}$$

$E_b = V_a - I_a R_a = 146.42 - 30 \times 0.25 = 138.92$ V;

$$N = \frac{E_b}{K_a\phi} = \frac{138.92}{0.172} = 807.67 \text{ rpm.}$$

(iii)   The armature current is constant

$\therefore$ the supply current is rectangular wave as shown below

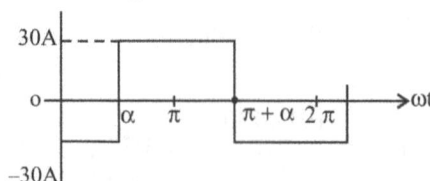

rms value of supply current = 30 A

rms value of supply voltage = 230 V

$\therefore$ Apparent power = $V_s I_s$ = 230 × 30 = 6900 VA.

Real power supplied = $V_a I_a$ = 146.42 × 30 = 4392.4 W

$\therefore$ Power factor = $\dfrac{4392.4}{6900}$ = 0.64 lagging

Example 1.10

The speed of a 220 V, 3.73 KW, 1000 rpm dc separately excited motor is controlled by a single phase full converter. The ac supply is 240 V, 50 Hz. A very large inductance is connected in series with the armature. Assume the motor and converter to be lossless. The motor emf constant is 1.9 V/rad/sec. For a speed of 1000 rpm and rated motor current, determine.

(i)   the firing angle of the converter

(ii)   the rms value of the supply current and thyristor current

(iii)   the supply power factor

Solution

(i)   Speed = 1000 rpm = $\dfrac{1000 \times 2\pi}{60}$ = 104.7 rad/sec

Back emf E_b = 104.7 × 1.9 = 198.96 V

Since the motor is lossless $E_b = E_a$ = 198.96 V

For 1ϕ fully controlled converter average output voltage

$$E_a = \frac{2\sqrt{2}V}{\pi} \cos \alpha.$$

$\therefore$

$$\frac{2\sqrt{2} \times 240}{\pi} \cos \alpha = 198.96$$

$$\cos \alpha = 0.920 .$$

$\therefore$

$$\alpha = 22.9° \text{ A}$$

(ii)   RMS value of supply current = Average armature current

$$= \frac{3730}{220} = 16.95 \text{ A}$$

RMS value of the thyrister current = $\dfrac{16.95}{\sqrt{2}}$ = 11.98 A

(iii)   Supply power factor at 1000 rpm and rated current.

Active power = $E_a I_a$ = 198.96 × 16.95 W

Apparent power = $V_s I_s$ = 240 × 16.95 VA

$$\therefore \text{ Power factor} = \frac{198.96 \times 16.95}{240 \times 16.95} = 0.829 \text{ lag.}$$

1.5   Operation of dc Series Motor Connected to 1φ Controlled Rectifier

In this motor the field winding is connected in series with the armature and is designed to carry the armature current. The equivalent circuit of the series motor is shown in the Fig. 1.9.

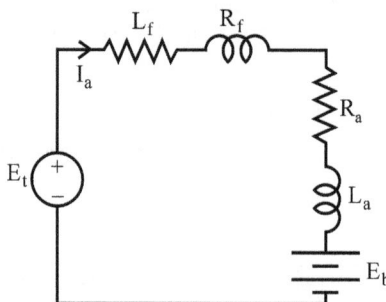

Fig. 1.9 Equivalent circuit of dc series motor connected to controlled rectifier.

E_t is the output voltage of rectifier and it is the voltage applied to the series motor.

From the equivalent circuit the following equations are written for steady state condition.

$$E_t = E_b + I_a (R_a + R_f) \qquad .....(1.35)$$

where E_t is the voltage applied to the motor, E_b is the back emf, I_a is the current through armature and field winding R_f is the resistance of series field winding and R_a is the armature resistance.

$$E_b = K_e \phi \omega \text{ Volts} \qquad .....(1.36)$$

$$T_m = K_e \phi I \text{ N-m} \qquad .....(1.37)$$

where K_e is called emf constant.

In series motor the flux is a function of armature current. If the magnetic circuit is not saturated the flux is proportional to the armature current. Hence $\phi = K_f I_a$

where K_f is a constant.

$$\therefore \qquad E_b = K_e K_f I_a \omega \qquad .....(1.38)$$

$$T_m = K_e K_f I_a^2 \qquad .....(1.39)$$

From eqn. (1.35) and (1.38)

$$K_e K_f I_a \omega = E_t - I_a (R_a + R_f) \qquad .....(1.40)$$

$$\omega = \frac{E_t}{K_e K_f I_a} - \frac{R_a + R_f}{K_e K_f}$$

From eqn. (1.39) $\quad I_a = \dfrac{\sqrt{T_m}}{\sqrt{K_e K_f}}$

From eqn. (1.40) $\quad \therefore \omega = \dfrac{E_t}{\sqrt{T_m}\sqrt{K_e K_m}} - \dfrac{R_a + R_f}{K_e K_f}$ .....(1.41)

This equation gives relationship between torque and speed. The torque-speed characteristic and armature current-torque characteristic are shown in Fig. 1.10 for a certain E_t applied to the motor

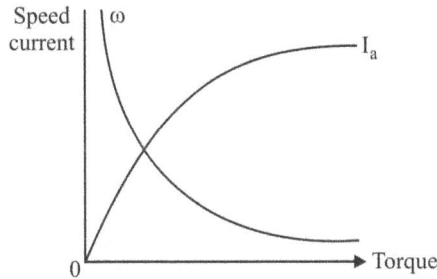

Fig. 1.10 Torque speed characteristic.

It can be seen from eqn. (1.41) that the speed of DC series motor can be controlled by varying E_t the voltage applied to the motor. The voltage E_t can be controlled by controlling the firing angle of rectifier. The torque speed characteristics for different values of E_t are shown in Fig. 1.11.

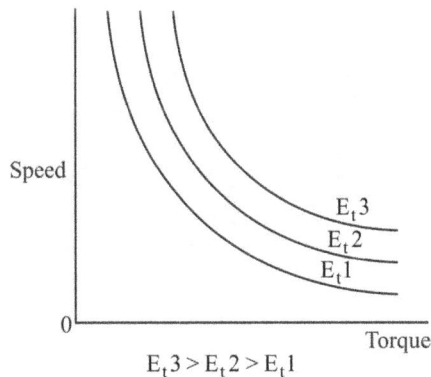

$$E_t 3 > E_t 2 > E_t 1$$

Fig. 1.11 Torque speed characteristics of series motor.

The controlled rectifier can be:

> Single phase half controlled converter
>
> Single phase fully controlled converter

Three phase half controlled bridge converter

Three phase fully controlled converter.

1.5.1 DC Series Motor Connected to Single Phase Half Controlled Converter

The circuit diagram is shown in Fig 1.12

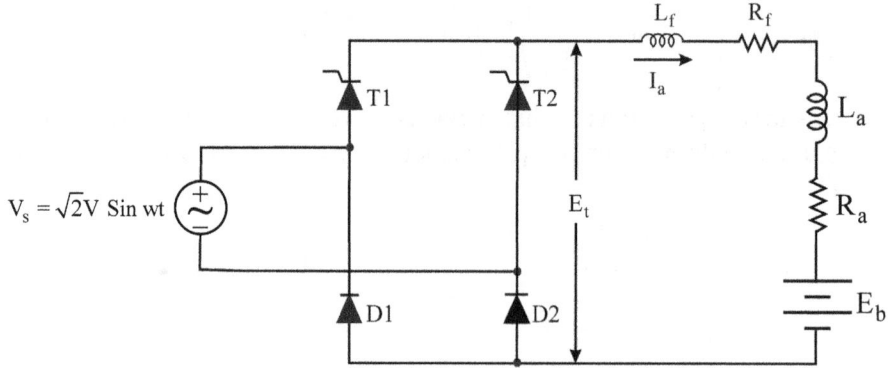

Fig. 1.12 DC series motor connected to 1-Φ half controlled bridge rectifier.

The waveforms of Vs, E_t, I_a are shown in the Fig. 1.13 for firing angle α.

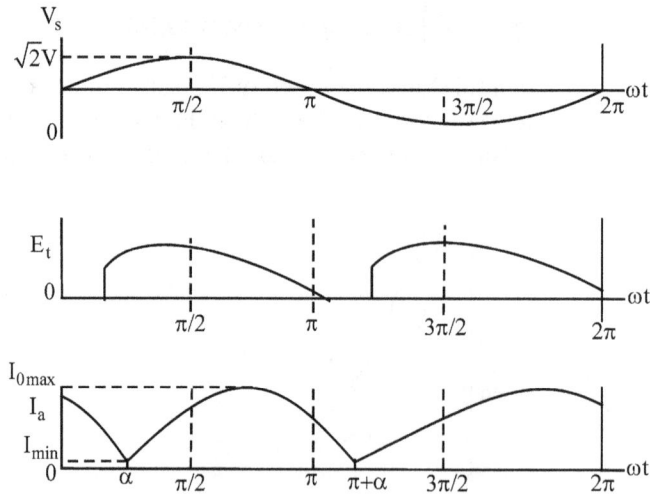

Fig. 1.13 HCR Waveforms of supply and load voltages and load current
(continuous current).

The armature current i_a is continuous. The back emf E_b varies as i_a varies although the speed is assumed as constant over one cycle of the input voltage. This is so because the flux changes as the armature current i.e., series field current changes.

The operation of the motor is given by the following equations :

For the time interval $\alpha < \omega t < \pi$

$$(R_a + R_f)\, ia + (L_a + L_f)\, \frac{di_a}{dt} + E_b = \sqrt{2}\ V \sin \omega t \qquad\qquad .....(1.42)$$

For the interval $\pi < \omega t < \pi + \alpha$

$$(R_a + R_f)\, i_a + (L_a + L_f)\, \frac{di_a}{dt} + E_b = 0 \qquad\qquad .....(1.43)$$

$$E_b = K_e\, \phi\, \omega = K_e\, f(i_a)\, \omega \qquad\qquad .....(1.44)$$

If $K_e\, f(i_a)\, \omega$ is substituted for E_b in eqns. (1.42) and (1.43) they become nonlinear equations and the solution for i_a can be obtained only by numerical method. A simple method of analysis is obtained when E_b is replaced by its average value. Thus $E_b = K\, I_a\omega$, where I_a is the average value of current.

Average value of applied voltage E_t to the motor at its terminals is equal to the average output voltage of the single phase half controlled converter and is equal to $\left[\dfrac{\sqrt{2V}}{\pi}(1+\cos\alpha)\right]$ where α is the firing angle. Thus

$$E_t = \frac{\sqrt{V}}{\pi}\ (1 + \cos \alpha).$$

Substituting this value of E_t in eqn. (1.41), the relation between torque and speed in terms of firing angle α is obtained as

$$\omega = \frac{\sqrt{2}V(1+\cos\alpha)}{\pi\sqrt{T}\sqrt{K}} - \frac{R_a + R_f}{K} \qquad\qquad .....(1.45)$$

where $\qquad K = K_e\, K_f$

Speed torque curves of series motor for different firing angles are shown in Fig. 1.14.

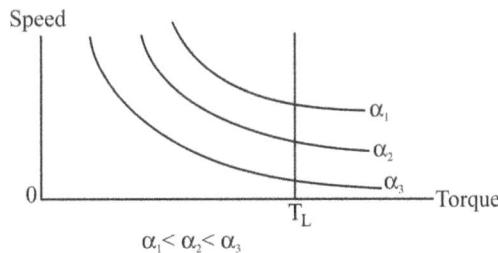

Fig. 1.14 Speed torque characteristics for different firing angles.

It can be seen from the curves that the speed of the motor can be controlled for any given load torque T_L by changing the firing angle α. If α is reduced the speed will increase.

For the half controlled converter the firing angle can be varied from 180^0 to 0^0 and the voltage E_t is always +ve. For series motor the torque developed by the motor is proportional to the square of the current. With the circuit arrangement shown, the direction of rotation of motor cannot be changed even though the polarity of voltage applied to the motor is changed.

1.5.2 dc Series Motor Connected to Single Phase Fully Controlled Converter

The circuit diagram is shown in Fig. 1.15.

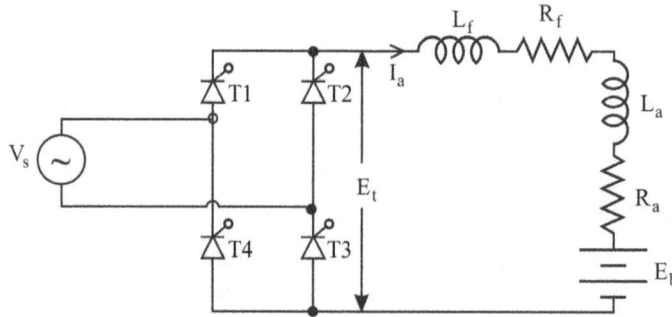

Fig. 1.15 Single phase f.c. converter connected to DC. series motor.

T1 and T2 are trigged at $\omega t = \alpha$ and T3 and T4 are triggered at $\omega t = \pi + \alpha$. The current is continuous. The waveforms of different voltages and currents are shown in Fig. 1.16.

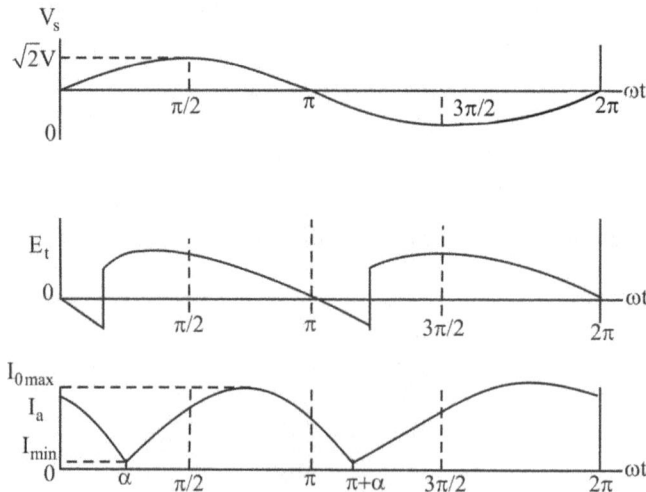

Fig. 1.16 Waveforms of voltages and current.

The average value of the output voltage

$$E_t = \frac{1}{\pi} \int_{\alpha}^{\pi+\alpha} \sqrt{2}V \sin \omega t\, d(\omega t) = \frac{2\sqrt{2}\ V}{\pi} \cos \alpha \qquad .....(1.46)$$

For $0 < \alpha < 90^0$ the E_t will be positive and
$90^0 < \alpha < 180^0$ the E_t will be negative.

For α less than 90^0 the equivalent circuit under steady state is as shown in Fig. 1.17.

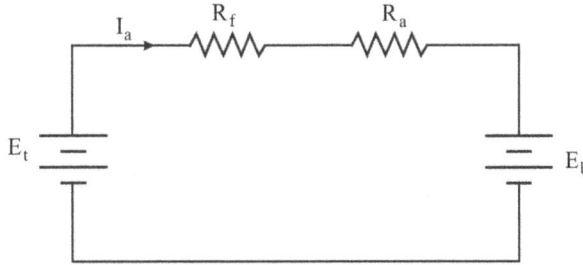

Fig. 1.17 Equivalent circuit for $0 < \alpha < 90^0$.

The converter works as Rectifier and Electrical energy is converted to mechanical energy and drives the motor.

For α greater than 90^0 the polarity of E_t changes. The current direction from the source will not change because of SCRs. The circuit will not work if the direction of E_b remains as it is. But if the direction of generated voltage E_b changes the mechanical energy of machine (now acting as generator) can be converted into electrical energy and fed back into the electrical system. Then the converter is said to be acting as inverter. The equivalent circuit is as shown in Fig. 1.18.

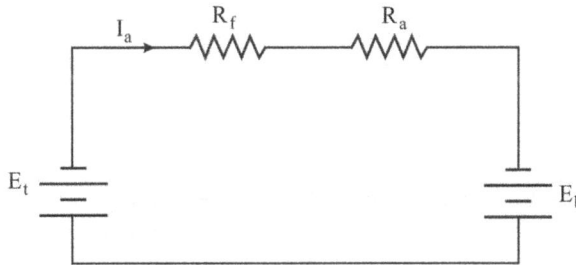

Fig. 1.18 Equivalent circuit for $90^0 < \alpha < 180^0$.

Thus the operation of dc series machine connected to 1ϕ fully controlled converter can be as follows :

As long as $0 < \alpha < 90^0$ the converter works as rectifier and the dc machine works as motor and as long as $90^0 < \alpha < 180^0$ the converter works as inverter and the dc machine works as generator.

The equations of motion will be as follows :

For $\qquad\qquad 0 < \alpha < 90^0$

$$\frac{2\sqrt{2}V\cos\alpha}{\pi} = E_b + I_a(R_a + R_f) \qquad\qquad .....(1.47)$$

$E_b = K\,I_a\,\omega$ (Assuming that flux is proportional to current I_a)

$$T_m = K\,I_a^2\,;\ I_a = \sqrt{\frac{T_m}{K}}$$

Substituting for E_b and I_a in eqn. (1.47), we get/

$$\frac{2\sqrt{2}V\cos\alpha}{\pi} = K\sqrt{\frac{T_m}{K}}\,\omega + \sqrt{\frac{T_m}{K}}(R_a + R_f) \qquad .....(1.48)$$

$$\omega = \frac{2\sqrt{2}V\cos\alpha}{\pi\sqrt{K}\sqrt{T_m}} - \frac{R_a + R_f}{K} \qquad\qquad .....(1.49)$$

The torque-speed characteristics of dc series motor connected to fully controlled converter for firing angles less than 90^0 is shown in Fig. 1.19.

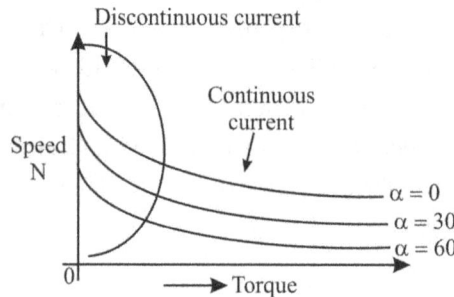

Fig. 1.19 Torque speed characteristics for both continuous and discontinuous modes of operations.

For $90^0 < \alpha < 180^0$ the polarities of generated voltage in the armature and terminal voltage at converter are as shown in Fig. 1.18. For the system to work properly magnitude of E_b should be greater than the magnitude of the terminal voltage E_t of converter. Thus

$$|E_b| = \left|\frac{2\sqrt{2}\ V\cos\alpha}{\pi}\right| + I_a(R_a + R_f) \qquad\qquad .....(1.50)$$

This operation is known as Regenerative braking operation.

Example **1.11**

A 230 V, 750 rpm 25 A dc series motor is driving at rated conditions a load whose torque is proportional to speed squared. The combined resistance of armature and field is one ohm. Calculate the motor terminal voltage and current for a speed of 400 rpm. State the assumption made for solving this problem,

Solution:

Rated condition : $\quad V_1 = 230$, $N_1 = 750$, $I_{a1} = 25$ A

$$T_1 = K \, \phi \, I_{a1} = K \, K_f \, I_{a1}^2$$

This shows that torque is proportional to square of current. But it is given that torque is proportional to square of speed. Hence the armature current I_a is proportional to speed N.

It is given that torque is proportional to speed squared.

Hence $T_1 = K_T \omega_1^2 = K_{T1} N_1^2$. This shows that current is proportional to speed.

$$\therefore \qquad \frac{I_{a2}}{I_{a1}} = \frac{N_2}{N_1} \, ;$$

The current I_{a2} at 400 rpm = (400 × 25)/750 = 13.33 A.

$$E_{b1} = K \, \phi_1 \, \omega_1 \; = K \, K_f \, I_{a1} \omega_1$$

We know that $\quad E_b = V - I_a (R_a + R_f)$

$$\therefore \qquad E_{b1} = 230 - 25 \times 1 = 205 \text{ V.}$$

At speed of 750 rpm back emf E_{b1} is 205 Volts.

Let E_{b2} be the back emf at 400 rpm.

Then $\qquad \dfrac{E_{b2}}{E_{b1}} = \dfrac{N_2 I_{a2}}{N_1 I_{a1}}$

$$E_{b2} = \frac{400 \times 13.33 \times 205}{750 \times 25} = 58.31 \text{ V}$$

$\therefore$ Applied voltage at 400 rpm = $E_{b2} + I_{a2} (R_a + R_f)$

$$= 58.31 + 13.33 \times 1 = \mathbf{71.64 \text{ Volts}}$$

Assumption : The flux is proportional to field current.

Example 1.12

If the motor in the above example 1.11 is connected to 1 ϕ half controlled converter ac supply of 400 V, what should be the firing angle for speed of 750 rpm and for a speed of 400 rpm.?

At 750 rpm $\qquad V_a = 230V$

$$\therefore \qquad \frac{\sqrt{2} \times 400}{\pi}(1 + \cos\alpha) = 230 \text{ V}$$

$$1 + \cos\alpha_1 = \frac{230 \times \pi}{\sqrt{2} \times 400} = 1.277$$

$$\cos\ \alpha_1 = 0.277$$

$$\alpha_1 = \mathbf{73.91^0}$$

At 400 rpm applied voltage

$$V_a = 71.64 \text{ volts}$$

$$\frac{\sqrt{2} \times 400}{\pi}(1 + \cos\alpha_2) = 71.64$$

$$1 + \cos\alpha_2 = 0.398$$

$$\cos\alpha_2 = 0.398 - 1 = -0.601$$

$$\boldsymbol{\alpha_2 = 126.94^0.}$$

Example 1.13

A 220 V dc series motor runs at 1000 rpm (clockwise) and takes an armature current of 100 A when driving a load with constant torque. Resistance of the armature and field windings are 0.05 Ω each. Find the magnitude and direction of the motor speed and armature current if the motor terminal voltage is reversed and the number of turns in the field winding is reduced to 80%. Assume linear magnetic circuits.

Solution :

Torque is proportional to product of flux and current. As magnetic circuit is linear flux is proportional to field current.

$$\text{Torque} = K_e \, \phi \, I_a$$

$$T_1 = K_e \, K_f \, I_{a1}^2$$

I_{a1} is the current at speed of 100 rpm.

$$T_2 = K_e \, \phi_2 \, I_{a2}$$

$$\phi_2 = K_f \, I_a$$

Since field turns are reduced to 80%.

$$\phi_2 = 0.8 \, K_f \, I_{a2}$$

∴ $$T_2 = K_e \, 0.8 \, K_f \, I_{a2}$$

$$I_{a2} = 0.8 \, K_e \, K_f \, I_{a2}^2$$

Since torque is constant

$$T_1 = T_2 \;\Rightarrow\; 0.8 \, K_e \, K_f \, I_{a2}^2 = K_e \, K_f \, I_{a1}^2$$

∴ $$I_{a2} = \frac{I_{a1}}{\sqrt{0.8}} = \frac{100}{\sqrt{0.8}} = 111.8 \text{ A}$$

By changing the polarity of supply voltage direction of both armature current and flux will change (T α φ I$_a$). Therefore direction of torque will not change. Hence it rotates in the same direction. Hence the motor continues to rotate in clockwise direction.

Back emf at 1000 rpm is E_{b1}

$$E_{b1} = 220 - 100 \, (0.05 + 0.05) = 210 \text{ V}$$

$$E_{b2} = 220 - 111.8 \, (0.05 + 0.8 \times 0.05)$$

$$= 220 - 111.8 \times 0.09 = 209.94 \text{ V}$$

$$N_2 = ?$$

$$E_{b2} = K \, \phi_2 \, N_2 = K0.8 \, K_f \, I_{a2} N_2 = K \, K_f \, 0.8 \times 111.8 \times N_2$$

$$E_{b1} = K \, K_f \, I_{a1} \, N_1 = K \, K_f \, 100 \times 1000$$

Taking ratio of above two equations we get,

$$\frac{E_{b2}}{E_{b1}} = \frac{0.8 \times 111.8 \times N_2}{100 \times 1000} = \frac{209.94}{210}$$

$$\therefore \qquad N_2 = \mathbf{1117.7 \text{ rpm}}$$

Example 1.14

A series motor drive is connected to 1 ϕ semi converter, which is supplied from 230 V, 50 Hz. The field and armature resistances are 1ohm and 1.5 ohm respectively. The torque constant M_{af} is 0.3 H and the load torque is 20 N-m. The damping coefficient f is 0.15 N-m-sec/rad. Calculate the average armature current and speed when the SCRs are triggered at an angle of 35^0. The current is continuous.

Solution :

Current is continuous. Hence the average voltage applied to motor

$$= \frac{\sqrt{2} \times 230}{\pi}(1 + \cos 35^0) = 188.35 \text{ V}$$

Let I_a be the average current and ω be the speed in radians per second.

Developed Torque $T_d = K \, I_a \, \phi \; = \; \dfrac{K \phi I_a^2}{I_a} \; = K_T \, I_a^2 \; = 0.3 \, I_a^2$

where $\qquad K_T = \dfrac{K\phi}{I_a} = 0.3 \text{H}$

Developed torque is equal to sum of load torque and frictional torque.

$$T_d = T_L + f\omega$$

where T_L is the load torque in N-m, f is the damping coefficient.

$\therefore$ $\qquad$ $0.3 \ I_a^2 = 20 + 0.15\omega$ $\qquad$ .....(i)

Equating the electrical power $E_b I_a$ to mechanical power output $T\omega$

$$T\omega = E_b \ I_a = [V_a - I_a \ (R_a + R_f)]I_a$$

$$0.3 \ I_a^2 \ \omega = [188.5 - (1 + 1.5) \ I_a \ ] \ I_a$$

$$\Rightarrow 0.3 \ I_a\omega = 188.5 - 2.5 \ I_a$$

$$\omega = \frac{188.5}{0.3I_a} - \frac{2.5}{0.3}$$

Substituting for ω in (i), we get

$$0.3 \ I_a^2 = 20 + 0.15 \left[\frac{188.5}{0.3I_a} - \frac{2.5}{0.3} \right]$$

$$= 20 + \frac{94.245}{I_a} - 1.25$$

$$= 18.75 + \frac{94.245}{I_a}$$

$$0.3 \ I_a^3 - 18.75 \ I_a = 94.245$$

By trial and error we solve for I_a

$\therefore$ $\qquad$ $I_a = 9.74$ A

Torque developed $= 0.3 \times 9.74^2 = 28.46$ N- m.

$\qquad$ $20 + 0.15 \ \omega = 28.46$

$\qquad$ $0.15\omega = 8.46$

$\qquad$ $\omega = 56.4$ rad/sec

$\qquad$ **N = 538.58 rpm**

Check: $\qquad$ $E_b = V_a - I_a \ (R_a + R_f)$

$\qquad$ $= 188.5 - 9.74 \times 2.5 = 164$ V

$\qquad$ $E_b \ I_a = 164 \times 9.74 = 1597$ W

$\qquad$ $T\omega = 28.46 \times 56.4 = 1605.14$ W

$\qquad$ $E_b \ I_a$ is thus approximately equal to $T\omega$

Problems

P1.1 (a) Derive an expression for the average output voltage of a 1ϕ full converter and draw the waveforms of output voltage for a firing angle α.

(b) Explain what is meant by rectification mode and inversion mode?

P1.2 (a) What is a full converter? Draw two full converter circuits.

(b) Why is the power factor of semi converters better than that of full converters?

(c) What is the principle of phase control?

P1.3 (a) Explain how the speed of a dc series motor is controlled using thyristors.

(b) A series motor is supplied from a rectified 1ϕ supply of 230 Vrms, 50 Hz frequency. The armature and field resistance together equal 2 Ω. The torque constant M_{af} is 0.23H and the load torque is 20 Nm. Neglect damping and find the average armature current and speed.

P1.4 (a) A fully controlled 1ϕ bridge connected to AC supply of 230 V rms and 50 Hz is used for speed control of DC motor with separate field excitation. The full load average current is 10 A and the converter operates at a firing angle of α = π/4. Neglecting the inductance and resistance of both armature and source, calculate the minimum value of series inductance L_d to provide continuous conduction.

(b) Distinguish between continuous and discontinuous conduction.

P1.5 Derive an expression for the average output voltage of a 1ϕ semi converter. Assuming a very highly inductive load, draw the waveforms of output voltage, load current and voltage across thyristors.

P1.6 (a) Explain the concept of constant torque control and constant power control.

(b) Explain how the speed control of dc motor is achieved illustrating the triggering circuits of the thyristors.

P1.7 Explain what is meant by constant torque and constant HP operation.

P1.8 (a) What are the assumptions made while doing the steady-state performance of the converter fed dc drives. Justify your answers.

(b) Explain the use of free wheeling diode in the converter fed dc drives. Take an example of 1-phase fully controlled converter for explanation. How it is going to affect the machine performance?

P1.9 A 1 ϕ, half controlled converter is fed from a 120 V rms, 60 Hz supply and provides a variable dc voltage at the terminals of a dc motor. The thyristor is triggered continuously by a dc signal. The resistance of armature circuit is 10 Ω and because of fixed motor excitation and high inertia, the motor speed is considered constant so that the back emf is 60 V. Find the average value of the armature current neglecting armature inductance.

P1.10   A dc series motor has $R_s = 3\,\Omega$ $R_s = 3\,\Omega$ and $M_{af} = 0.15$ H. The motor speed is varied by a phase –controlled bridge. The firing angle is $\pi/4$ and the average speed of the motor is 1450 rpm. The applied ac voltage to the bridge is $330\sin\omega t$. Assuming continuous motor current find the steady state average motor current and torque. Sketch the waveforms for output voltage, current and gating signals.

P1.11 (a)   A DC shunt motor operating from a 1ϕ half controlled bridge at a speed of 1450 rpm has an input voltage $330\sin 314t$ and a back emf 75 V. The SCRs are fired symmetrically at $\alpha = \pi/3$ in every half cycle and the armature has a resistance of $5\,\Omega$. Neglecting armature inductance, find the average armature current and the torque.

   (b)   Sketch the speed-torque characteristics.

P1.12   Two independent single – phase semi-converters are supplying the armature and field circuits of the separately excited dc motor for controlling its speed. The firing angle of the converter, supplying the field, is adjusted such that maximum field current flows. The machine parameters are : Armature resistance of $0.25\,\Omega$, field circuit resistance of $147\,\Omega$, Motor voltage constant $K_v = 0.7032$V/A-rad/s. The load torque is $T = 45$ N.m at 1000 rpm. The converters are fed from a 208 V, 50 Hz ac supply. The friction and windage losses are neglected. The inductance of the field and armature circuits is sufficient enough to make the armature and field currents continuous and ripple free. Determine

   (a)   the field current,

   (b)   the delay angle of the armature converter, input power factor of the armature circuit converter

P1.13 (a)   Explain how the speed of a dc series motor is controlled using converters.

   (b)   A series motor is supplied from a full converter whose $\alpha = 65^0$, 1ϕ supply of 230 V rms, 50 Hz frequency. The armature and field resistance together equal $2\,\Omega$. The torque constant M_{af} is 0.23 H and the load torque is 20 Nm. Neglect damping and find the average armature current and speed.

P1.14   A single –phase fully controlled thyristor converter is supplying a dc separately excited dc motor. Draw the neat waveform diagrams and explain various operating modes of the drive both motoring and regenerative braking for

   (a) $\lambda < \alpha$,   (b) $\lambda > \alpha$,

   where α : is the firing angle, λ is the angle at which the source voltage equal to the motor back emf. Assume the armature of the separately excited dc motor can be replaced by simple R-L and back emf load.

Objective Type Questions

1. A 1-φ, half –controlled rectifier connected to 220 V, 50 Hz ac supply operating with firing angle of 90^0 and supplying continuous current to a separately excited dc motor. The average voltage supplied to the motor armature is

   (a)   220 V   (b)   110 V

   (c)   99 V   (d)   89 V   [   ]

2. A 1- φ half –controlled converter connected to 220 V, 50 Hz ac supply is supplying continuous current to a separately excited dc motor. The firing angle is 60^0. The average current drawn by the armature is 20 A.
   The power input to the armature is

   (a)   4400W   (b)   2200W

   (c)   2971 W   (d)   2791 W   [   ]

3. A 1-φ half controlled converter connected to 220 V, 50 Hz ac supply is operating with firing angle of 45^0. It is supplying continuous current of 20 A to a separately excited dc motor. The armature resistance is 0.1 Ω. The back emf is

   (a)   167.05 V   (b)   150.5 V

   (c)   169.05 V   (d)   171.05 V   [   ]

4. A separately excited dc motor connected to 220 V dc supply is developing a torque of 40 N-m, when running at a speed of 400 rpm and drawing a current of 20 A. If the current drawn is 40 A, the torque developed is

   (a)   40 N-m   (b)   20 N-m

   (c)   80 N-m   (d)   160 N-m   [   ]

5. A separately excited dc motor is driving a centrifugal pump whose torque is proportional to the square of the speed. The armature current is 20 A at speed of 500 rpm. The armature current at the speed of 200 rpm is

   (a)   6.4 A   (b)   3.2 A

   (c)   9.6 A   (d)   8 A   [   ]

6. A 1-φ half controlled converter connected to 220 V, 50 Hz ac supply is used for controlling the speed of a separately excited dc motor whose armature resistance is negligible. When the firing angle is 60^0 the motor is rotating at a speed of 800 rpm. The armature is coupled to a constant torque load. The firing angle for a speed of 600 rpm is

   (a) 70.1^0 (b) 80.3^0

   (c) 82.82^0 (d) 83.1^0 [   ]

7.   A 220 V, dc series motor has armature resistance of 0.2 Ω & field resistance of 0.5 Ω. It runs at a speed of 1000 rpm when the armature current is 10 A. When the armature current is 20 A the speed is

(a)   500 rpm   (b)   2000 rpm

(c)   241.75 rpm   (d)   483.5 rpm   [   ]

8.   A 1-φ fully controlled converter connected to 220 V, 50 Hz, ac supply is supplying power to a dc series motor. The armature resistance is 0.3 Ω and field resistance is 0.4 Ω. The firing angle of converter is 30^0. The back emf is 160 V. The average current drawn by the motor is

(a)   17.14 A   (b)   16.47 A

(c)   14.7 A   (d)   18 A   [   ]

9.   The speed of 220 V, 3.73 KW, 1000 rpm dc motor (Separately excited) is controlled by 1-φ Full converter. The ac supply is 240 V, 50 Hz. Assume constant armature current and the motor and converter as loss less. The motor emf constant is 1.9 V /rad/sec. For a speed of 800 rpm and rated motor current the firing angle 'α' is equal to

(a) 40^0 (b) 37^0

(c) 42.55^0 (d) 38.48^0 [   ]

10.   The speed of 220 V, 3.73 KW, 1000 rpm dc motor (separately excited) is controlled by 1-φ Full converter. The ac supply is 240 V, 50 Hz. Assume constant armature current and the motor and converter as loss less. The motor e.m.f constant is 1.9 V /rad/sec. For a speed of 800 rpm and rated motor current the rms value of supply current is

(a)   16 A   (b)   16.95 A

(c)   18 A   (d)   15.95 A   [   ]

11.   The speed of 220 V, 3.73 KW, 1000 rpm dc motor (separately excited) is controlled by 1-φ Full converter. The ac supply is 240 V, 50 Hz. Assume constant armature current and the motor and converter as loss less. The motor emf constant is 1.9 V /rad/sec. For a speed of 800 rpm and rated motor current rms value of Thyristor current is

(a)   12.9 A   (b)   11.98 A

(c)   10 A   (d)   5 A   [   ]

12. The speed of 220 V, 3.73 KW, 1000 rpm dc motor (separately excited) is controlled by 1-ϕ Full converter. The ac supply is 240 V, 50 Hz. Assume constant armature current and the motor and converter as loss less. The motor e.m.f constant is 1.9 V /rad/sec. For a speed of 800 rpm and rated motor current the supply power factor is

    (a)   0.7 Lag   (b)   0.663 Lag

    (c)   0.633 lead   (d)   0.5 Lag   [   ]

13. A 1-ϕ fully controlled converter connected to 220 V, 50 Hz, AC supply is delivering average current of 20 A with firing angle of 120^0. The active power drawn from the source is

    (a)   1980 W   (b)   −1980 W

    (c)   2000 W   (d)   − 2000 W   [   ]

14. A 220 V, 800 rpm dc series motor develops 100 N-m of torque at full load current of 100 A. The torque developed when it draws 80 amps is equal to

    (a)   80 N-m   (b)   100 N-m

    (c)   64 N-m   (d)   156.5 N-m   [   ]

State Whether True or False

15. The power factor at which half controlled converter operates is less then that of fully controlled converter for a given load.   [   ]

16. The cost of smoothing choke required for a single phase converter is less than that of 3-ϕ converter.   [   ]

17. A single phase half controlled converter is operating in discontinuous current mode of operation. The firing angle is 120^0 and the extinction angle 210^0. The free wheeling duration is 30^0 [   ]

18. A single phase fully controlled converter is a two quadrant converter.   [   ]

19. The speed of separately excited dc motor above its rated speed is controlled by varying the field current.   [   ]

20. A single phase half controlled converter supplying discontinuous current to the armature of separately excited dc motor has three modes of operation.   [   ]

21. If displacement factor and distortion factor of line commutated converter are 0.9 and 0.8 respectively. The power factor = ...

22. Give two methods of improving the power factor of line commutated converters with out using forced commutation.

23. Draw the equivalent circuit of a dc separately excited motor connected to single phase fully controlled converter.

24. Draw the schematic diagram of separately excited dc motor connected to single phase half controlled converter.

25. Draw the power circuit of single phase fully controlled converter driving a dc series motor.

26   Draw the speed torque characteristics of a separately excited dc motor connected to 1-ϕ controlled rectifier for two firing angles α_1 & α_2 where $\alpha_2 > \alpha_1$.

27. Draw the speed torque characteristics of a series dc motor connected to 1-ϕ controlled rectifier for two firing angles α_1 & α_2 where $\alpha_2 > \alpha_1$.

2

Control of DC Motors by Three Phase Converters

Control of DC Motors by three phase converters : Three phase semi and fully controlled converters connected to dc separately excited and dc series motors –output voltage and current waveforms – speed and torque expressions – speed-torque characteristics – problems.

2.1 Separately Excited DC Motor connected to Three Phase Converters

There are four types of converters :
1. Three phase half wave converter.
2. Three phase half-controlled bridge converter.
3. Three phase fully-controlled bridge converter.
4. Three phase dual converter

2.1.1 Separately Excited dc Motor connected to Three Phase Half Wave Converter

The circuit diagram is shown in Fig. 2.1.

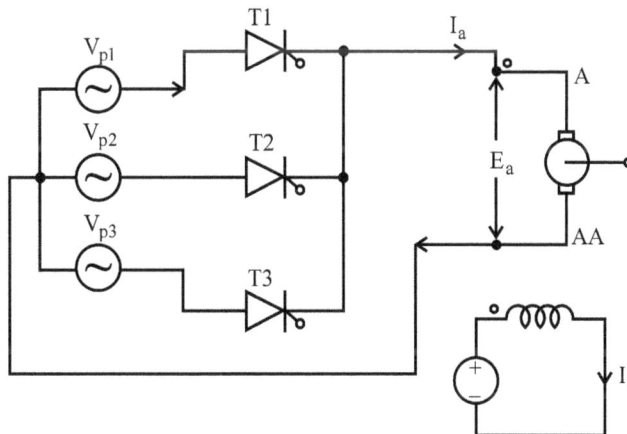

Fig. 2.1 Three phase half-wave converter connected to separately excited dc motor.

This converter is a three pulse converter. The current in the armature is from A to AA. This circuit is not generally used because of the dc components inherent in the line currents that tend to saturate the transformer from which currents are drawn. Assuming continuous conduction the average voltage applied to the armature is given by the equation

$$V_a = 0.675\ V_L \cos \alpha \qquad \qquad .....(2.1)$$

where V_L is line voltage

The waveforms of three phase voltages V_{p1}, V_{p2}, and V_{p3}, voltage applied to armature E_a and current through armature I_a and line current I_{L1} drawn from phase-1 are shown in Fig. 2.2.

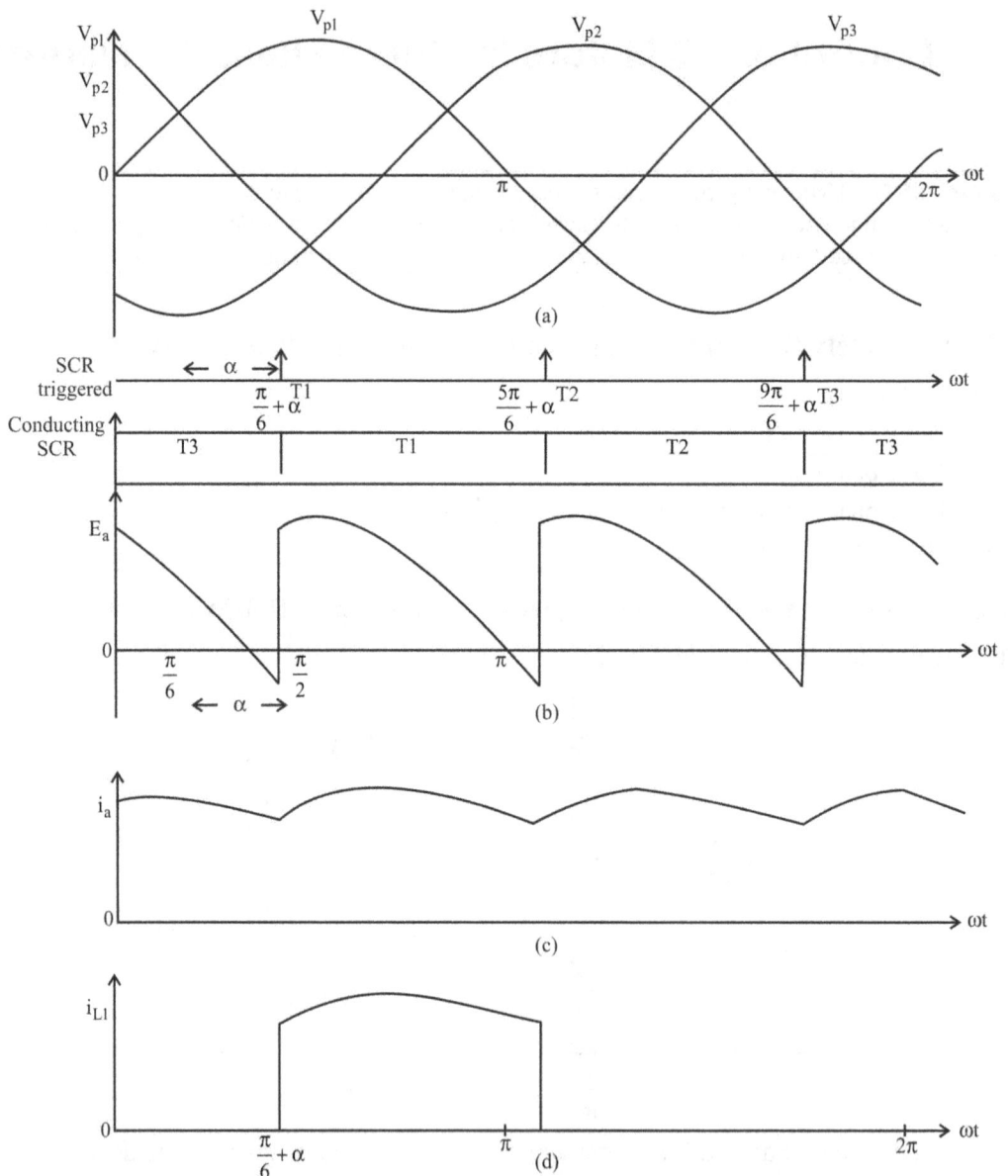

Fig. 2.2 Waveforms of (a) phase voltages, (b) armature voltage, (c) armature current and (d) line current.

2.1.2 Three Phase Half Controlled Bridge Converter Connected to Separately Excited DC Motor

The circuit diagram is shown in Fig. 2.3. The SCRs T1, T3, and T5 are triggered at 120^0 intervals in each cycle of input voltage. The waveforms of three Phase voltages V_{p1}, V_{p2} and V_{p3}, armature voltage E_a, armature current I_a and current I_{L1} drawn from line1 are shown in Fig. 2.4(a) for continuous current

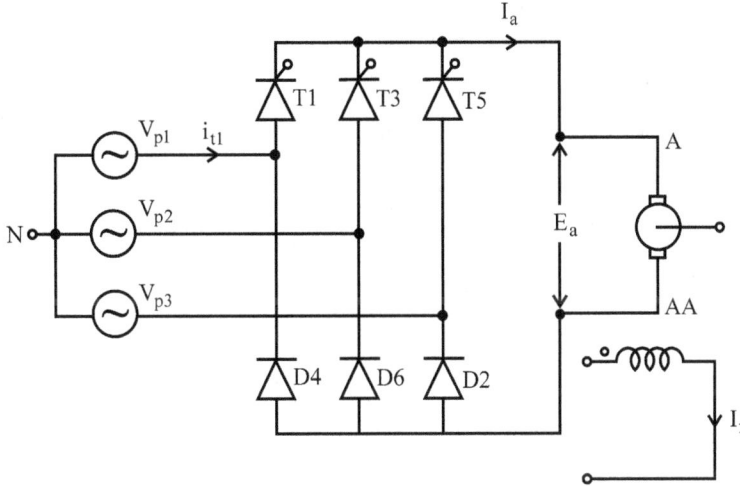

Fig. 2.3 Three phase half controlled bridge rectifier or semiconverter.

Assuming continuous conduction the average voltage applied to the armature E_a is derived as follows:

The three input phase voltages V_{p1}, V_{p2} and V_{p3} are assumed as

$$V_{p1} = \sqrt{2}\ V \sin \omega t$$
$$V_{p2} = \sqrt{2}\ V \sin(\omega t - 120^0)$$
$$V_{p3} = \sqrt{2}\ V \sin(\omega t - 240^0)$$

The output voltage has three symmetrical pulses of 120^0 duration.

One pulse is from $\omega t = \left(\dfrac{\pi}{6} + \alpha\right)$ to $\omega t = \left(\dfrac{5\pi}{6} + \alpha\right)$

From $\dfrac{\pi}{6} + \alpha$ to $\dfrac{\pi}{2}$ output voltage E_a is equal to $(V_{p1} - V_{p2})$.

From $\dfrac{\pi}{2}$ to $\dfrac{5\pi}{6} + \alpha$ output voltage E_a is equal to $(V_{p1} - V_{p3})$.

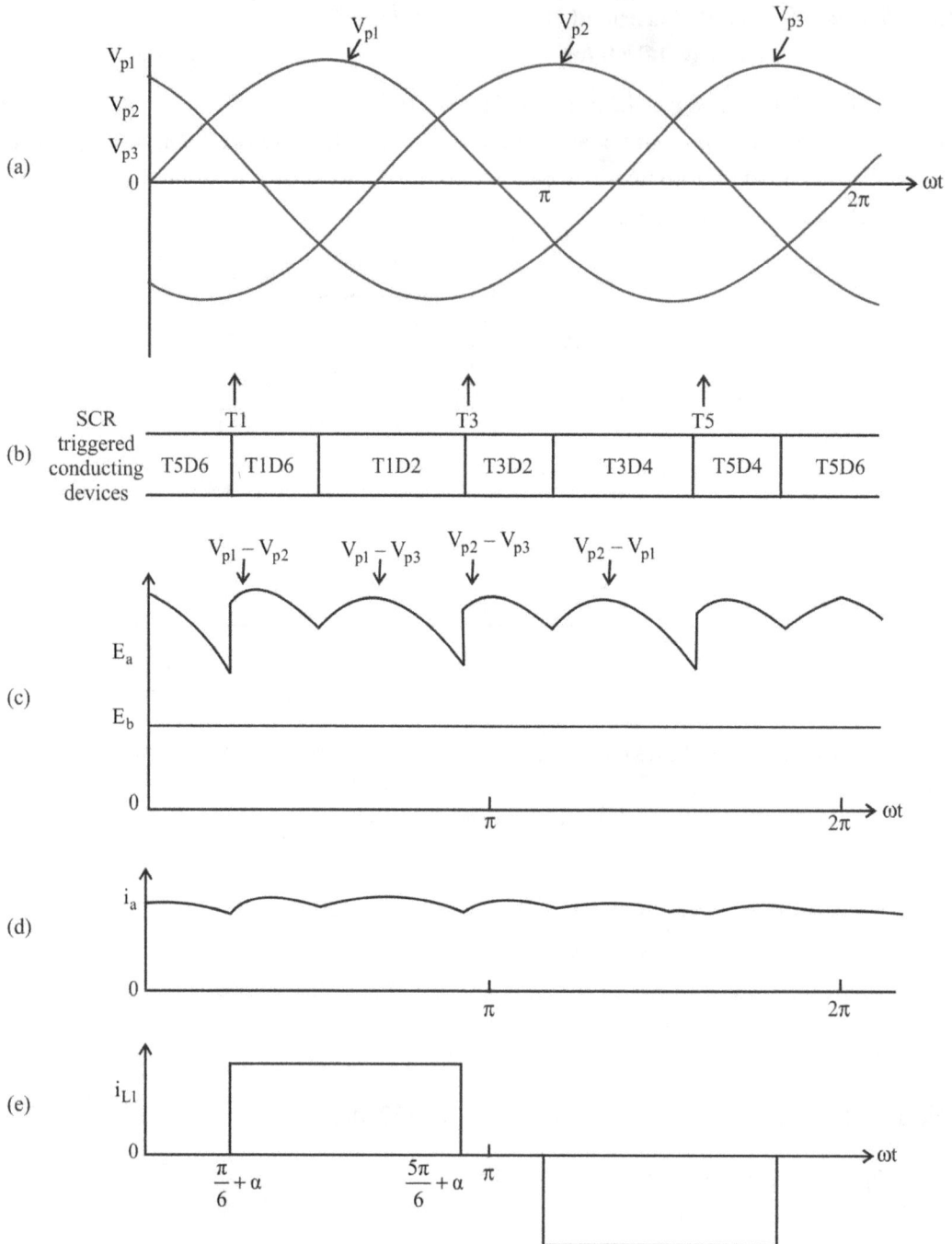

Fig. 2.4(a) Waveforms of continuous conduction (a) 3 phase voltages (b) conducting intervals of devices (c) voltage applied to armature & back emf (d) Armature current (e) line current.

The average value of this voltage pulse is taken.

$$E_a = \frac{3}{2\pi}[\int_{\frac{\pi}{6}+\alpha}^{\frac{\pi}{2}} \{\sqrt{2}\ V(\sin\omega t - \sin(\omega t - 120^0))d\omega t\} +$$

$$\int_{\frac{\pi}{2}}^{\frac{5\pi}{6}+\alpha} \{\sqrt{2}\ V(\sin\omega t - \sin(\omega t - 240^0))d\omega t\}]$$

$$= \frac{3\sqrt{2}\sqrt{3}V}{2\pi}(1+\cos\alpha)$$

$$= \frac{1.35V_L(1+\cos\alpha)}{2}$$

$$E_a = 0.675V_L(1+\cos\alpha) \qquad\qquad .....(2.2)$$

where V_L is line voltage and α is the firing angle

The current may be discontinuous depending on firing angle, back emf etc. The waveforms are shown in Fig. 2.4 (b). The average voltage E_a applied to armature can be calculated from firing angle and extinction angle.

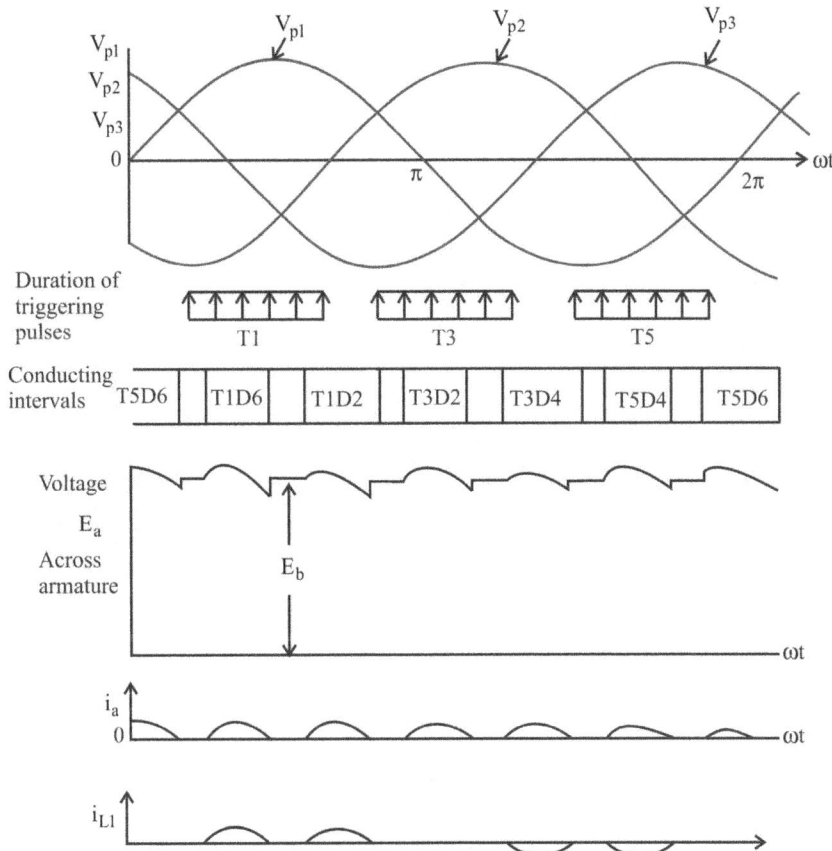

Fig. 2.4(b) Waveforms for discontinuous current mode of operation.

Speed-Torque expression for continuous current case

Under steady state condition the following equations hold good. The armature current is assumed as continuous.

$$0.675V_L(1 + \cos\alpha) = E_b + I_a R_a$$

$$E_b = K_1\omega\phi = K\omega$$

$$T_m = K_1 I_a \phi = KI_a$$

$$0.675V_L(1 + \cos\alpha) = K\omega + T_m R_a/K \qquad\qquad .....(2.3)$$

$$\omega = \frac{0.675V_L(1+\cos\alpha)}{K} - \frac{T_m R_a}{K^2} \qquad\qquad .....(2.4)$$

where ω is speed in radians per second and T_M is torque developed in Newton-meters. The torque–speed characteristics are shown in Fig. 2.5.

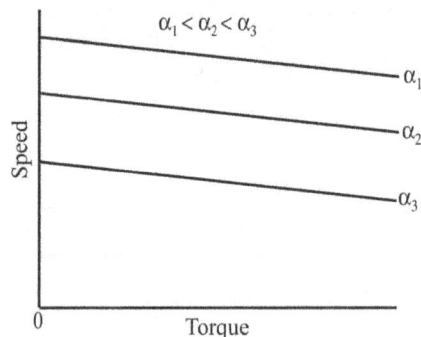

Fig. 2.5 Torque speed characteristics for different firing angles.

2.1.3 Three Phase Fully Controlled Bridge Converter Connected to Separately Excited dc Motor

The circuit diagram is shown in Fig. 2.6.

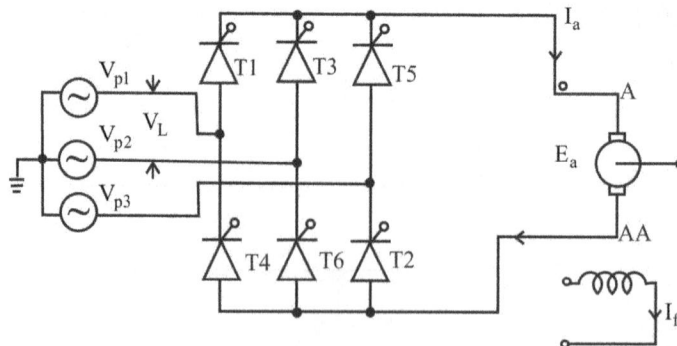

Fig. 2.6 Three phase fully controlled bridge converter connected to separately excited dc motor.

The thyristors are triggered at 60^0 intervals in serial order T1 to T6 in each cycle of input voltage. The waveforms of three phase voltages V_{p1}, V_{p2}, V_{p3}, voltage applied to armature E_a, armature current I_a, and ac line current I_{L1} are shown in Fig. 2.7 for a firing angle α. The armature current I_a is assumed as continuous.

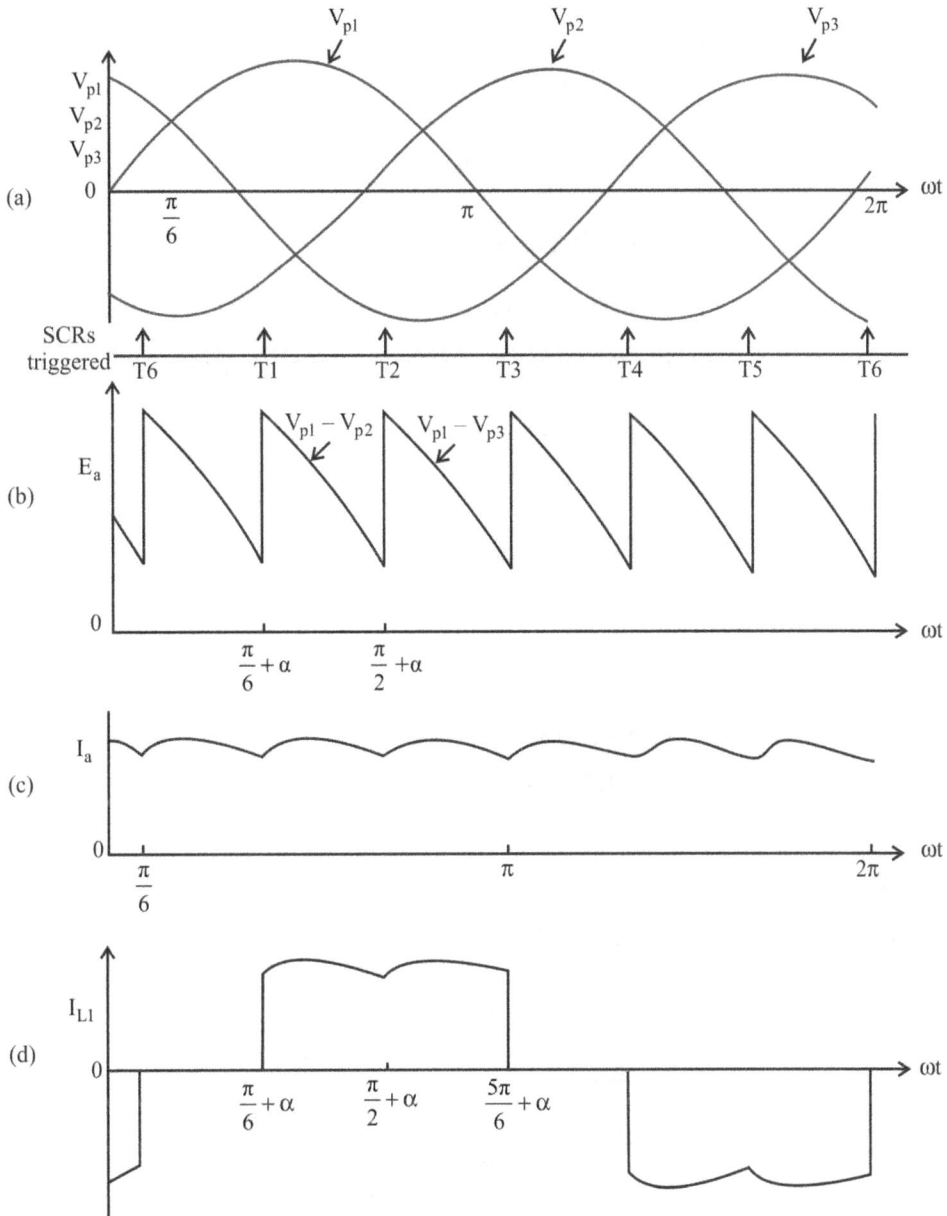

Fig. 2.7 Waveforms of (a) phase voltages (b) motor terminal voltage (c) armature current and (d) line current.

Assuming continuous conduction the average voltage applied to armature is obtained by taking the average value of output voltage over the period from $\left(\dfrac{\pi}{6}+\alpha\right)$ to $\left(\dfrac{\pi}{2}+\alpha\right)$.

$$Ea = \frac{3}{\pi}\int_{\frac{\pi}{6}+\alpha}^{\frac{\pi}{2}+\alpha} \sqrt{2}\ V[\sin\omega t - \sin(\omega t - 120^0)]d(\omega t),$$

$$= \frac{3\sqrt{2}\ V_L}{\pi\Pi}\cos\alpha$$

$$= 1.35 V_L\cos\alpha$$

where V_L is rms value of line voltage   .....(2.5)

If I_a is the average armature current, rms value of line current is

$$I_{Lrms} = 0.8615\ I_a \qquad\qquad .....(2.6)$$

$$\text{Power factor} = E_a I_a/\sqrt{3}V_L I_{Ll} = 0.9546\cos\alpha \qquad\qquad .....(2.7)$$

$$1.35 V_L\cos\alpha = E_b + I_a R_a \qquad\qquad .....(2.8)$$

$$E_b = K_1\omega\phi = K\omega$$

$$T_m = K_1 I_a\phi = K I_a$$

$$1.35\ V_L\cos\alpha = K\omega + T_m R_a/K \qquad\qquad .....(2.9)$$

$$\omega = \frac{1.35 V_L\cos\alpha}{K} - \frac{T_m R_a}{K^2} \qquad\qquad .....(2.10)$$

The torque speed curves for different values of firing angles are shown in Fig. 2.8. The firing angle can be varied from 90^0 to 0^0 only for motor operation.

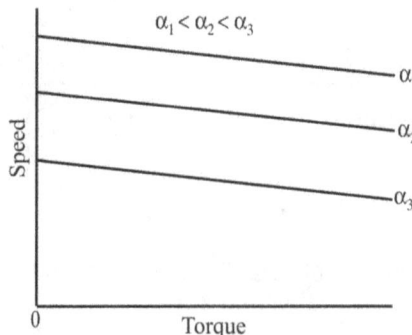

Fig. 2.8 Torque speed characteristics.

2.2.1 dc Series Motor Connected to 3φ Half-controlled Bridge Rectifier

The circuit diagram is shown in Fig. 2.9. S1-S2 is series field connected in series with armature A-AA. T1 is triggered at firing angle α. T1, T3 and T5 are triggered in order at 120^0 intervals in each cycle of input voltage.

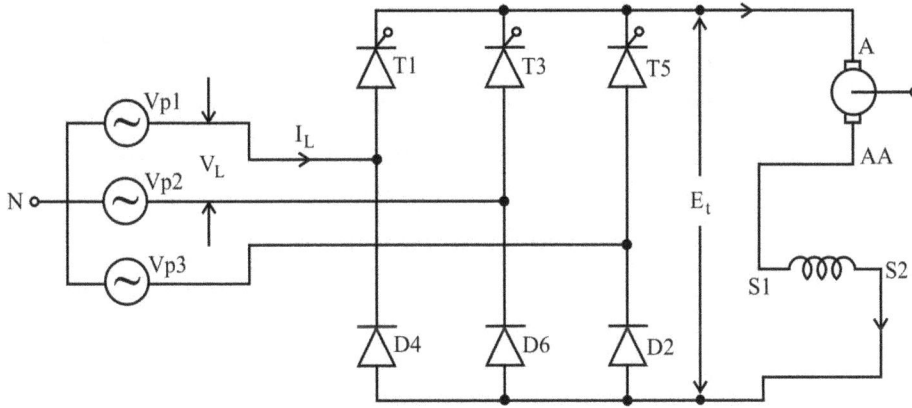

Fig. 2.9 Three phase half controlled bridge rectifier connected to D.C series motor.

The waveforms of voltages and currents are shown in Fig. 2.10. E_t is the voltage applied to motor.

The equivalent circuit is shown in Fig. 2.11. The current is assumed as ripple free. Under steady state condition the following equations hold good.

$$E_t = \frac{1.35V_L}{2}(1 + \cos \alpha) \qquad \qquad .....(2.11)$$

$$E_b + I_a(R_a + R_{se}) = E_t \qquad \qquad .....(2.12)$$

$$T_m = K\,I_a^{\,2} \qquad \qquad .....(2.13)$$

$$E_b = KI_a\omega \qquad \qquad .....(2.14)$$

$$\therefore \qquad \omega = \frac{1.35V_L(1 + \cos\alpha)}{2\sqrt{KT_m}} - \frac{R_a + R_{se}}{K} \qquad \qquad .....(2.15)$$

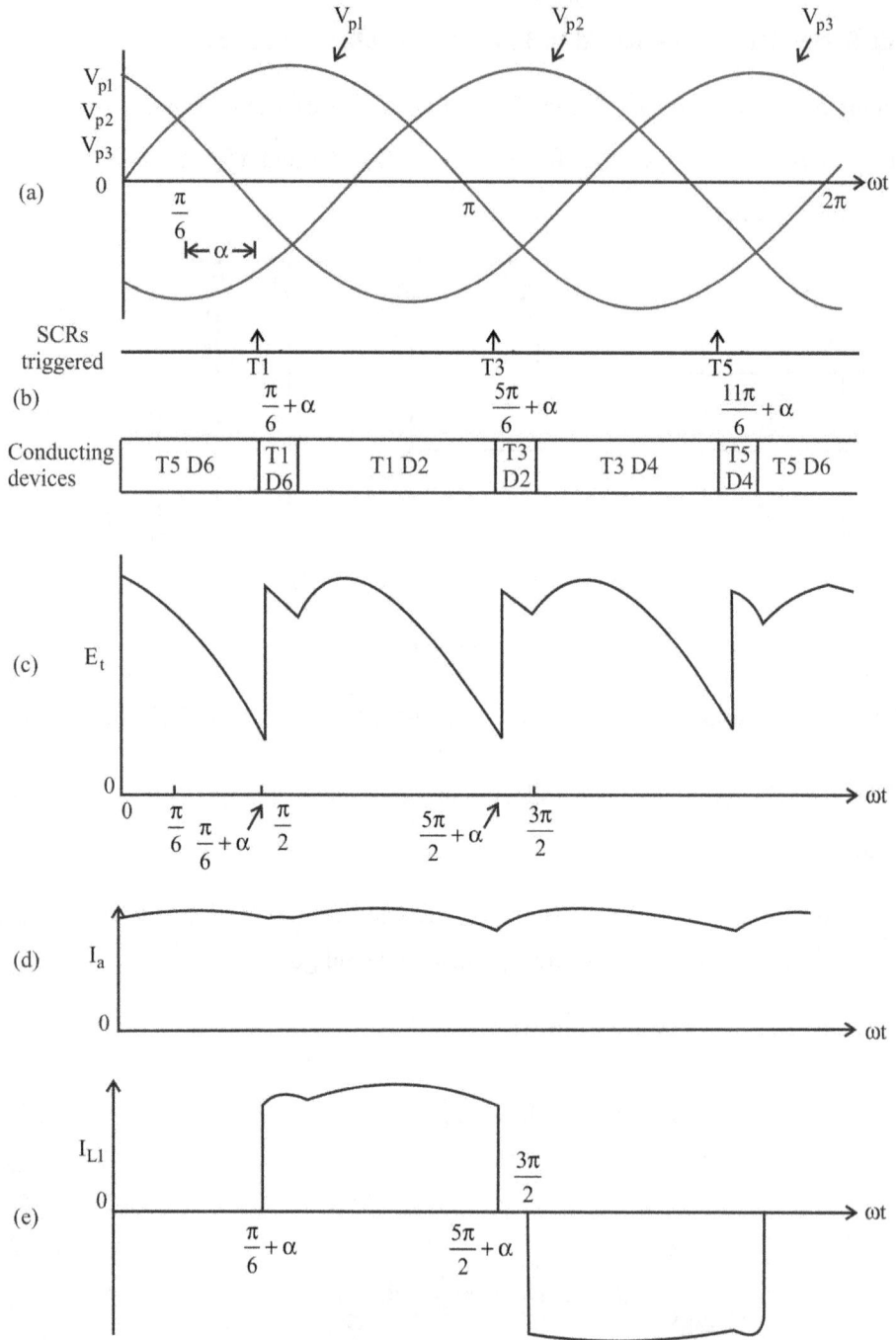

Fig. 2.10 Wave forms of (a) supply phase voltages (b) conducting intervals of devices (c) voltage applied to motor (d) armature current and (e) line current.

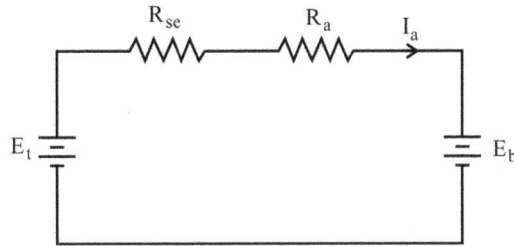

Fig. 2.11 Equivalent circuit.

The speed torque characteristics are shown in Fig. 2.12.

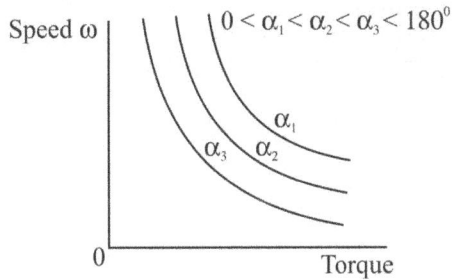

Fig. 2.12 Torque speed characteristics for different firing angles.

The firing angle α can be changed from 180^0 to 0^0 for controlling the speed of the motor.

2.2.2 DC Series Motor Connected to 3ϕ Fully Controlled Bridge Rectifier

The circuit diagram is shown in Fig. 2.13.

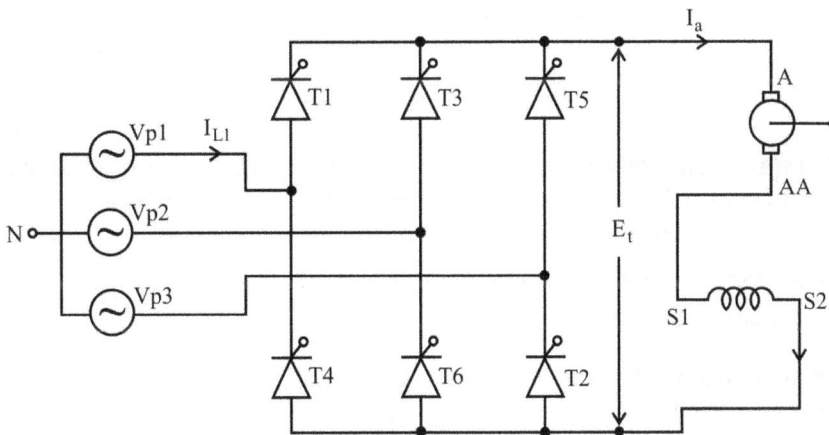

Fig. 2.13 Three Phase fully controlled bridge converter connected to dc series motor.

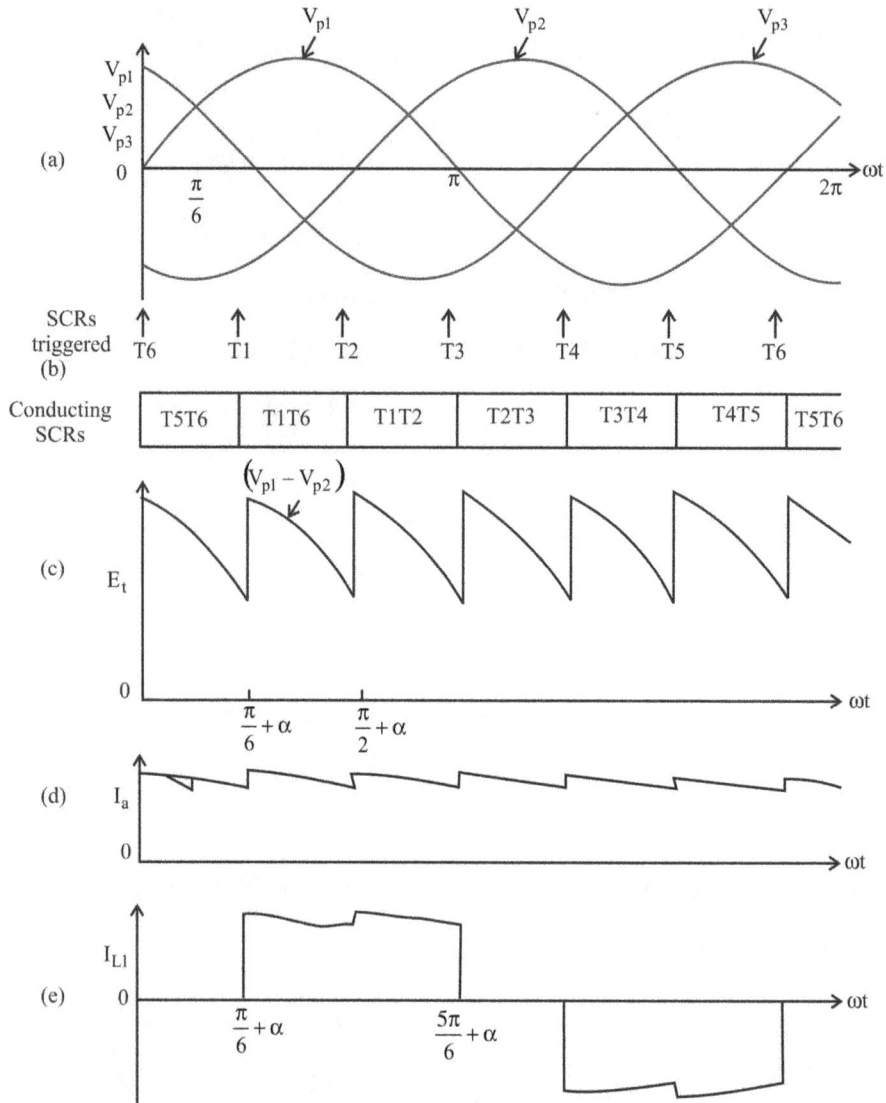

Fig. 2.14(a) Waveforms of 3ϕ supply voltages (b) conducting intervals of scrs (c) waveforms of voltage E_f applied to motor (d) waveforms of current supplied to motor (e) waveform of line current drawn from ac source.

The waveforms of voltages and currents for $\alpha = 30^0$ are shown in Fig. 2.14.

The equivalent circuit is shown in Fig. 2.15.

The waveforms of voltages and current are shown in Fig. 2.14 for $0 < \alpha < 90^0$. The current is assumed as constant. The average voltage applied to the motor terminals is given by $E_t = 1.35\ V_L \cos \alpha$, where V_L is the rms value of the line voltage of 3-phase supply.

The voltage is positive and the current is positive. The electrical energy is converted into mechanical energy and the dc machine acts as motor.

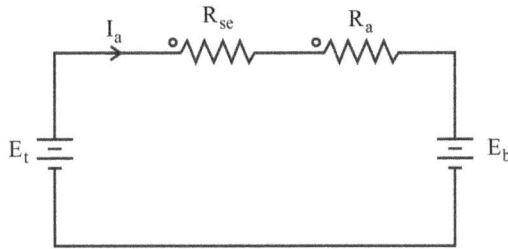

Fig. 2.15 Equivalent circuit steady state.

Under steady state condition the following equations hold good.

$$1.35\ V_L \cos \alpha = E_b + I_a (R_{se} + R_a) \qquad\qquad .....(2.16)$$

$$E_b = KI_a\omega \qquad\qquad .....(2.17)$$

$$T_m = KI_a^2 \qquad\qquad .....(2.18)$$

$$KI_a\omega + I_a (R_{se} + R_a) = 1.35\ V_L \cos \alpha.$$

$$\omega = [1.35\ V_L \cos \alpha - I_a(R_{se}+R_a)]/KI_a$$
$$= \frac{1.35 V_L \cos \alpha}{\sqrt{T_m K}} - \frac{R_{se} + R_a}{K} \qquad\qquad .....(2.19)$$

The torque speed characteristics for different values of α are shown in Fig. 2.16.

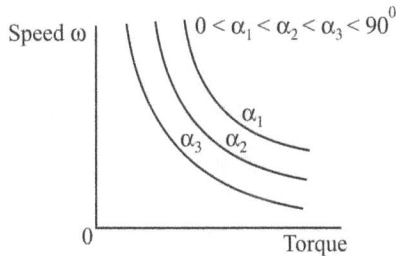

Fig. 2.16 Torque speed curves for different firing angles.

If the firing angle is greater than 90^0 and the current is assumed as constant the average output voltage is negative. Since the voltage is negative and current is positive, the power flow should be into the ac mains. This can happen only if the dc machine acts as generator and generates voltage greater than the terminal voltage V_t of the converter. This action is known as regenerative braking.

2.3   Power Factor Improvement

A Line commutated converter driving a dc motor is a Non-linear Load because the current drawn from the source contains harmonics.

Power factor = Displacement factor × Distortion factor.

For improving the pf of non-linear load both, displacement factor and distortion factor is to be improved.

Displacement factor can be improved by filtering out harmonics in the input current and voltage of non-linear loads.

Let us consider two examples for further clarification of these points.

1. A diode bridge rectifier always operates at unity displacement factor but its power factor is low due to low distortion factor. Therefore filters are connected at its input terminals to filter out harmonics.

2. A line commutated thyristor converter operates at low displacement factor and a low distortion factor especially at low loads. When its power rating is large, a static VAR Compensator is employed to improve displacement factor and filters are employed to improve distortion factor.

Harmonics produced by non-linear loads, apart from reducing power factor create the following.

(i) They interfere with other loads on the same line.

(ii) Produce Electromagnetic Interference.

(iii) Saturate and overheat transformers and motors on the same line. Increase Noise.

(iv) Overheating of capacitors leading to their failures.

(v) Malfunction of electronic equipment and Protective devices.

(vi) Increase the line losses.

Phase controlled converters (1Φ & 3Φ) are widely used because they are simple, less expensive, reliable and do not require any commutation circuit.

But the supply power factor in these converters is low when the out put voltage is low i.e., as the firing angle increases the power factor will become very low.

The various techniques for improving the power factor are:

1. Phase angle control (PAC)

2. Semi converter operation of full converter

3. Asymmetrical firing.

The above three methods use natural commutation. We can use forced commutation converters for improving the pf. Each thyristar is provided with its own commutation circuitry and therefore can be commutated at any desired instant. The following are the various control schemes.

1. Extinction angle control (EAC)

2. Symmetrical angle control (SAC)

3. Pulse width modulation (PWM)

Worked Out Examples

Example 2.1

A separately excited dc motor rated at 10 kW, 300 V, 1000 rpm is supplied with power from 3-phase half controlled bridge rectifier. The ideal 3 phase power supply is rated at 220 V, 50 Hz. This motor has armature resistance R_a = 0.2 Ω and sufficient added inductance to maintain continuous conduction. It delivers rated power at rated speed when α = 0°. If the firing angle α is retarded to 30°, calculate the speed, power factor and efficiency of operation if (a) the load torque is constant and (b) the load torque is proportional to speed.

Solution :

(a) Current is continuous

Hence when α = 0, the average output voltage of the converter

$$= \frac{1.35V_L}{2}(1+\cos\alpha)$$

$$= \frac{1.35 \times 220}{2}(1 + \cos 0°)$$

$$= 297 \text{ V}$$

Rated output = 10000 W

∴ $\quad\quad\quad (V - I_a R_a) I_a = 10000$

$\quad\quad\quad (297 - 0.2 I_a) I_a = 10000$

∴ $\quad\quad\quad I_a = 34.5 \text{ A}$

Full load armature current = 34.5 A

Rated speed = 1000 rpm = 104.72 rad/sec

Torque developed $= \dfrac{10000}{104.72} = 95.5 \text{ N-m}$

Now α = 30°

∴ Average voltage $= \dfrac{1.35 \times 220}{2}(1 + \cos 30°)$

$$= 277.1 \text{ V}$$

Since the torque is held constant $I_a = 34.5$ α

Back emf $E_b = V - I_a R_a = 277.1 - 34.5 \times 0.2$

$$= 270.19 \text{ V}$$

Back emf at N = 1000 rpm and α = 0° is

$$E_b = 297 - 34.5 \times 0.2 = 290.1$$

∴ New speed $= \dfrac{1000 \times 270.19}{290.1} = 931.36 \text{ rpm}$

Power factor :

RMS value of line current $I_L = I_a \sqrt{\dfrac{2}{3}}$

$$= 34.5 \sqrt{\dfrac{2}{3}}$$

$$= 28.16 \text{ A}$$

Apparent power input $= \sqrt{3} \, V_L I_L$

$$= \sqrt{3} \times 220 \times 28.16$$

$$= 10733 \text{ VA}$$

Active power input $= V_a I_a$

$$= 277.1 \times 34.5$$

$$= 9560 \text{ W}$$

$\therefore$ Power factor $= \dfrac{9560}{10733} = 0.89$ lag.

Efficiency :

$$\text{Efficiency} = \dfrac{\text{Motor power output}}{\text{Power input to motor}}$$

$$= \dfrac{E_b I_a}{V_a I_a} = \dfrac{270.19}{277.1}$$

$$= 0.9747 = 97.47 \text{ %}$$

Let N_3 be the speed, E_{b3} the back emf and I_{a3} be the armature current.

$$\dfrac{T_3}{T_1} = \dfrac{N_3}{N_1} = \dfrac{I_{a3}}{I_{a1}} = \dfrac{E_{b3}}{E_{b1}}$$

$N_1 = 1000$ rpm ; $I_{a1} = 34.5$; $T_1 = 95.5$ N.m

$E_{b1} = 290.1$ V; $E_{b3} = 277.1 - 0.2 \, I_3$

$\therefore$ $\dfrac{277.1 - 0.2 I_{a3}}{290.1} = \dfrac{I_{a3}}{34.5}$

$\therefore$ $I_{a3} = 32.19$

But $\dfrac{N_3}{N_1} = \dfrac{I_{a3}}{I_{a1}}$

$\therefore$ $\dfrac{1000 \times 32.19}{34.5} = 933$ rpm

Power factor :

$$I_L = I_a \sqrt{\frac{2}{3}} = 32.19 \times \sqrt{\frac{2}{3}} = 26.28A$$

Apparent power input $= \sqrt{3} \times 220 \times 26.28 = 10015$ VA

Active power input $= V_a I_a = 277.1 \times 32.19 = 8916.63$

$\therefore$ Power factor $= \dfrac{8916.63}{10015} = 0.89$ log.

$$\text{Efficiency} = \frac{E_b I_a}{V_a I_a} = \frac{E_b}{V_a} = \frac{277.1 - 0.2 \times 32.19}{277.1} = 0.9771$$

$$= 97.71\%$$

Example 2.2

A 80 kW, 440 V, 800 rpm dc motor is operating at 600 rpm developing 75% rated torque is controlled by 3φ, 6 pulse thyristor converter. If the back emf at rated speed is 410 V, determine the triggering angle of the converter. The input to the converter is 3 φ, 415 V, 50 Hz. ac supply.

Solution :

Rated armature current $I_a = \dfrac{80 \times 1000}{440} = 181.8A.$

Back emf E_{b1} at rated speed $= 410$ V

$\therefore$ $440 - I_a R_a = 410$ V

$\therefore$ $I_a R_a = 30v$

$$R_a = \frac{30}{181.8} = 0.165 \ \Omega$$

Motor is operating at 600 rpm.

Back emf E_{b2} at 600 rpm $= \dfrac{410 \times 600}{800} = 307.5$ V

At 75% of rated torque armature current $= 0.75 \times 181.8$

$$= 136.35$$

Voltage applied to armature $= 307.5 + 136.35 \times 0.165$

$$= 330 \text{ V}.$$

$\therefore$ $1.35 V_L \cos\alpha = 330$

$$\cos\alpha = \frac{330}{1.35 \times 415} = 0.589$$

$$\alpha = 53.91°$$

Example 2.3

A 220 V, 600 rpm, 500 A, separately excited dc motor has armature and field resistances of 0.02 Ω and 10 Ω respectively. Armature is fed from a 3 phase fully controlled rectifier and field from half-controlled single phase rectifier. A three phase three wire ac source with a line voltage of 440 V is available. Armature rectifier is fed from a three phase transformer with Y– Δ connection and field from a single phase transformer.

(a)   output voltages of transformers must be such that for zero firing angles rated voltages are maintained across the motor armature and field. Calculate the transformer turns rations.

(b)   with the transformer turns ratio as in (a) calculate firing angles of the armature rectifier for rated torque and field and 400 rpm. Assume continuous conduction.

Solution :

The circuit diagram is given below

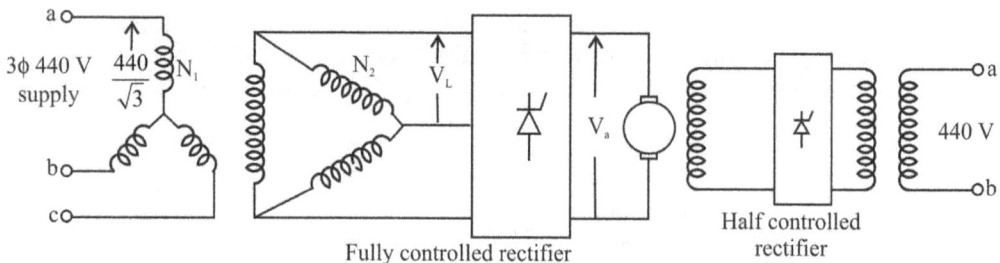

Fully controlled rectifier   Half controlled rectifier

Rated dc voltage of armature V_a = 220 V

This we should get for α = 0

We know that for 3ϕ, f.c. converter $V_a = 1.35 V_L \cos α$.

∴ $1.35 V_L \cos 0° = 220$

$V_L = 163$ V

Since the transformer is Δ connected on secondary, line voltage = phase voltage = 163 V.

$$\text{Phase voltage on the primary} = \frac{440}{\sqrt{3}} = 254 \text{ V.}$$

$$\text{Turns ratio} = \frac{\sec \text{ondary turns}}{\text{primary turns}} = \frac{163}{254} = 0.64$$

Field circuit

(a)   Rated voltage = 220 V

For single phase half controlled converter

$$V_{dc} = \frac{\sqrt{2}V}{\pi} (1 + \cos α)$$

$$220 = \frac{\sqrt{2} \times 2V}{\pi}$$

$$V = \frac{220 \times \pi}{2\sqrt{2}} = 244 \text{ V.}$$

$$\text{Turns ratio} = \frac{244}{440} = 0.554$$

(b)　Rated voltage = 220 V

Rated speed = 600 rpm

Rated current = 500 A

Back emf at rated speed of 600 rpm = 220 – 500 × 0.02 = 210 V.

$$\text{Back emf at 400 rpm} = \frac{210 \times 400}{600} = 140 \text{ V.}$$

Armature voltage at this speed and rated torque

$$V_a = 140 + 500 \times 0.02 = 150 \text{ V.}$$

$$\cos \alpha = \frac{150}{1.35 \times 163} = 0.681$$

$$\alpha = 47.02°$$

Example 2.4

A separately excited dc motor rated at 10 kw, 300 V, 1000 rpm is supplied with power from a fully controlled three phase bridge rectifier. The ideal 3φ power supply is rated at 220 V, 50 Hz. This motor has an armature resistance $R_a = 0.2 \ \Omega$ and sufficient added inductance to maintain continuous conduction. It delivers rated power at rated speed at $\alpha = 0°$. If the firing angle is retarded to $\alpha = 30°$, calculate the speed, power factor and efficiency of operation if the load torque is constant.

Solution :

Average output voltage at $\alpha = 0°$ is given by

$$E_a = 1.35 V_L \cos 0° = 1.35 \times 220 = 297 \text{ volts.}$$

Power output = 10,000 w

$$E_b I_a = 10000$$

$$(E_a - I_a R_a) I_a = 10000$$

$$297 \ I_a - 0.2 I_a^2 = 10000$$

$$0.2I_a^2 - 297I_a + 1000 = 0$$

$$\therefore \qquad I_a = 34.5 \text{ A}$$

$$E_b = 297 - 0.2 \times 34.5 = 290.1 \text{V}$$

Firing angle $\alpha = 30°$

Average voltage applied to armature $= 1.35 \times 220 \times \cos 30°$

$$= 257.2 \text{ V}$$

Since the torque is constant $I_a = 34.5$ A

New back e.m.f $= 257.2 - 0.2 \times 34.5$

$$= 250.3 \text{ V}$$

Back emf is proportional to speed.

$$\therefore \qquad \text{speed} = \frac{1000 \times 250.3}{290.1}$$

$$= 862.8 \text{ rpm.}$$

The wave form of line current is given below. It will be rectangular wave with height of 34.5A

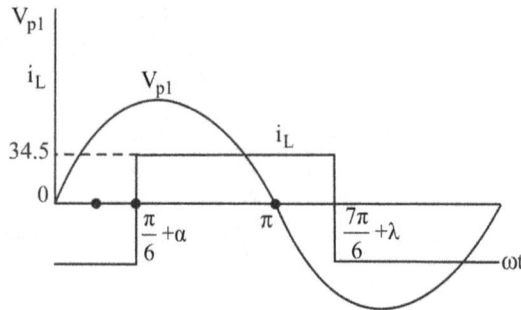

rms value of line current $= 34.5$ A

line voltage $= 220$ V

Apparent power input $= \sqrt{3} \times 220 \times 34.5$

$$= 13146 \text{ VA}$$

Active power input $= 257.2 \times 34.5$

$$= 8873\text{w.}$$

$$\therefore \qquad \text{power factor} = \frac{8873}{13146} = 0.675 \text{ lag}$$

$$\% \text{ Efficiency} = \frac{\text{power output}}{\text{power input}} \times 100 = \frac{E_b I_a}{E_a I_a} \times 100$$

$$= \frac{E_b}{E_a} \times 100 = \frac{250.3}{257.2} \times 100 = 97.3\%$$

Example 2.5

A three phase full converter is used to control the speed of a 220 V, 3.73 kw, 1200 rpm dc separately excited motor. The ac supply is 240 V, 50 Hz. The motor emf constant is 1.7 Vs/rad. The armature resistance is 1.5 ohms. For $\alpha = 60°$, the motor speed is 800 rpm, determine,

    (i)   the average value of motor current, assuming it to be ripple free.

    (ii)   the rms value of the thyristor current and supply line current and

    (iii) the supply power factor

Solution :

The circuit diagram is shown below.

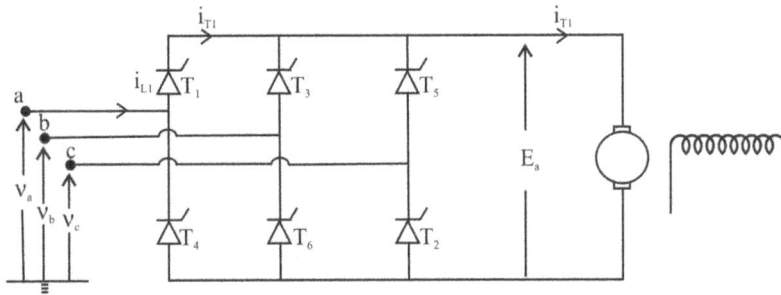

$$V_{ab} = V_{bc} = V_{ca} = 240v$$

$$T_1 \text{ is triggered at wt} = \frac{\pi}{6} + \alpha$$

$$= 30 + 60 = 90°$$

    (i)   For $\alpha = 60°$, the average voltage E_a applied to the armature is given by

$$E_a = 1.35 V_L \cos \alpha = 1.35 \times 240 \times \cos 60° = 162 \text{ V}$$

Back emf at 800 rpm $E_b = \dfrac{800 \times 2\pi \times 1.7}{60} = 142.41 \text{ V}$

Armature current $I_a = \dfrac{E_a - E_b}{R_a} = \dfrac{162 - 142.41}{1.5}$

$$= 13.06 \text{ A}$$

    (ii)   For full converter each thyristor conducts for $\dfrac{2\pi}{3}$ rad in each cycle of 2π radius.

The current through thyristor is equal to I_a.

$$\text{RMS value of thyristor current } I_T = I_a \sqrt{\frac{2\pi}{3} \frac{1}{2\pi}}$$

$$= \frac{13.06}{\sqrt{3}} = 7.5 \text{ A}$$

Supply line current will be a.c with rectangular waveforms as shown in figure.

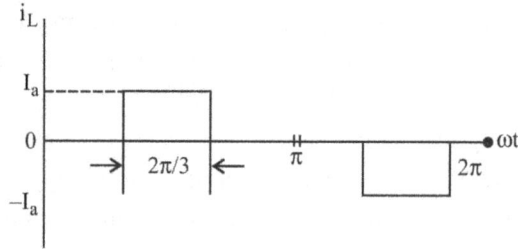

$$\text{RMS value of line current} = I_a \sqrt{\frac{2\pi}{3} \frac{1}{\pi}} = I_a \sqrt{\frac{2}{3}}$$

$$= 13.06 \sqrt{\frac{2}{3}} = 10.61 \text{Amps.}$$

(iii) Apparent power input $= \sqrt{3} \, V_L I_L$

$$= \sqrt{3} \times 240 \times 10.61$$

$$= 4410.5 \text{VA}$$

Active power input $= E_a I_a = 162 \times 13.06 = 2115.7$ VA

$$\therefore \text{ Power factor} = \frac{2115.7}{4410.5} = 0.48 \text{ lag}$$

***Example* 2.6**

The speed of a 25 h.p, 380 V, 1800 rpm separately excited dc motor is controlled by a 3 phase full converter and the field current is set to the maximum possible value. The ac input is a 3ϕ star connected 210 V, 50 Hz supply. The armature and field parameters are $R_a = 0.15 \ \Omega$, $R_f = 250 \ \Omega$ and $K_a = 1.2$ V/A rad/sec. Assume that the armature and field currents are continuous and ripple free. By neglecting the system losses determine the following :

(a) the firing angle of the armature converter α , if the motor supplies the rated power at rated speed.

(b) the no load speed, if the firing angles are the same as in part (a) and the armature current at no load is 10% of the rated value and

(c) the speed regulation.

Solution :

(a) The motor supplies rated power. i.e., 25 h.p.

$\therefore$ power output $= 25 \times 746 = 18650$ w.

$E_b I_a = 18650$; speed $= 1800$ rpm $= 188.49$ rad/sec.

Applied voltage to field circuit $= 1.35 \times 210 \times \cos 0^\circ$

$$= 283.5 \text{ V}$$

$$\text{Field current} = \frac{V_f}{R_f} = \frac{283.5}{250} = 1.134 \text{ A.}$$

$K_a = 1.2$ V/A rad/sec; $E_b = K_a I_f \omega$

$\therefore$ Back emf $E_b = 1.2 \times 1.134 \times 188.49$

$= 256.5$ V

$E_b I_a = 18650$

$\therefore$ $I_a = \dfrac{18650}{256.5} = 72.7$ A

Voltage applied to armature $E_a = E_b + I_a R_a$

$$= 256.5 + 72.7 \times 0.15$$

$$= 267.4 \text{ V}$$

$1.35 V_L \cos \alpha = E_a$

$$\cos \alpha = \frac{267.4}{1.35 \times 210} = 0.943$$

$\alpha = 19.4^\circ$

(b) No load armature current $= \dfrac{10}{100} \times 72.7 = 7.27$A

now back emf $E_b = E_a - I_a R_a$

$$= 267.4 - 7.27 \times 0.15$$

$$= 266.3$$

$$\text{Speed} = \frac{266.3}{1.2 \times 1.134} \times \frac{60}{2\pi} = 1867.73 \text{ rpm.}$$

(c) Speed regulation $= \dfrac{\text{No load speed} - \text{full load speed}}{\text{full load speed}} \times 100$

$$= \frac{1867.73 - 1800}{1800} \times 100$$

$$= 3.84\%$$

Example **2.7**

The speed of a 150 h.p, 650 V, 1750 rpm separately excited dc motor is controlled by 3 phase full converter. The converter is operating from 3 phase 460 V, 50 Hz, supply. The rated armature current of the motor is 170 A. The motor parameters are R_a = 0.099 Ω, L_a = 0.73 mH and $K_a\phi$ = 0.33 V/rpm. Neglect losses in the converter system. Determine.

(a) No load speeds at firing angles α = 30°. Assume that at no-load, the armature current is 10% of the rated current and is continuous.

(b) The firing angle to obtain rated speed of 1750 rpm at rated motor current. Also compute the supply power factor

(c) The speed regulation for the firing angle obtained in part (b).

Solution :

(a) *No load condition* :

No load current = 0.1×170

$$= 17 \text{ A}$$

When $\alpha = 0°$; the voltage applied to armature E_a

$$E_a = 1.35 \times 460 \times \cos 0° = 621.22 \text{ V}$$

$$E_b = E_a - I_a R_a = 621.22 - 17 \times 0.099 = 619.5 \text{ V}$$

$$\text{Speed} = \frac{E_b}{K_a\phi} = \frac{619.5}{0.33} = 1877.4 \text{ rpm}$$

For $\alpha = 30°$ $E_a = 1.35 \times 460 \times \cos 30° = 537.99 \text{ V}$

$$E_b = E_a - I_a R_a = 537.99 - 17 \times 0.099 = 536.3 \text{ V}$$

$$\text{No load speed} = \frac{E_b}{K\phi} = \frac{536.3}{0.33} = 1625.17 \text{ rpm}$$

(b) Full load condition :

Speed = 1750 rpm

$$E_b = K_a\phi N = 0.33 \times 1750 = 577.5 \text{ V}$$

Rated motor current = 170 Amps

$\therefore$ $E_a = E_b + I_a R_a = 577.5 + 170 \times 0.099 = 594.33 \text{V}$

$$1.35 V_L \cos \alpha = E_a$$

$$\cos \alpha = \frac{594.33}{1.35 \times 460} = 0.957$$

$\therefore$ $\alpha = 16.9°$

rms value of line current $= I_a \sqrt{\dfrac{2}{3}}$

$$= 170 \sqrt{\dfrac{2}{3}} = 138.8 \text{ A}$$

Apparent power input $= \sqrt{3} \, V_L \, I_L$

$$= \sqrt{3} \times 460 \times 138.8 = 11.591$$

Active power input $= E_a I_a = 594.33 \times 170$

Supply power factor $= \dfrac{E_a I_a}{\sqrt{3} V_L I_L} = \dfrac{594.33 \times 170}{\sqrt{3} \times 460 \times 138.8}$

$$= 0.91 \text{ lag.}$$

(c) **Speed regulation :** At full load condition $I_a = 170$ A and speed $= 1750$ rpm. If the load on the motor is thrown off the I_a becomes 17 A. For the same firing angle of $16.92°$ the terminal voltage remains at 594.3 V.

The back emf will be $E_b = 594.33 - 17 \times 0.099 = 592.65$

$\therefore$ The no load speed $= \dfrac{592.65}{0.33} = 1795.9$ rpm

Speed regulation $= \dfrac{1795.9 - 1750}{1750} \times 100$

$$= 2.62\%$$

Example 2.8

The speed of a separately excited dc motor is controlled by means of a 3 phase semi converter from a 3 phase, 415V, 50 Hz supply. The motor constants are inductance 10 mH, resistance 0.9 ohms and armature constant 1.5V/rad/sec (Nm/A). Calculate the speed of this motor at a torque of 50 N-m when the converter is fired at 45^0. Neglect losses in the converter

Solution :

We know that $E_b = K_m \omega$

$$T = K_m I_a$$

Armature current $I_a = \dfrac{50}{1.5} = 33.33$ A

For $\alpha = 45°$ the average voltage applied to armature E_a

$$E_a = \dfrac{1.35 V_L}{2} (1 + \cos \alpha) = \dfrac{1.35 \times 415}{2} (1 + \cos 45°)$$

$$= 478.2 \text{ volts.}$$

$$E_b = E_a - I_a R_a = 478.2 - 33.3 \times 0.9 = 448.23 \text{ V}$$

Speed

$$\omega = \frac{E_b}{K_m} = \frac{448.23}{1.5} = 298.8 \text{ rad/sec}$$

$$N = \frac{298.8 \times 60}{2\pi} = 2853.5 \text{ rpm}$$

Problems

P.2.1   A 600V, 1500 rpm, 80A separately excited dc motor is fed through a three-phase semiconverter from a 3 phase 400 V, 50 Hz, supply. Armature resistance $R_a = 1 \ \Omega$. Armature current is assumed as constant.

For a firing angle of 45° at 1200 rpm compute the rms values of source and thyristor currents, average value of thrysitor current and input supply power factor.

> *Ans :*
>
> > Rms value of source current = 36.8 A
> >
> > Rms value of thyristor current = 26 A
> >
> > Average value of thyristor current = 15.02 A
> >
> > Input supply p.f = 0.815 lag.

P.2.2   A 100 KW, 500V, 200 rpm separately excited dcmotor is energized from 400V, 50 Hz, 3 phase source through a 3ϕ full converter. The voltage drop in conducting thyristors is 2 V. The dc motor parameters are $R_a = 0.1 \ \Omega$, $K_m = 1.6$ V-S/rad, $L_a = 8$ mH. Rated armature current = 210 A. No load armature current is 10% of rated current. Armature current is continuous and ripple free.

Find the no load speed at firing angle of 30°.

Find the firing angle for a speed of 2000 rpm at rated armature current. Determine also the supply power factor.

Find the speed regulation for the firing angle obtained in (b).

> *Ans :*   (a) 2767 rpm   (b) 48.47°; 0.633 lag (c) 5.64%

P2.3   A 230 V, 10 Kw, 1000 rpm separately excited dcmotor has its armature resistance of 0.3Ω and field resistance of 300 Ω. The speed of this motor is controlled by two 3 phase full conveter one in the armature circuit and the other in the field circuit and both are fed from 400 V, 50 Hz source. The motor   constant is 1.1 V-S/rad. Armature and field currents are ripple free.

(a)   With field converter setting to maximum field current calculate firing angle for the armature converter for load torque of 60 Nm at rated speed.

(b)   With the load torque as in part (a) and zero degree firing angle for armature converter, speed is to be raised to 3000 rpm. Determine the firing angle of the field converter.

> *Ans :* (a) 66.37°   (b) 31.4°

P2.4   (a)   Derive an expression for the average output voltage of a 3 ϕ, half wave converter and draw the Wave forms of output voltage for a firing angle α.

        (b)   What is the effect of source inductance on the average output voltage.

P2.5   Explain the principle of working of a 3 ϕ full converter supplying a highly inductive load with a neat circuit diagram and waveforms of output voltage and output current.

P2.6   (a)   Derive an expression for the average output voltage of a 3 ϕ full converter

        (b)   What is the frequency of the lowest order harmonic in the 3-ϕ full converters?

P2.7   A 3 ϕ half-wave bridge comprising three thyristors is fed from a 277 rms, line to neutral, 60 HZ supply and provides an adjustable dc voltage at the terminals of a separately excited dc motor. The motor specifications are R_a = 0.02 Ω. L_a = 0.01H, E_a = 1.2 ωm and full load I_a = 500A. Find the firing angle α so that the motor operates at full load current and at the rated speed of 200 rad/sec.

Assume continuous conduction and neglect thyristor forward voltage drop.

Objective Type Questions

1.   A 3-ϕ half controlled converter is connected to 400 V, 3-ϕ, 50 Hz ac supply is delivering 10 A to dc motor. If the firing angle α = 30^0 the power input to the motor is

    (a)   6000W   (b)   5100W

    (c)   5020W   (d)   5038W   [   ]

2.   A 3-ϕ fully controlled bridge rectifier is connected to 440V ac supply. The load current is continuous. If the firing angle α = 60^0, the average output voltage is ....

    (a)   400 V   (b)   300 V

    (c)   297 V   (d)   279V   [   ]

3.   A 3-ϕ half controlled converter connected to 400 V, 3-ϕ supply is delivering 20 A continuous current with firing angle α = 60^0 .

   The rms value of the line current is

    (a)   20 A   (b)   15 A

    (c)   16.33 A   (d)   8.16 A   [   ]

4.   A 3-ϕ half wave converter connected to 3-ϕ supply with the line voltage of V_L and operating with firing angle α is delivering continuous current to a dc load. The average voltage is

    (a) $1.35 V_L \cos \alpha$ (b) $\dfrac{1.35}{1.352} V_L \cos \alpha$

    (c) $1.35 V_L (1 + \cos\alpha)$ (d) $\dfrac{1.352}{2} (1 + \cos\alpha)$ [   ]

5. A 3-ϕ half controlled bridge rectifier is

    (a)   Full converter   (b)   3 Pulse Converter

    (c)   Having 6 SCRs   (d)   Two quadrant converter   [   ]

6. A 3-ϕ fully controlled converter is

    (a)   Single quadrant converter   (b)   Six pulse converter

    (c)   3 Pulse converter   (d)   Having 3 SCRs & 3 diodes [   ]

7. A Separately excited dc motor rated at 10 KW, 300 V, 1000 rpm is supplied with power from 3-ϕ, half controlled bridge rectifier which is connected to ideal 3-ϕ, 220 V, 50 Hz supply. The motor has armature resistance of 0.2 Ω. The motor draws continuous current and delivers rated power at rated speed when $\alpha = 0^0$. What is the armature current?

    (a)   1450   (b)   34.47

    (c)   40.5   (d)   30.5   [   ]

8. A 220 V, 1500 rpm, 50 A separately excited dc motor is fed from a 3-ϕ, fully controlled bridge rectifier. Available A.C source has line voltage of 440 V, 50 Hz. A star-delta connected transformer is used to feed the armature so that motor terminal voltage equals rated voltage when converter-firing angle is zero. The transformer ratio should be

    (a)   1   (b)   0.5

    (c)   0.641   (d)   0.461   [   ]

9. A line commutated converter driving a dc motor can be consider as

    (a)   Linear load

    (b)   Non-linear load

    (c)   Linear load at low speeds & Non –linear at high speeds

    (d)   Linear load at high torques & Non linear at low torques.   [   ]

3

Four Quadrant Operation of DC Drives

Four quadrant operation of DC drives : Introduction to four quadrant operation –Motoring operations, Electric braking:- Plugging, Dynamic and Regenerative braking operations – Four quadrant operation of DC motors by dual converter– Closed loop operation of DC motor (Block diagram only)

3.1 Introduction to Four Quadrant Operation

The operation of drive by convention is shown in four quadrants as in Fig. 3.1. If the motor is rotating, say, in clockwise direction its operation is in I and II quadrants and if the rotation is in anticlockwise direction its operation is in III and IV quadrants. X-axis is the torque and Y axis is speed. In the first quadrant and third quadrant both torque and speed are in the same direction and the electric machine is operating as motor. In second and fourth quadrants the torque is opposing motion and this operation is called electric braking. Second quadrant operation is called forward braking and fourth quadrant operation is known as reverse braking.

When operating in second and fourth quadrants the machine is operating as generator, converting mechanical energy into electrical energy.

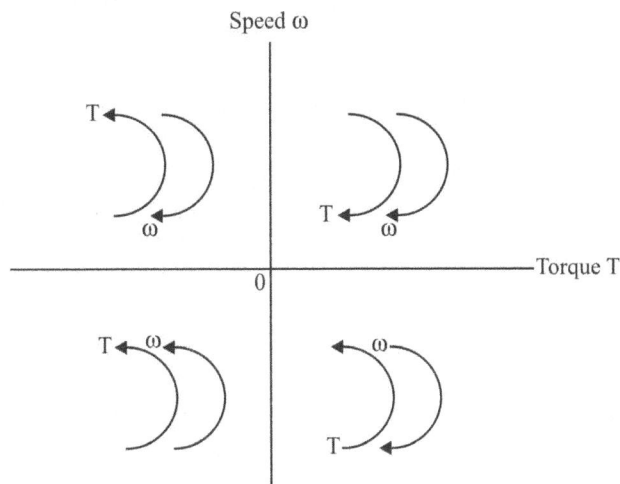

Fig. 3.1 Torque-speed characteristic of DC motor.

If a DC machine is considered for four quadrant operation generally separately excited machine is used. Field current is kept constant and armature voltage is varied.

(a)   First quadrant operation is called forward motoring.

(b)   Second quadrant operation is called forward braking.

(c)   Third quadrant operation is called reverse motoring.

(d)   Fourth quadrant operation is called reverse braking.

3.2   Electrical Braking of Motors

There are three types of electrical braking. They are:

(a)   Dynamic or Rheostatic braking.

(b)   Plugging or counter current braking.

(c)   Regenerative braking.

(a)   ***Dynamic or Rheostatic braking :*** In this type of braking the supply to the armature is disconnected and an external resistance is connected across the armature. The field current is held constant. The generated voltage in the armature circulates the current through the external resistance. The direction of current through armature changes. Electromagnetic torque developed by the motor is in opposition to the direction of rotation. Operation is in second quadrant. Thus the kinetic energy of the moving system is converted into electrical energy and dissipated in the resistance of the armature and the external resistance. The braking torque is proportional to the product of field current and armature current. The armature current now depends on the generated voltage. Generated voltage depends on speed of armature. When the speed is zero there is no braking torque.

(b)   ***Plugging or Counter Current braking :*** In this type of braking, the supply to the armature is reversed thus forcing the current to flow in opposite direction in the armature. The direction of the field current is not   changed. Thus generated torque opposes the direction of rotation. The supply voltage plus generated voltage will force the current. To limit this current an external resistance is to be connected in series with the armature. The kinetic energy plus the energy supplied by the dc source is dissipated as heat in the external resistance and armature resistance. By this method the speed can be reduced quickly to zero. But when the speed comes to zero the dc supply has to be removed to make the braking current zero, otherwise the motor will rotate in opposite direction.

(c)   ***Regenerative braking :*** In dynamic braking or plugging the kinetic energy is converted into electrical energy and wasted as heat in the resistance. But in regenerative braking, the kinetic energy is converted into electrical energy and fed back into the electrical system for use by other devices connected to the system. Regenerative braking is possible only if the generated voltage in the dc machine driving the load is greater than the supply voltage so that the direction of current in the armature is reversed generating a braking torque. The generated voltage will be made higher than supply voltage by either increased speed, or by increasing the field flux or by reducing the supply voltage as in controlled rectifiers.

The methods of applying all the above types of braking to dc motors, using power electronics will be discussed.

3.3 Dynamic Braking

3.3.1 Dynamic Braking Application to a Separately Excited DC Motor

The schematic diagram given in Fig. 3.2 can be used for Dynamic Braking a separately excited dc motor. The supply voltage E_d can be obtained from controlled rectifier. E_d can be changed by changing firing angle.

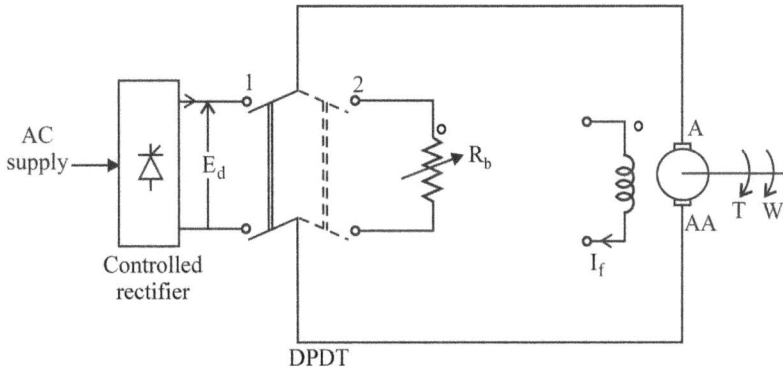

Fig. 3.2 Dynamic braking.

Initially the DPDT switch is in position (1). The dc machine is working as motor driving the load at speed ω, and the armature current enters at terminal A and its magnitude is given by equation

$$I_a = \frac{E_d - E_b}{R_a} \qquad\qquad .....(3.1)$$

where E_d is the supply voltage

and E_b is the generated voltage (or back emf)

The equivalent circuit is shown in Fig. 3.3.

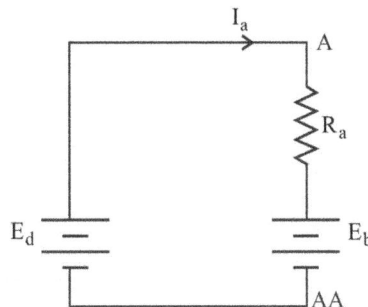

Fig. 3.3 Equivalent circuit during motoring.

The motor is operating in the first quadrant at point (1) shown in Fig. 3.5.

To apply braking DPDT switch is thrown to position (2).This will disconnect the DC supply and connects the braking resistance R_b across the armature. Now the current direction in armature is reversed, coming out from terminal A. The equivalent circuit is as shown in Fig. 3.4.

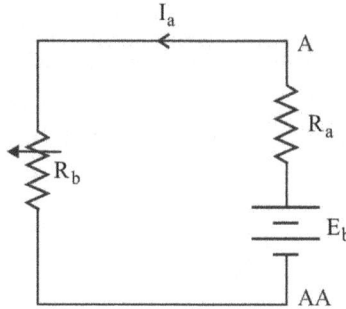

Fig. 3.4 Equivalent circuit during dynamic braking.

$$I_a = \frac{E_b}{R_a + R_b}$$ .....(3.2)

Braking torque is proportional to the product of armature current and field flux.

Now the operating point shifts to point (2) in second quadrant. Since the torque acts in opposite direction the speed decreases and I_a also decreases, decreasing the braking torque. The deceleration decreases. To keep the braking torque at desired level, the I_a has to be kept high. The braking resistance R_b can be reduced manually to keep the braking current high. The operating point shifts as shown in the Fig. 3.5 and ultimately the speed comes down to zero making the braking current and torque zero. The operating point shifts to origin.

A family of dynamic braking characteristics for different values of external resistance is shown in the Fig. 3.5. This method of braking is applied in non-reversible drives.

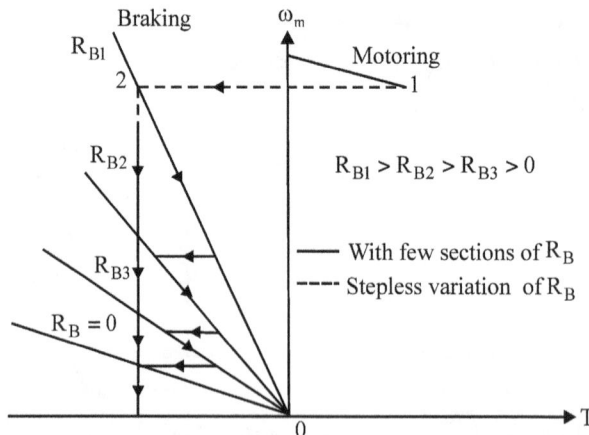

Fig. 3.5 Dynamic braking speed-torque curves with variable braking resistance.

The braking torque can be held constant at all speeds by continuously varying the braking resistance. This is achieved by placing an electronic switch (SCR) in series with the braking resistance and operating the switch at very high speed. When the switch is ON the current is $I_a = \dfrac{E_b}{R_a + R_b}$ and when the switch is OFF the current is zero. By changing the duty cycle i.e., ratio of ON period to ON period plus OFF period, the average braking current is controlled and hence the braking torque. The Fig. 3.6 shows the method of applying dynamic braking using a chopper in series with braking resistance.

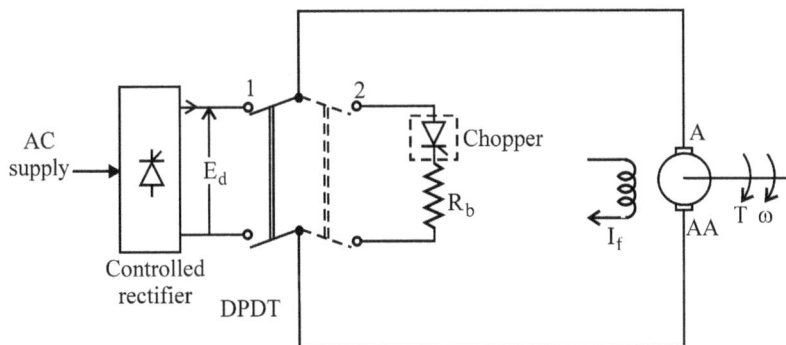

Fig. 3.6 Dynamic braking of separately excited dc motor using a chopper.

Example 3.1

A. DC separately excited motor has armature resistance of 0.04 Ω. Motor is coupled to an over hauling load with a torque of 400 N-m drawing a current of 200 A. Motor is braked by dynamic braking with braking resistance of 1 Ω.

(a)   At what speed the motor will hold the load?

(b)   What should be the braking resistance for keeping the speed at 600 rpm.

Solution:

(a) $E_b = k \, \phi \, \omega$

$T = k \, \phi \, I_a$

$400 = k \, \phi \times 200$

∴ $k \, \phi = 2$

$E_b = I_a (R_a + R_b) = I_a (1 + 0.04)$

E_b under dynamic braking $= 200 \times 1.04 = 208$ V

$\text{Speed} = \dfrac{208}{k\phi} = \dfrac{208}{2} = 104 \text{ rad/sec}$

$= \dfrac{104 \times 60}{2\pi} = \textbf{993 rpm}$

(b)
$$E_b \text{ at } 600 \text{ rpm} = \frac{2 \times 600 \times 2\pi}{60}$$

$$= 125.66 \text{ V}$$

$$0.04 + R_b = \frac{E_b}{I_a} = \frac{125.66}{200} = 0.628$$

$$R_b = 0.628 - 0.04 = \mathbf{0.588\ \Omega}$$

Example 3.2

A dc shunt motor has the armature resistance of 0.04 Ω and the field winding resistance of 10 Ω. Motor is coupled to an over-hauling load with a torque of 400 N-m. Following magnetization curve was measured at 600 rpm.

Field current A	2.5	5	7.5	10	12.5	15	17.5	20	22.5	25
Back emf V	25	50	73.5	90	102.5	110	116	121	125	129

Motor is braked by self-excited dynamic braking with a braking resistance of 1 Ω. At what speed motor will hold the load?

Solution :

$$E_b = K_e\ \phi\ \omega \qquad T = K_e\ \phi\ I_a$$

The equivalent circuit during dynamic braking of shunt machine is shown below.

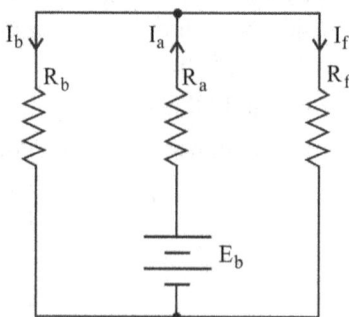

$$I_f = I_a\ \frac{R_b}{R_b + R_f} = I_a\ \frac{1}{1 + 10} = \frac{I_a}{11}$$

$$K_e\ \phi = \frac{E_b}{\omega} = \frac{E_b \times 60}{2\pi \times 600} = \frac{E_b}{20\pi}$$

$$I_a = 11\ I_f \quad \therefore T = K_e\ \phi\ I_a = 11\ K_e\ \phi\ I_f$$

From the magnetization characteristic we calculate $K_e\ \phi$ and T at different values of field current.

I_f	12.5	15	17.5	20	22.5	25
E_b	102.5	110	116	121	125	129
$K_e\phi = \dfrac{E_b}{20\pi}$	1.63	1.75	1.846	1.925	1.989	2.05
$T = 11k_e\phi\, I_f$	224	288.75	355.355	423.5	492.2	563.75

From the above data of I_f verses T, we get the value of I_f for T = 400 N-m; I_f should lie between 17.5 and 20 A

Assuming linear variation between 17.5 & 20,

$$\text{Slope} = \frac{20 - 17.5}{423.5 - 355.35} = 0.03668$$

Therefore the field current corresponding to T = 400 N-m is

$$17.5 + (400\text{-}355) - (0.03668) = \mathbf{19.15\ A}$$

Now for I_f = 19.5 A

$$I_a = 11 \times 19.15 = 210.65\ A$$

$$E_b = V + I_a R_a = I_f R_f + I_a R_a = 19.15 \times 10 + 210.65 \times 0.04$$

$$= 191.5 + 8.426 = 199.926\ V$$

From the plot $k_e\phi$ for I_f of 19.15 is 1.898.

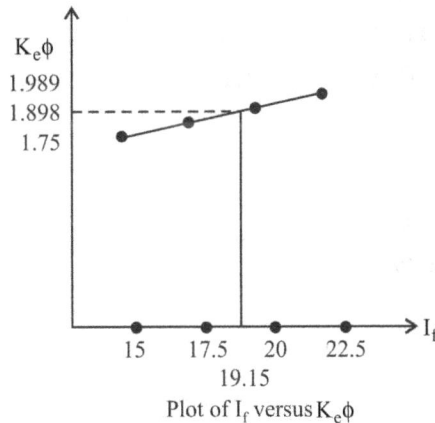

Plot of I_f versus $K_e\phi$

But

$$E_b = K_e\phi\, \omega$$

∴

$$\omega = \frac{E_b}{K_{e\varphi}} = \frac{203.58}{1.898} = 105.335$$

$$N = \frac{105.335 \times 60}{2\pi} = 1005\ rpm$$

Example 3.3

A dc shunt motor has the armature resistance of 0.04 Ω and the field winding resistance of 10 Ω. Motor is coupled to an over hauling load with a torque of 400 N-m. Following magnetization curve was measured at 600 rpm.

Field current	A	2.5	5	7.5	10	12.5	15	17.5	20	22.5	25
Back emf	V	25	50.0	73.5	90	102.5	110	116	121	125	129

Calculate the value of R_b when the motor is required to hold over-hauling load at 1200 rpm.

Solution :

$$T = K\phi\, I_a; \quad I_a = \frac{400}{k\phi}$$

$$E_b = K\phi\omega$$

$$T = K\phi\, I_a$$

$$\omega = \frac{1200 \times 2\pi}{60} = 40\pi; \quad K\phi = \frac{E_b}{40\pi}$$

$$I_a = \frac{T}{K\phi};$$

voltage across armature $V = E_b - I_a R_a$.

From the given data $K\phi$, I_a

V, and $I_f R_f$ are calculated and tabulated.

Under steady state $I_f R_f$ should be equal to V. This happens for value of I_f that lies between 2.5 and 5.

	I_f	2.5	5	7.5	10	12.5	15	17.5	20	22.5	25
At 600 rpm	E_b	25	50	100	147	180	205	220	232	242	250
	$K\phi$	0.3 97	0.79 6	1.17	1.432	1.631	1.75	1.846	1.926	1.989	
	I_a	10 07	502. 5	341.9	278	245.2	228.6	216.68	207.68	201.1	
At 1200 rpm	E_b	50	100	200	294						
	V	9.7 2	79.9	133.36	168	195					
	$I_f R_f$	25	50	75	100	125					

$$E_b - I_a R_a = I_f R_f = 10\, I_f$$

V at 1200 rpm = $E_b - I_a R_a = 50 - 1007 \times 0.04 = 9.72$

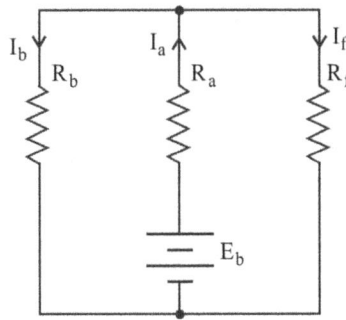

V will be equal to $I_f R_f$ for I_f that lies between 2.5 & 5 A. This is calculated as given below. This can also be obtained graphically.

$$9.72 + \left(\frac{79.9 - 9.72}{2.5}\right)(x - 2.5)$$

$$= 50 + \left(\frac{25 - 50}{2.5}\right)(5 - x)$$

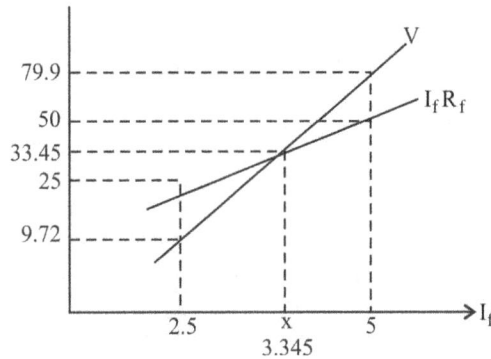

At $I_f = x$; $E_b - I_a R = 10\,I_f$

x will be between 2.5 and 5A

$$9.72 + 28.07\,(x - 2.5) = 50 - 10\,(5 - x)$$
$$28.07\,x - 60.45 = 10\,x$$
$$18.07\,x = 60.45$$
$$x = \frac{60.45}{18.07} = 3.345$$

At $I_f = 3.345$ A $V = I_f R_f = 33.\,45$ V

I_a is calculated for I_f of 3.345 A assuming linear relation between 2.5 and 5 A.

∴ $I_a = 1007$ at $I_f = 2.5$
                      $I_a = 502.5$ $I_f = 5$
                      $y = m\,x + c$
                      $1007 = 2.5m + c$ .....(i)

$$502.5 = 5m + c \qquad \qquad \text{.....(ii)}$$

from (i) – (ii), $504.5 - 2.5\,m$

∴ $m = -201.8$

$c = 1007 + 2.5 \times 201.8 = 1511.5$

$y = -201.8\,x + 1511.5$

At $x = 3.345\ y = -201.8 \times 3.345 + 1511.5 = 836.5.$

$I_b = I_a - I_f = 836.5 - 3.345 = 833.134\ A$

$I_b R_b$ is equal to V.

$$R_b \times 833.13 = 3.345 \times 10 = 33.45$$

$$R_b = \frac{33.45}{833.13} = \textbf{0.0401}\ \Omega$$

3.3.2 Dynamic Braking of DC Series Motor

There are certain difficulties in applying dynamic braking to series motor. They are explained below. Fig 3.7 shows the connection for dynamic braking. When switch is in position (1) it operates as motor. The direction of rotation is clockwise & direction of torque developed is clock wise. The field current enters at S1 and leaves at S2. Flux ϕ is positive. The armature current enters at A and leaves at AA. I_a is positive. Torque is positive. If the supply is disconnected by opening the switch the field and the armature currents will be zero. The flux will be very small and equal to residual flux. The speed of motor is not changed. At this instant there will be small back emf due to residual magnetism (flux ϕ).

$$E_b = K\,\phi\,\omega.$$

For dynamic braking the switch is thrown to position (2); in position (2) braking resistance R_b is connected across the motor. A small current flows in the opposite direction due to back emf generated. This current will destroy the residual magnetism and the E_b will be zero and the machine ceases to generate any emf. Thus no torque and no braking effect.

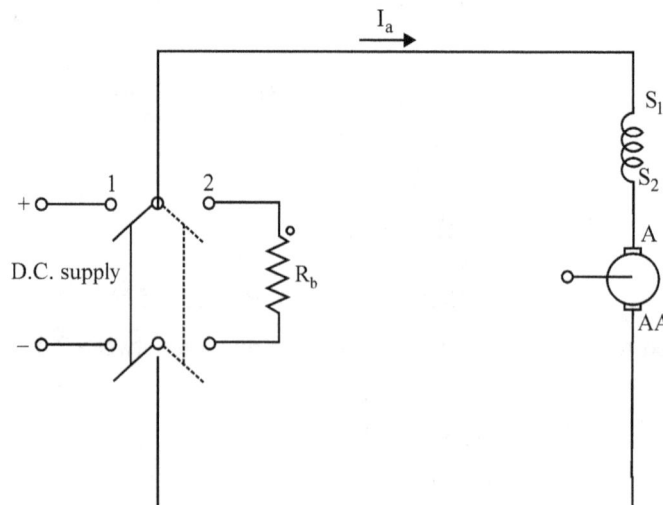

Fig. 3.7 Dynamic braking of series motor.

To obviate the above difficulty, the circuit has to be modified so that the direction of field current is not changed even though the direction of armature current changes. The modified circuit is shown in Fig. 3.8. The series field is connected across diode bridge as shown.

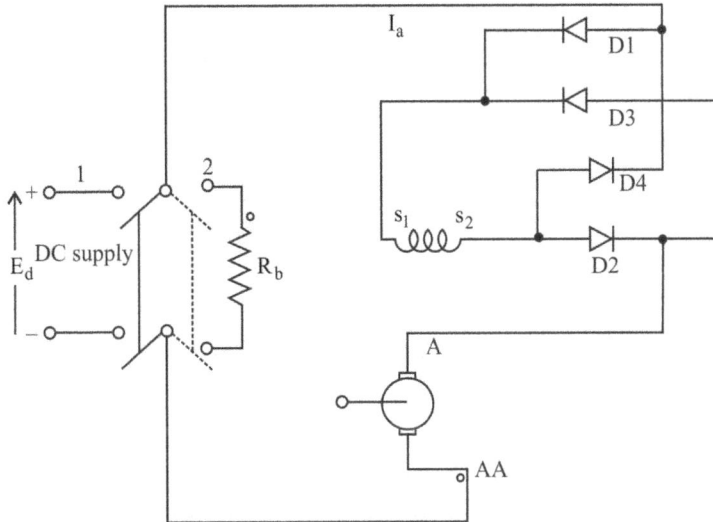

Fig. 3.8 Dynamic braking of dc series motor.

When the switch is in position 1, normal motor action takes place. The equivalent circuit is shown in Fig. 3.8.1. DC supply of E_d volts is applied to series motor. Current enters at S1 of series field and terminal A of armature. Torque is positive. The following equations hold good.

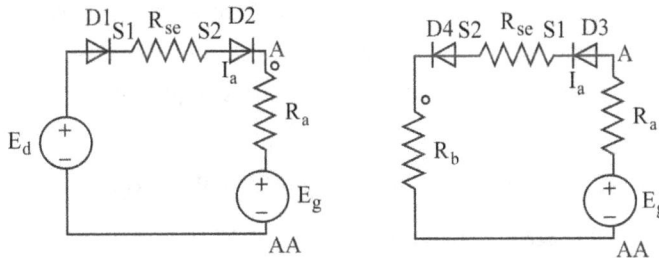

Fig. 3.8.1 Switch in position 1–Motoring.   **Fig. 3.8.2** Switch in position 2–Dynamic braking.

$$\text{Torque } T = kI_a\phi \qquad\qquad .....(3.3)$$

$$E_d = E_g + I_a (R_a + R_{se}) = K\;\phi\omega + \frac{T}{K\phi}(R_a + R_{se}) \qquad\qquad .....(3.4)$$

$$\omega = \frac{E_d}{K\phi} - \frac{T}{(K\phi)^2}(R_a + R_{se}) \qquad\qquad .....(3.5)$$

$$\phi = K_f I_a \qquad\qquad .....(3.6)$$

$$T = K\phi I_a = KK_f I_a^2; \quad I_a = \sqrt{\frac{T}{KK_f}} \qquad \qquad .....(3.7)$$

Substituting for I_a and ϕ in eqns. 3.5 and 3.7,

$$\omega = \frac{E_d}{KK_f I_a} - \frac{T(R_a + R_{se})}{(KK_f I_a)^2} = \frac{E_d}{\sqrt{KK_f T}} - \frac{(R_a + R_{se})}{KK_f} \qquad .....(3.8)$$

The torque speed characteristic during dynamic braking is shown in Fig. 3.9.

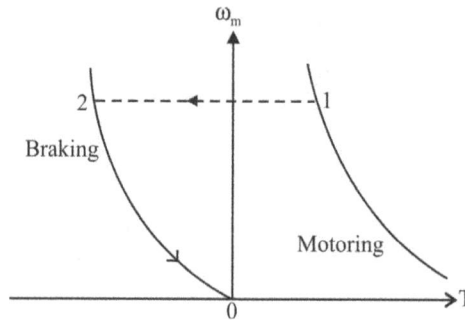

Fig. 3.9 Torque speed characteristic.

For applying dynamic brake, the switch is thrown to position 2. The dc supply is disconnected and braking resistance R_b is connected across the motor. Now generated voltage E_g forces the current to leave the terminal A and enter the terminal S1 of the series field. The current in armature is reversed but direction of current in series field remains same. Therefore the torque is negative and the motor starts decelerating and speed decreases. As speed decreases E_g also decreases, current decreases. Torque decreases. Ultimately speed comes down to zero. The following equations hold good for braking operation.

$$I_a = E_g/(R_a + R_{se} + R_b) = K\phi \omega/(R_a + R_{se} + R_b) \qquad .....(3.10)$$

$$T = K\phi\, I_a = (K\phi)^2\, \omega/(R_a + R_{se} + R_b) \qquad .....(3.11)$$

The torque speed characteristic during braking lies in second quadrant and shown in Fig. 3.9.

***Example* 3.4**

Calculate the resistance to be connected across a DC series motor used in a crane, when the supply is cut off, and dynamic braking is applied to limit the speed to 500 rpm, if the descending load exerts a constant load torque of 200 N-m. The magnetization curve of the motor running at 750 rpm is a straight line given by $E_g = (5.7\ I_a + 228.6)$ V, between $I_a = 30$ A to 50 A. The total motor resistance including the series field resistance is 1.1 Ω.

Solution :

Descending speed = 500 rpm

$$= 500 \times 2\pi /60 \quad \text{rad/sec}$$

$$= 52.36 \text{ rad/sec.}$$

$$T\omega = E_g I_a$$

If I_a is the armature current to develop torque of 200 N-m, then

$$E_g I_a /\omega = 200 \text{ N-m.}$$

Mechanical Power Input = $E_g I_a$

$$= 200 \times 52.36$$

$$= 10. 472 \text{ Kw}$$

E_g at 500 rpm = $(5.7 I_a + 228.6)$ 500/750

$$= 3.8 I_a + 152.4 \text{ Volts.}$$

Output voltage or Voltage across external resistance

$$= 3.8 I_a + 152.4 -1.1 I_a$$

$$= 2.7 I_a +152.4 \text{ V}$$

Output Power developed = VI_a

$$(2.7 I_a + 152.4) I_a = 200 \times 52.36 = 10472 \text{W}$$

$$2.7 I_a^2 + 152.4 I_a - 10472 = 0$$

$$I_a = (-152.4 \pm \sqrt{(152.4^2 + 4 \times 2.7 \times 10472)})/ 2 \times 2.7$$

$$= (-152.4 \pm 364) / 5.4$$

$$= 40.07 \text{ A}$$

Output voltage V = $2.7 I_a +152.4$

$$= (2.7 \times 40.07) + 152.4$$

$$= 260.6 \text{ V}$$

∴ External resistance, $R_{ex} = 260.6/40.07$

$$= 6.5 \text{ } \Omega.$$

3.4 Plugging

3.4.1 *Plugging of Separately excited dc Motor*

For plugging, the supply voltage to the armature is reversed so that it assists the back emf in forcing armature current in reverse direction. An arrangement for plugging separately excited dc motor is shown in Fig. 3.10.

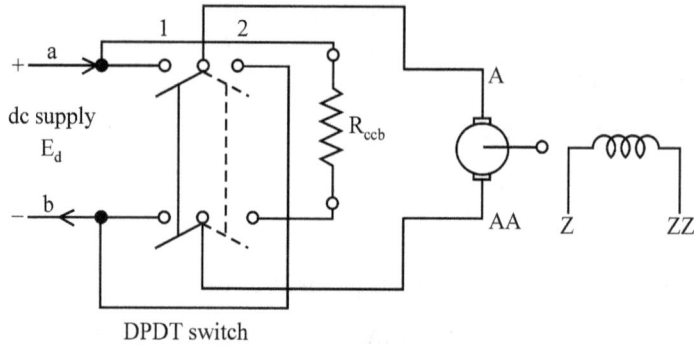

Fig. 3.10 Plugging of separately excited dc motor.

When switch is in position 1 supply voltage opposes the back emf and motor is operating at certain speed in first quadrant. The equivalent circuit is as shown in Fig. 3.10(a).

$$I_a = (E_d - E_b)/R_a \qquad \qquad \ldots\ldots(3.12)$$

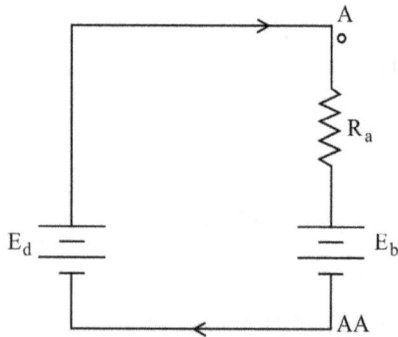

Fig. 3.10 (a) Motoring

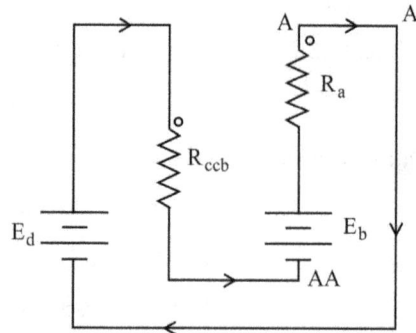

Fig. 3.10 (b) Plugging.

To apply plugging the switch is thrown to position 2. The supply voltage is in the same direction as that of back emf. A resistance R_{ccb} is connected in series with the armature to limit armature current. The equivalent circuit is as shown in Fig. 3.10(b). The armature current is reversed, while field current is not. Hence torque is developed in opposite direction making the motor to decelerate

$$I_a = (E_d + E_b)/(R_a + R_{ccb}) \qquad \qquad \ldots\ldots(3.13)$$

We know that $E_b = k\phi\omega$ and Torque $T = k\phi I_a$. Substituting for E_b and I_a in eqn. (3.13) and rearranging we get relation between torque and speed as

$$\omega = -\frac{T(R_a + R_b)}{(k\phi)^2} - \frac{E_d}{(k\phi)} \qquad \qquad \ldots\ldots(3.14)$$

The torque –speed characteristic during motoring and braking are shown in Fig. 3.11.

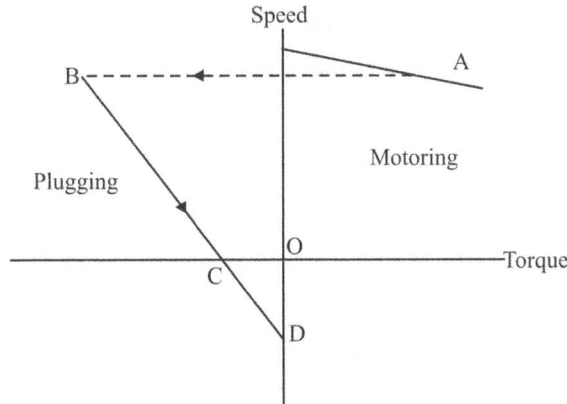

Fig. 3.11 Torque speed characteristics during plugging.

Let us assume that the motor is running at speed corresponding to point A in first quadrant. DPDT switch is in position (1). For braking, the switch is thrown to position (2). The operating point shifts to point B in second quadrant on the characteristic line drawn as per eqn. 3.14. Since torque is opposite to direction of rotation, speed decreases. Hence back emf decreases and armature current decreases and hence torque decreases. The operating point moves along the line BCD. When the operating point reaches C, the speed is zero. Back emf is zero. But the supply voltage is still there. At this instant the supply is to be cut off, otherwise motor starts moving in opposite direction because of torque produced even at zero speed. If the supply is not switched off, the motor accelerates in opposite direction in third quadrant. Since the direction of rotation is changed the direction of back emf also changes, opposing the supply voltage and thus acting as motor rotating in opposite direction. It settles down at a speed corresponding to load torque. Plugging gives fast braking due to high average torque even with one section of braking resistance. Plugging is highly inefficient because in addition to power generated by the motor, power supplied by the source is also wasted in resistance.

Example 3.5

A 220 V, 970 rpm, 100 A dc separately excited motor has an armature resistance of 0.05 Ω. It is braked by plugging from an initial speed of 1000 rpm. Calculate

  (a) Resistance to be placed in armature circuit to limit braking current to twice the full load value,

  (b) Braking torque, and

  (c) Torque when the speed has fallen to zero.

Solution :

At 970 rpm back emf $E_b = 220 - 100 \times .05 = 215$ V

At 1000 rpm $E_b = \dfrac{215 \times 1000}{970} = 221.65$ V

(a)   Plugging

$$V_a + E_b = I_a (R_a + R_b)$$

$$220 + 221.65 = 200 (0.05 + R_b)$$

$$R_b + 0.05 = \frac{441.65}{200} = 2.208$$

$$R_b = 2.208 - 0.05 = 2.158 \ \Omega$$

(b)

$$T \ \omega_m = E_b \ I_a = 221.65 \times 200 = 44330$$

$$\omega_m = \frac{1000 \times 2\pi}{60} = 104.71 \ \text{rad/sec.}$$

$$T = \frac{44330}{104.71} = 423.335 \ \text{N-m}$$

(c)   When the speed is zero back emf is zero

$$\therefore \qquad I_a = \frac{V_a + 0}{R_a + R_b} = \frac{220}{2.208} = 99.63 \ \text{A}$$

$$\text{Torque at this current} = \frac{99.63}{200} \times 423.35 = 210.9 \ \text{N-m}$$

3.4.2 *Plugging of dc series motor (counter current braking)*

Since the field winding is in series with armature we have to change the direction of current either in field or armature to get braking torque. Armature current reversal is preferred as its electrical time constant is less than that of field. An arrangement for plugging series motor is shown in Fig. 3.12.

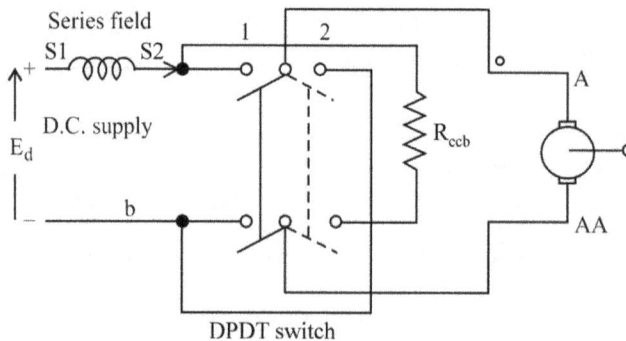

Fig. 3.12 Plugging of dc series motor.

When the DPDT switch is in position (1) the current flows from S_1 to S_2 in series field and from A to AA in armature. The induced emf in armature opposes the applied voltage. Torque is positive and motor rotates, say, in clockwise direction. Let its operating point be A in first quadrant. The equivalent circuit is shown in Fig. 3.12(a).

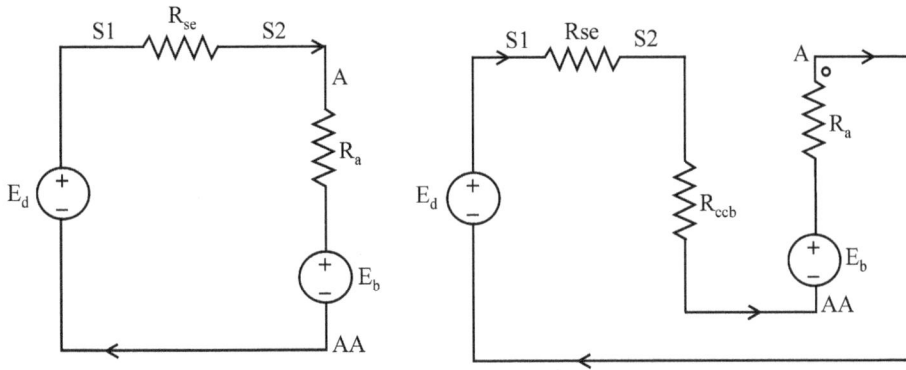

Fig. 3.12(a) Equivalent circuit for motoring.   **Fig. 3.12(b)** Equivalent circuit plugging.

$$E_d = E_b + I_a(R_a + R_{se}) = KI_a \omega + I_a(R_a + R_{se}) \qquad .....(3.15)$$

And torque is proportional to square of current.

To apply brake, the switch is thrown to position (2). The current flows from S1 to S2 in series field and from AA to A in armature. Since the direction of armature current is changed keeping direction of field current as it is, the direction of torque produced is in opposition to direction of rotation. The generated voltage in armature is in the same direction as that of supply voltage. The equivalent circuit is shown in Fig. 3.12(b).

$$E_d = -E_b + I_a(R_a + R_{se} + R_{ccb}) = -KI_a \omega + I_a (R_a + R_{se} + R_{ccb}) \quad .....(3.16)$$

The moment the switch is thrown to position (2), operating point shifts to point B in II quadrant as shown in Fig. 3.13. The braking torque depends on product of I_a and flux. Flux direction has not changed but armature current direction has changed hence torque direction is changed. Therefore T is proportional to $-I_a^2$. I_a depends on applied voltage E_a, E_b and on R_{ccb}. As the speed decreases E_b decreases. I_a decreases and hence the braking torque decreases. When the motor comes to rest the generated voltage is zero i.e., E_b is zero. But the supply voltage forces the motor to rotate in opposite direction. Hence the moment speed becomes zero, the supply has to be disconnected and mechanical brakes are to be applied. The torque speed curves during motoring and braking are shown in Fig. 3.13.

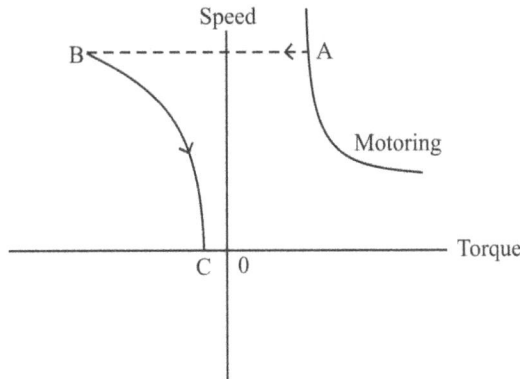

Fig. 3.13 Torque speed characteristics during plugging of DC series motor.

The following equations hold good for motoring action :

$$E_b = KI_a\omega$$

$$T = KI_a^2; \quad I_a = \sqrt{\frac{T}{K}}$$

Motoring $\qquad E_b = E_d - I_a(R_a + R_{se})$ .....(3.17)

$$KI_a\omega = E_d - \sqrt{\frac{T}{K}}(R_a + R_{se})$$

Substituting for I_a, , we get

$$\omega\sqrt{KT} = E_d - \frac{\sqrt{T}}{\sqrt{K}}(R_a + R_{se})$$

$$\omega = \frac{E_d}{\sqrt{T}\sqrt{K}} - \frac{\sqrt{T}}{\sqrt{K}\sqrt{K}\sqrt{T}}(R_a + R_{se})$$

$$= \frac{E_d}{\sqrt{T}\sqrt{K}} - \frac{(R_a + R_{se})}{K}$$ .....(3.18)

The following equations hold for plugging :

$$-I_a = \frac{E_d + E_b}{(R_a + R_{se} + R_{ccb})}$$ .....(3.19)

We know that $T = KI_a I_f = -K(I_a)^2$ and $E_b = KI_f\omega$. As I_f and ω are positive E_b continues to have same direction as before. But the torque which is proportional to I_a and I_f is negative

$$T = K\left[\frac{E_d + E_b}{R_a + R_{se} + R_{ccb}}\right]^2$$ .....(3.20)

Braking torque $\qquad T_b = -T \quad$ Let $R = R_a + R_{se} + R_{ccb}$

From eqn. 3.20 $\qquad E_d + E_b = R\frac{\sqrt{T_b}}{\sqrt{K}}$

$$E_d + KI_f\omega = R\frac{\sqrt{T_b}}{\sqrt{K}} \Rightarrow E_d + K\frac{\sqrt{T_b}}{\sqrt{K}}\omega = R\frac{\sqrt{T_b}}{\sqrt{K}}$$ .....(3.21)

Rearranging eqn. 3.21, we get the relation between torque and speed as

$$\omega = \frac{R}{K} - \frac{E_d}{\sqrt{KT_b}}$$ .....(3.22)

Example 3.6

A DC series motor has the following parameters:

$$P = 15 \text{ Kw}, V = 220 \text{ V}, r_{se} = 0.08 \ \Omega, r_{sh} = 0.17 \ \Omega$$
$$I = 80 \text{ A}, N = 350 \text{ rpm}.$$

What resistance must be connected in series with the armature at the instant of plugging to limit the current to 120 A? Calculate the braking torque at the instant of plugging. Also find the braking torque when the speed of the motor has fallen to 75% of its rated speed. The magnetization curve of the motor is a straight line given by $E_g = (1.07 \ I_a + 115)$ V between $I_a = 60$ A to 120 A.

Solution :

When motoring, $\qquad \omega = 350 \times 2 \ \pi \ / \ 60$
$$= 36.65 \text{ rad/sec.}$$
$$\text{Back emf} = 220 - 80 \ (0.08 + 0.17)$$
$$= 200 \text{ V}$$

At the instant of plugging, total voltage $= 220 + 200$
$$= 420 \text{ V}$$

Current is to be limited to 120 A

Total resistance $= 420/120 = 3.5 \ \Omega$

Additional resistance to be connected $= 3.5 - (0.17 + 0.08) = 3.25 \ \Omega$

When the current is 120 A, $E_g = (1.07 \ I_a + 115)$ V
$$= (1.07 \times 120) + 115$$
$$= 243.4 \text{ V}$$

Power developed $= E_g I_a$
$$= 243.4 \times 120$$
$$= 29.2 \text{ KW}$$

$\omega = 36.65$ rad/sec; Torque, $T = P/\omega = 29208/36.65 = 796.9$ N-m

When the speed of the motor has fallen to 75%, the expression for E_g is
$$E_g = 0.75(1.07 \ I_a + 115) \text{ V}$$
$$= (0.8025 \ I_a + 86.25) \text{ V}$$

Total Voltage $= 220 + (0.8025 \ I_a + 86.25)$

This is equal to the drop in the 3.5 Ω resistance

$\therefore \qquad 3.5 \ I_a = 220 + (0.8025 \ I_a + 86.25)$
$$2.6975 \ I_a = 306.25$$
$$I_a = 113.53 \text{ A}$$
$$E_g = (0.8025 \times 113.53) + 86.25$$
$$= 177.35 \text{ V}$$

Power developed = $E_g I_a$

$= 177.35 \times 113.53$

$= 20135$ W

Speed $= 0.75 \times 350 = 262.5$ rpm.

$\omega = 262.5 \times 2\pi / 60 = 27.48$ rad/sec.

Torque, $T = E_g I_a / \omega = 20135/27.48 = 732.71$ N-m.

3.5   Regenerative Braking

In this type of braking the kinetic energy of moving system is converted into electrical energy and fed back into electrical system. This type of braking is possible if the generated voltage in the dc machine is greater than the supply voltage. The generated voltage is directly proportional to product of speed and field flux. If the speed is constant generated voltage can be increased by increasing the field current. If the dc supply is obtained from a controlled converter, the converter is to be operated in inversion mode by making firing angle greater than 90^0. The supply voltage is made less than the generated voltage of machine, so that power is taken from the machine and fed back to the mains.

3.5.1 Regenerative Braking of Separately Excited dc Motor

A schematic diagram for applying regenerative braking to separately excited dc motor is given in Fig. 3.14.

For motoring the switch is in position 1. Converter is working as rectifier. Firing angle less than 90^0. $E_a > E_b$. Current flows from A to AA in armature.

$$E_a = E_b + I_a R_a.$$

Power input to motor = $E_a I_a$

The equivalent circuit is shown in Fig. 3.14.a.

Electrical power output of converter = power input to motor = $E_a I_a$ W   .....(3.23)

Mechanical power developed by motor = $E_b I_a = T\omega$ W   .....(3.24)

Torque developed $T = K I_a$

Back emf $E_b = K\omega$

$$I_a = (E_a - E_b)/R_a$$

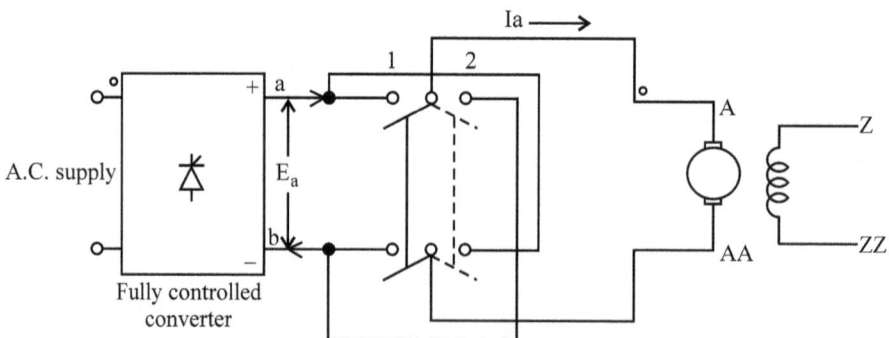

Fig. 3.14 Regenerative braking of separately excited dc motor.

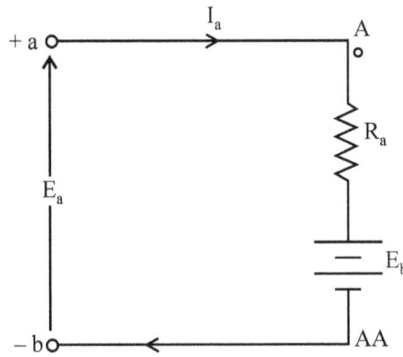

Fig. 3.14(a) Equivalent circuit during motoring operation.

Regenerative Braking – The switch is in position 2. The equivalent circuit is shown in Fig 3.14b. Converter is to be operated in inversion mode. Firing angle is greater than 90^0. $E_a < E_b$. Current flows from AA to A in the armature.

$$E_b = E_a + I_a R_a.$$

Power fed back to the mains $= E_a I_a$ .....(3.25)

Electrical power generated by DC

Machine now acting as generator $= E_b I_a$ .....(3.26)

$$I_a = (E_b - E_a)/R_a$$

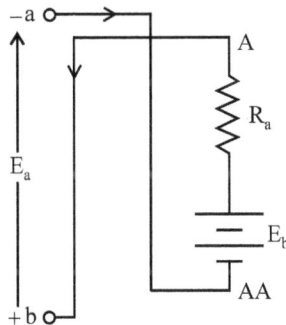

Fig. 3.14(b) Equivalent circuit during regenerative braking.

During motoring, the operation is in the first quadrant and during regenerating the operation is in the second quadrant. The braking torque depends on armature current. Armature current $I_a = (E_b - E_a)/R_a$. The braking torque can thus be controlled by controlling the firing angle of the converter which in turn controls E_a.

3.5.2 Regenerative Braking of Series Excited dc Motor

A schematic diagram for regenerative braking of DC series motor is shown in Fig. 3.15.

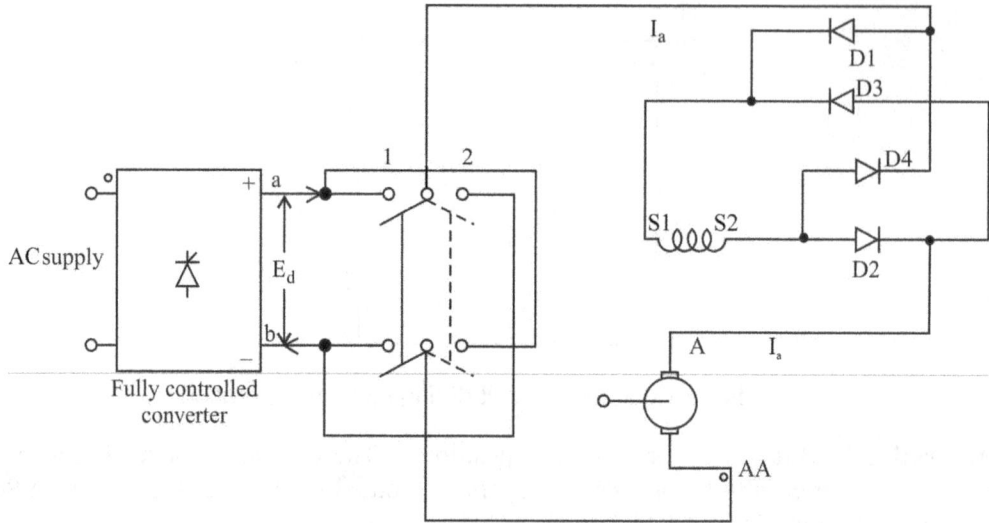

Fig. 3.15 Regenerative braking of series excited dc motor.

A diode bridge is connected across series field so that flux direction does not change even though the direction of armature current changes. When armature current enters at A and leaves at AA torque is positive and the machine operates in first quadrant. When armature current enters at AA and leaves at A, torque is negative and the machine operates in second quadrant.

Motoring : The DPDT switch is in position 1. The equivalent circuit is shown in Fig. 3.15(a). The converter firing angle is less than 90^0. Supply voltage is positive at upper terminal and current leaves from positive terminal of converter. Current in field winding is from S1 to S2. Flux is positive. Armature current is from A to AA and it is positive. Therefore torque is positive, speed is positive. The DC supply voltage to motor is E_s. $E_s > E_b$.

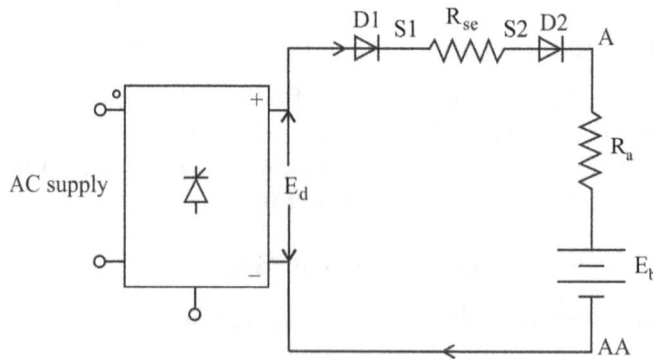

Fig. 3.15(a)

$$I_a = I_f = (E_d - E_b)/(R_a + R_{se}) \qquad \qquad .....(3.27)$$

$$T = KI_a I_f; \ E_b = KI_f \omega.$$

Mechanical power developed $= E_b I_a = T \omega \ W \qquad \qquad .....(3.28)$

Electrical power input to motor $= E_d I_a \ W \qquad \qquad .....(3.29)$

Regenerative braking : For regenerative braking DPDT is thrown to position 2. Firing angle of converter should be greater than 90^0. The polarity of converter output voltage will be positive at bottom terminal. The firing angle has to be adjusted such that $E_d < E_b$. The equivalent circuit is shown in Fig. 3.15(b).

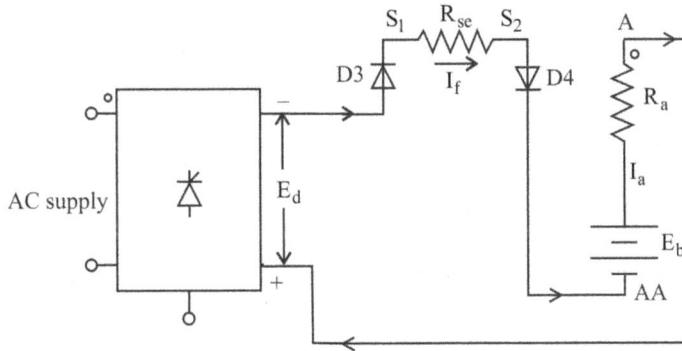

Fig. 3.15(b) Braking.

The direction of current in series field has not changed but that in armature has changed. Hence the torque will be negative and opposes the direction of rotation.

Now $\qquad \qquad I_a = -I_f = (E_b - E_d)/(R_a + R_{se}) \qquad \qquad .....(3.30)$

Braking current can be controlled by firing angle.

Electrical power generated by DC machine now acting as generator $= E_b I_a \qquad .....(3.31)$

Electrical power input to converter and fed back to mains $= E_d I_a \qquad .....(3.32)$

Braking torque developed $= KI_a^2 \qquad \qquad .....(3.33)$

Generated voltage in armature $= K I_f \omega. \qquad \qquad .....(3.34)$

3.6 Four Quadrant Operation of dc Motors by Dual Converters

A dual converter is made up of two fully controlled converters connected in anti parallel as shown in Fig. 3.16. For four quadrant operation of separately excited dc motor the dual converter is operated in circulating current mode. For this mode of operation circulating current limiting inductors are connected as shown in the figure. If α_P is the firing angle of positive converter and α_n is the firing angle of negative converter, $\alpha_P + \alpha_n = 180^0$. It can also be operated in circulating current free mode of operation. Operation of the motor in different quadrants will now be explained for circulating current mode of operation, with the help of equivalent circuits.

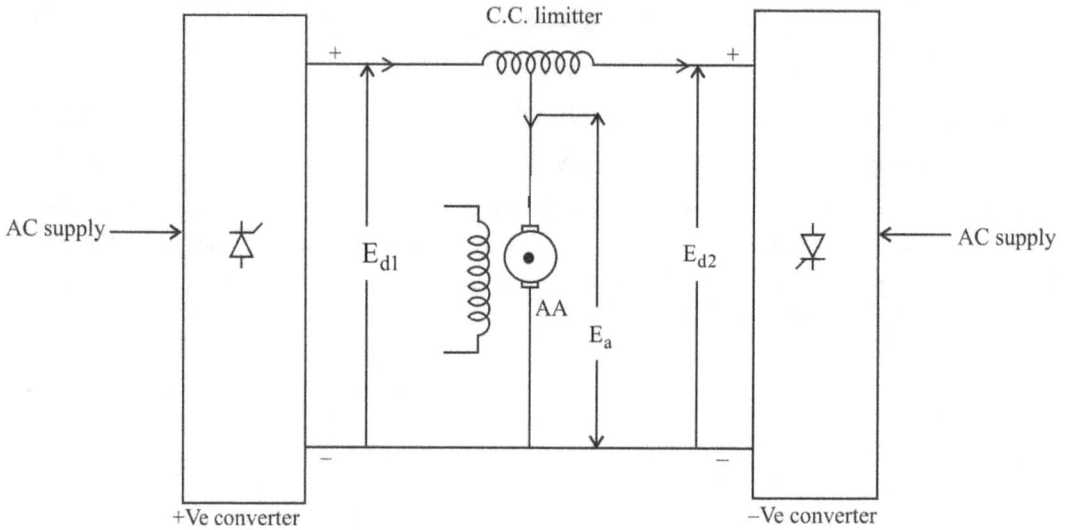

Fig. 3.16 Dual converter.

(a) *First quadrant operation or forward motoring :* Torque produced is positive. Motor is rotating in clock wise direction. Positive converter operating as rectifier and negative converter operating as inverter. Voltage applied to armature E_a is positive at terminal A. Current in armature is from A to AA, and is supplied by positive converter only. E_{d1} is equal to and E_{d2} and are greater than E_b. So that current flows from A to AA in armature. $E_a = E_b + I_a R_a$. The equivalent circuit is shown in Fig. 3.16(a).

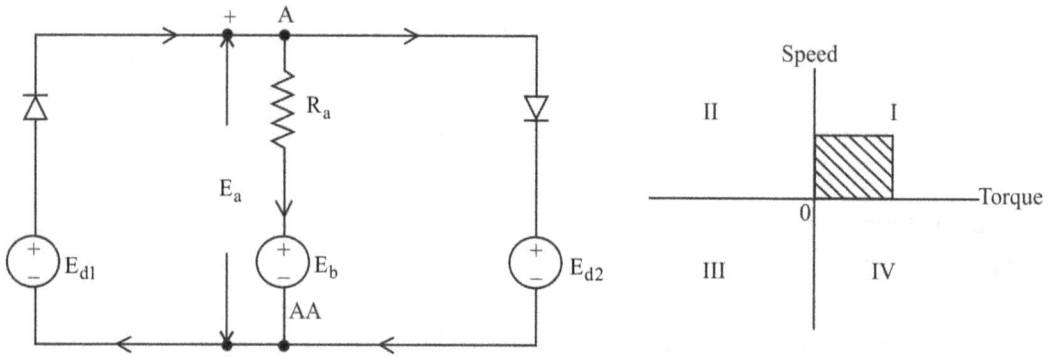

Fig. 3.16(a) First quadrant operation.

(b) *Second quadrant operation or forward braking :* Torque produced is negative. i. e., counter clockwise. Motor is rotating in clock wise direction. Positive converter operating as rectifier and negative converter operating as inverter. Voltage applied to armature E_a is positive at terminal A. E_{d1} and E_{d2} are made less than E_b so that current in armature is from AA to A, and is supplied by negative converter only. $E_a = E_b - I_a R_a$. Mechanical power is converted into electrical power and fed back to ac mains through converter 2. Refer Fig. 3.16(b). This is regenerative braking. As the speed decreases during braking E_{d2} & E_{d1} are to be reduced to make armature current constant at predetermined value.

We can also use plugging for operation in second quadrant with certain precautions mentioned already. For plugging converter 2 has to operate as rectifier and converter 1 has to operate as inverter so that E_{d2} will aid E_b.

$$I_a = (E_{d2} + E_b)/R_a$$

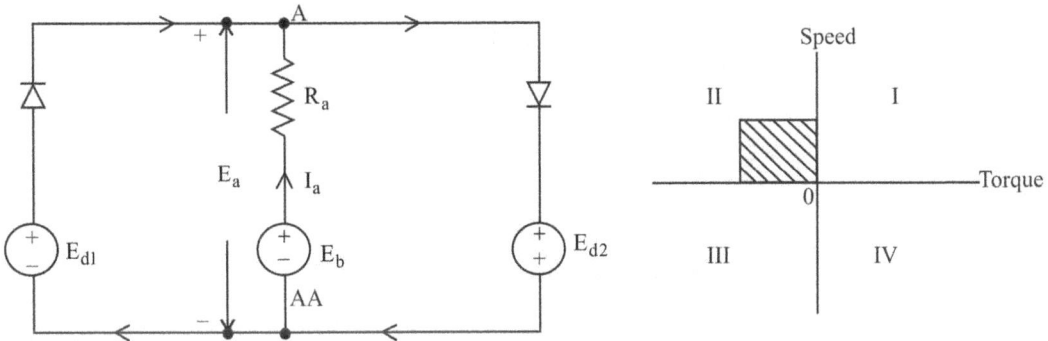

Fig. 3.16(b) Second quadrant operation. (Regenerative braking).

(c) *Third quadrant operation or reverse motoring :* Torque produced is negative i.e., counter clockwise. Motor is rotating in anti-clock wise direction. Positive converter operating as inverter and negative converter operating as rectifier. Voltage applied to armature E_a is positive at terminal AA. Current in armature is from AA to A, and is supplied by negative converter only. $E_b + I_aR_a = E_{d2}$

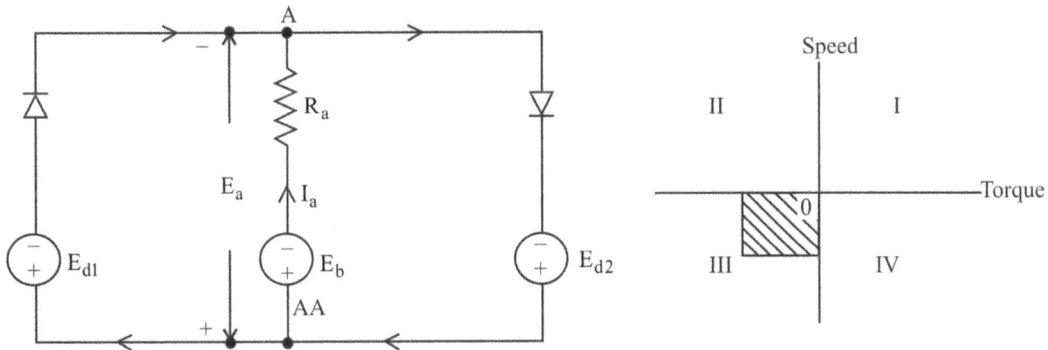

Fig. 3.16(c) Third quadrant operation.

E_{d1} & E_{d2} should be made greater than E_b, the back emf.

(d) *Fourth quadrant operation or reverse braking :* Torque produced is positive i.e., clockwise. Motor is rotating in anti-clock wise direction. Positive converter operating as inverter and negative converter operating as rectifier. Voltage applied to armature E_a is positive at terminal AA. Current in armature is from A to AA, and is supplied by converter 1 only. $E_a = E_b - I_a R_a$. Mechanical power is converted to electrical power and fed back to ac mains through converter 1. Refer to Fig 3.16(d). E_{d1} and E_{d2} are to be less than E_b.

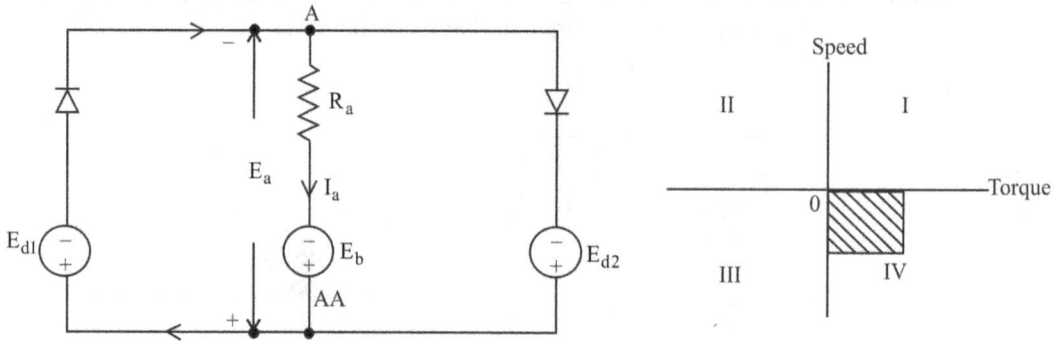

Fig. 3.16(d) Fourth quadrant operation.

Example 3.7

A 220 V, 750 rpm, 200 A separately excited motor has an armature resistance of 0.05 Ω. Armature is fed from a 3 ϕ, non circulating current mode dual converter, consisting of fully controlled rectifiers A and B. Rectifier A provides motoring operation in the forward direction and rectifier B in the reverse direction. Line voltage of ac source is 400 V. Calculate firing angle of rectifier for the motoring operation at rated torque and 600 rpm assuming continuous conduction.

Solution :

$$E_b \text{ at 750 rpm} = 220 - 200 \times 0.05$$
$$= 220 - 10 = 210V$$
$$E_b \text{ at 600 rpm} = \frac{210 \times 600}{750}$$
$$= 168 \text{ V}$$

I_a at rated torque = 200 A

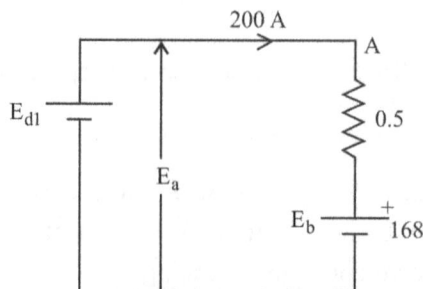

∴   Converter output

$$\text{Voltage } E_c = E_b + I_a R_a$$

$$= 168 + 200 \times 0.05$$

$$= 178 \text{ V}$$

∴ $$1.35 \times 400 \times \cos \alpha_1 = 178$$

$$\cos \alpha_1 = \frac{178}{1.35 \times 400} = 0.3296$$

$$\alpha_1 = 70.75°$$

Only rectifier A should be triggered at 70.75°. No triggering pulses are to be given to rectifier B.

Example 3.8

A 220 V, 1500 rpm, 50 A separately excited DC motor with armature resistance of 0.5 Ω is fed from a circulating current dual converter with 3 ϕ ac source voltage of 165 V (line). Determine converter firing angles for the following operating points:

(a)   Motoring operation at rated motor torque and 1000 rpm.

(b)   Braking operation at rated motor torque and –1000 rpm.

Solution :

(a)   Back emf E_b at rated torque and rated speed of 1500 rpm

$$= 220 - 50 \times 0.5 = 195 \text{ V}$$

$$\text{Back emf } E_b \text{ at 1000 rpm} = \frac{1000 \times 195}{1500} = 130 \text{ V}$$

Supply voltage to armature $E_a = 130 + 0.5 \times 50$

$$= 155 \text{ V}$$

For 3ϕ converter $E_a = 1.35 \text{ V}_L \cos\alpha$

$$1.35 \times 165 \times \cos\alpha = 155$$

$$\cos\alpha = \frac{155}{1.35 \times 165} = 0.695$$

$$\alpha_1 = 45.9° \qquad \alpha_2 = 134.1°$$

For motoring operation converter 1 will be rectifier and converter 2 will be inverter.

(b)   Braking operation at rated motor torque and –1000 rpm when rotating at   –
1000 rpm back emf polarity is reversed and operating in 3rd quadrant.

Equivalent circuit is as shown below.

Converter (1) will be operating as inverter and it receives the power. Converter 2 will not allow the current to flow through it.

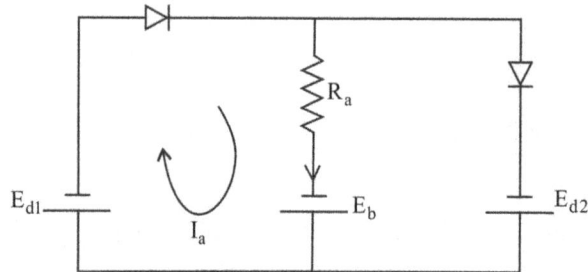

$$E_b \text{ at } 1000 \text{ rpm} = 130 \text{ V}$$

$$E_a = 130 - 50 \times 0.5 = 105 \text{ V}$$

$$1.35 \times 165 \times \cos\alpha_1 = -105$$

$$\cos\alpha_1 = -\frac{105}{1.35 \times 165} = -0471$$

$$\therefore \qquad \alpha_1 = 118.12°$$

$$\alpha_2 = 61.88°$$

Example 3.9

A dual converter 3 ϕ bridge circuit supplied power to a 540 V, 40 A separately excited DC motor with an armature resistance of 1.2 Ω. The voltage drops on the bridge thyristors are 1.2 V at rated motor current. Power is supplied by an ideal 3 ϕ source with rms line voltage of 400 V, 50 Hz. Find the necessary firing delay angles and motor back emf for :

   (a)   Motoring operation at rated load current with motor terminal voltage of 400 V.

   (b)   Regeneration operation at rated load current with motor terminal voltage of 400 V.

   (c)   Motor plugged at rated load current with a terminal voltage of 400 V and a current limiting resistor of 15 Ω.

Solution:

(a)   *Motoring operation*

In first quadrant.

Converter (1) $\alpha_1 < 90°$

Converter (2) $\alpha_2 > 90°$

Converter (1) supplies current.

Voltage drop in each thyristor is 1.2 V. Total voltage drop in converter will be $2 \times 1.2 = 2.4$ V. This is shown as drop in the diode shown in the diagram.

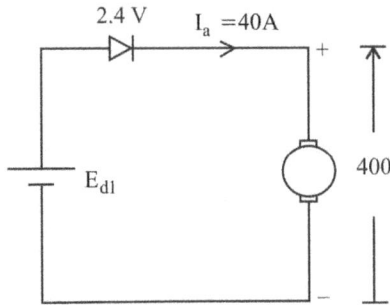

$\therefore$ $E_{dl} = 2.4 + 400$

$\therefore$ $E_{dl} = 402.4 \text{ V}$

             $1.35 \times 400 \times \cos\alpha_1 = 402.4;$

      $\cos$ $\alpha_1 = 0.744$

                $\alpha_1 = 41.88°$

$\therefore$ $\alpha_2 = 138.12°$

(b)   ***Regeneration :*** Operation will be in 2nd quadrant. Converter 2 receives the current. Converter (1) rectification and converter (2) inversion. The voltage across converters should be less than voltage across motor. The equivalent circuit is shown below.

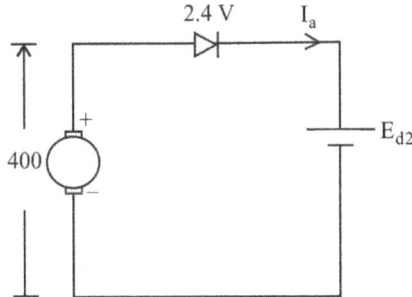

From the circuit it can be seen that

             $E_{d2} = 400 - 2.4$

                  $= 397.6 \text{ V}$

$\therefore$ $E_{d2} = -397.6$

             $1.35 \times 400 \times \cos\alpha_2 = -397.6$

             $\cos\alpha_2 = -\dfrac{397.6}{1.35 \times 400} = -0.736$

             $\alpha_2 = 137.4°$

             $\alpha_1 = 42.6°$

(c)   **Plugging :** Braking current can flow only through converter (2). The voltage generated by the converter (2) should aid the voltage across armature. The equivalent   circuit will be as shown below.

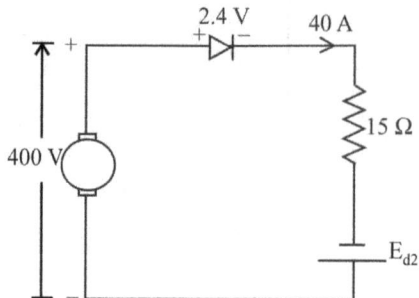

From the circuit $E_{d2} + 400 = 2.4 + 40 \times 15$

$$= 602.4$$

$$E_{d2} = 602.4 - 400 = 202.4 v.$$

$$1.35 \times 400 \times \cos\alpha_2 = 202.4$$

$$\cos \quad \alpha_2 = 0.3748$$

$$\alpha_2 = 67.98°$$

$$\alpha_1 = 112.02°$$

Example 3.10

A 3 phase dual converter feeds a 500 V, 60 A dc motor with separate excitation. The converter is fed from a 420 V, 50 Hz supply. Assuming voltage drop of 20 V in the converter, determine the firing angle and back emf for :

(a)   Motoring operation at full load current with motor terminal voltage of 450 V.

(b)   Regeneration operation at full current with terminal voltage of 450 V.

(c)   The motor is plugged at a terminal voltage of 400 V with a current limiting resistor of 10 Ω.

Solution :

(a)   First quadrant operation. Converter (1) supplies power.

$$E_{d1} = 450 + 20 = 470 \text{ V}; E_{d1} = 1.35 \text{ V}_L \cos \alpha_1$$

$$1.35 \times 420 \times \cos \alpha_1 = 470$$

$$\alpha_1 = 34.1°$$

$$\alpha_2 = 145.9°$$

Back emf $E_b = 450 - 60 \times 1.5$

$$= 360 \text{ V}$$

Regeneration : 2nd quadrant operation. Converter 2 receivers power.

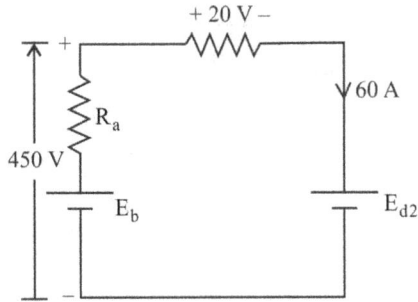

$$-E_{s2} = 450 - 20 = 430V$$

$$1.35 \times 420 \times \cos \alpha_2 = -430$$

$$\cos \alpha_2 = -0.758$$

$$\alpha_2 = 139.32°$$

$$\alpha_1 = 40.68^0$$

Back emf $E_b = 450 + 60 \times 1.5 = 540$ V

(c) *Plugging :* Converter (2) should supply power and it must aid back emf

$$E_{d2} + 400 = 20 + 60 \times 10$$

$$= 620$$

$$E_{d2} = 620 - 400 = 220$$

$$1.35 \times 420 \times \cos\alpha_2 = 220$$

$$\alpha_2 = 67.16°; \alpha_1 = 112.84°$$

$$E_b = 400 + 60 \times 1.5 = 490 \text{ V}$$

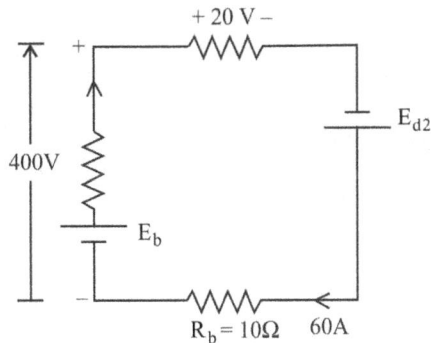

3.7   Closed Loop Operation of Separately Excited dc Motor Fed from Controlled Rectifier

The speed of motor can be controlled up to its rated speed by controlling the voltage applied to the armature keeping the field current at its rated maximum value. For controlling the speed above rated value the voltage applied to the armature is held at its maximum value and the field current is reduced. Of course the motor can rotate at speed greater than the rated speed at lesser torque. This mode of operation is known as constant horse power mode.

First the relation between the speed and armature voltage is derived. The open loop transfer function is obtained as given below. Refer to Fig. 3.17.

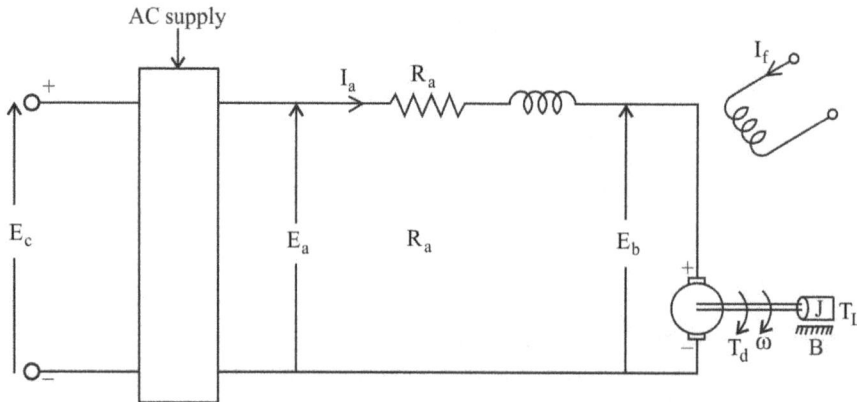

Fig. 3.17 Separately excited dc motor driven by controlled converter.

Let e_c = Voltage applied to converter to control the firing angled.

K_c = Gain of converter

e_a = Output voltage of converter and also the applied voltage to the armature.

R_a = Armature resistance

L_a = Armature inductance

i_a = Armature current

e_b = Back emf

T_d = Torque developed by the motor

ω = Speed of motor in rad/sec.

J = Moment of inertia of load & motor

B = Frictional co-efficient

T_L = Load torque in N-m

I_f = Field current

The following equations hold good.

$$e_a = K_c e_c$$

$$R_a i_a + L_a \frac{di_a}{dt} + e_b = e_a$$

$$e_b = K_a \omega \phi = K_a K_f \omega i_f = K \omega i_f$$

$$T_d = K_a i_a \phi = K_a K_f i_a i_f = K i_a i_f$$

$$J \frac{d\omega}{dt} + B\omega + T_L = T_d$$

Taking Laplace transform of above equations

$$E_a = K_c E_c \qquad\qquad .....(3.35)$$
$$(R_a + L_a s)\, I_a + E_b = E_a \qquad\qquad .....(3.36)$$
$$I_a = (E_a - E_b)/(R_a + L_a s) \qquad\qquad .....(3.37)$$
$$E_b = K \omega\, I_f \qquad\qquad .....(3.38)$$
$$T_d = K I_a I_f \qquad\qquad .....(3.39)$$
$$\omega = (T_d - T_L)/(Js + B) \qquad\qquad .....(3.40)$$

The above equations can be represented by block diagram as shown in Fig. 3.18.

Let $\dfrac{L_a}{R_a} = \tau_a$ armature time constant

$\dfrac{J}{B} = \tau_m$ mechanical time constant.

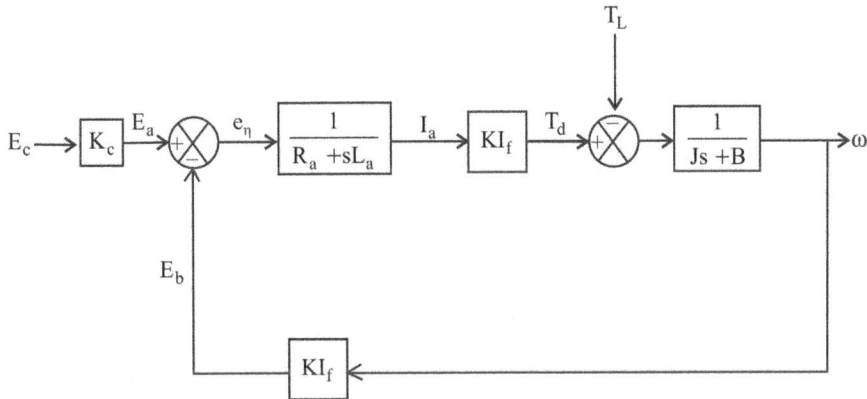

Fig. 3.18 Block diagram of open loop control system given in Fig. 3.17.

From the block diagram we get the following transfer functions.

At $\quad T_L = 0 \qquad \left.\dfrac{\omega}{E_c}\right|_{T_L=0} = \dfrac{K_c K I_f}{R_a B(1 + s\tau_a)(1 + s\tau_m) + K^2 I_f^2} \qquad .....(3.41)$

At $\qquad E_c = 0 \qquad \dfrac{\omega}{T_L}\bigg|_{EC=0} = \dfrac{R_a(1+s\tau_a)}{R_aB(1+s\tau_a)(1+s\tau_m)+K^2I_f^2}$ .....(3.42)

Steady state values of changes in speed $\Delta\omega$ due to changes in ΔE_a and ΔT_L are obtained by substituting s = 0 in the above equations.

$$\Delta\omega = \dfrac{K_cKI_f}{R_aB+K^2I_f^2}\,\Delta E_c \qquad\qquad .....(3.43)$$

$$\Delta\omega = \dfrac{R_a}{R_aB+K^2I_f^2}\,\Delta T_L \qquad\qquad .....(3.44)$$

If the motor is to be run at a specified speed we have to use closed loop control scheme. The basic block diagram of closed loop speed control system is given in Fig. 3.19.

In this system the actual speed ω is sensed by a speed sensor. The output of the speed sensor is a voltage e_ω. This voltage signal is compared with the reference signal. The difference signal $e = e_r - e_\omega$ is fed to the P-1 controller whose output signal e_c is the control voltage applied to the converter whose output is a dc voltage e_a applied to the armature of dc motor. The P-I controller change the voltage e_a till the motor runs at desired speed making the error signal e_n as zero. If there is a change in the load torque the speed will change which in turn changes e_ω generating the error signal. This error signal changes the armature voltage e_a which changes the speed. Thus the speed is held constant at desired value or set value.

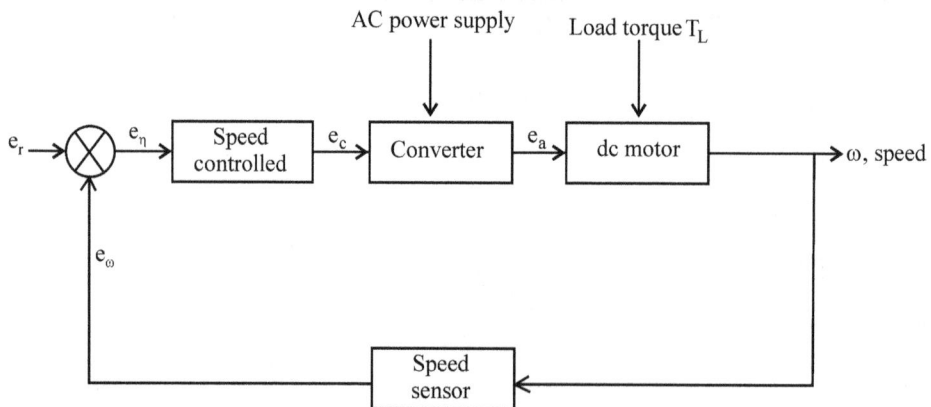

Fig. 3.19 Closed loop speed control.

There is a problem in this type of control. If there is a sudden increase in load torque, the speed drops and hence back emf decreases. This in turn increases the armature current to a large value of course the current will again come to study state value slowly. But during acceleration the current will be very large damaging the insulation of armature.

To overcome this problem we can have inner current loop to limit the armature current to its rated value even during acceleration.

The block diagram with inner current loop is shown in Fig. 3.20.

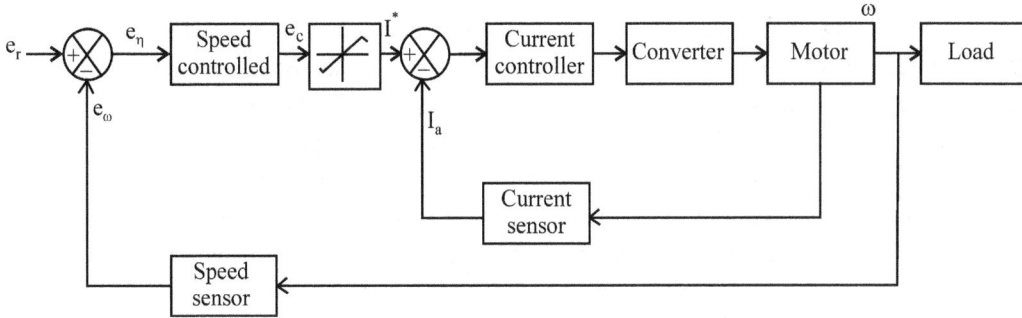

Fig. 3.20 Closed loop speed control with current limit.

I* is a voltage signal corresponding to the rated armature current of the motor. The actual current in the armature is sensed by current sensor and compared with the reference value I*. The current controller will see that the armature current will never be greater than I*. Whatever may be the difference in set speed and actual speed the current limiter output will never be greater than I*. Thus even during starting or due to sudden increase in load torque, the motor will accelerate at rated torque with rated armature current.

If we want to control the speed above and below rated speed the block diagram shown in Fig. 3.21 can be used.

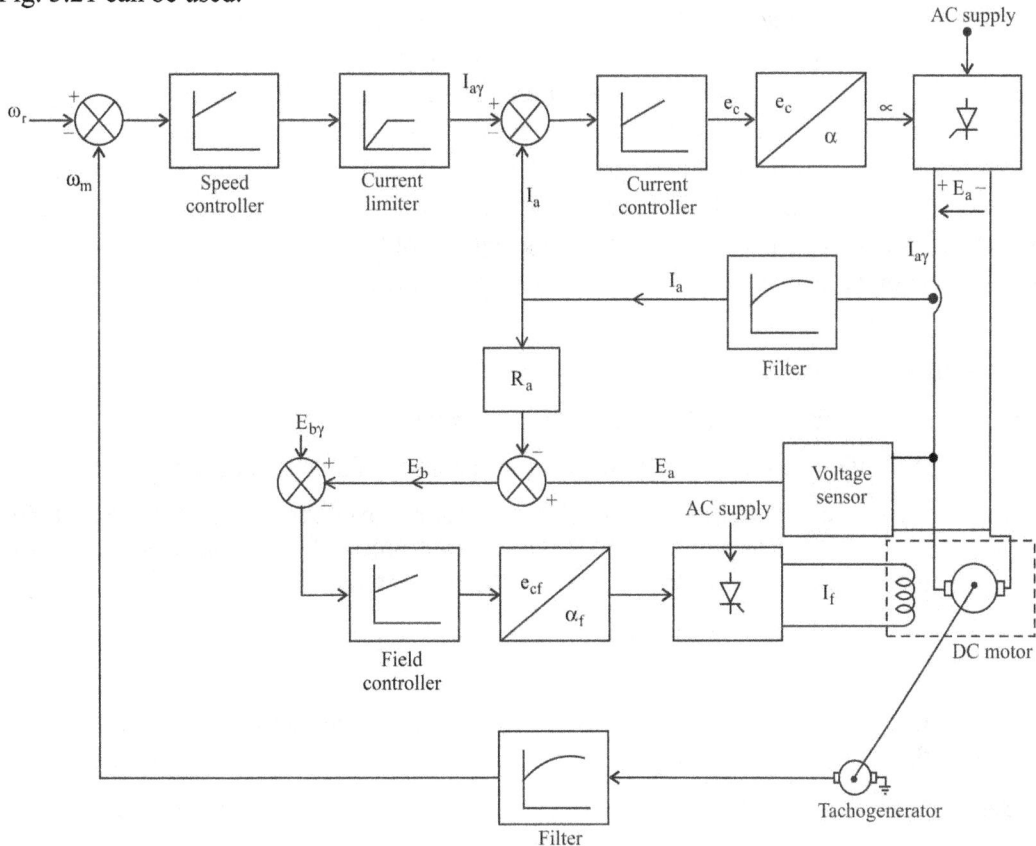

Fig. 3.21 Closed loop speed control with inner current loop and field weakening.

The working of this scheme is briefly given. ω_r is the reference speed signal. It can be either below or above the rated speed of the motor. ω_m is the signal corresponding to the actual speed of the motor. This signal is generated using a Tachogenerator coupled to the armature and a filter. The difference between ω_r and ω_m is given as input to the speed controller which is a p-I controller. The output of this p-I controller is given as input to current limiter. The maximum of this current limiter is set to be equal to the armature rated current I_{ar}. The actual current passing through the armature is sensed by a dc transformer, filtered and a signal I_a corresponding to the actual armature current is obtained and compared with reference signal $I_{a\gamma}$, the difference between $I_{a\gamma}$ and I_a is given as input to current controller. The output of this current controller is given to the firing circuit of the rectifier. The firing circuit output is the firing angle α which in turn changes the output voltage of the rectifier. Thus by the action of speed controller and current controller the load will be driven at the speed set by input signal ω_γ without exceeding the rated armature current. Torque developed by the motor is proportional to the product of armature current and field current. Thus there is a limitation on the load torque. The motor can drive the load of maximum torque at speeds less than or equal to the rated speed of the motor. If ω_γ is less than the rated speed the field current I_f is kept at its maximum value. But if ω_γ is greater than the rated speed the armature current is kept at its rated value and the field current has to be reduced. Back emf E_b is proportional to the product of speed and field current. But back emf is equal to armature voltage E_a minus armature resistance drop $I_a R_a$. Thus $E_b = E_a - I_a R_a$*. This E_b signal is obtained as indicated in the diagram. Actual E_b is compared with the reference value $E_{b\gamma}$ which will be set at 0.85 to 0.95 of the rated armature voltage. The difference between $E_{b\gamma}$ and E_b is used for changing the field current I_f. The field controller which is a p-I controller will change the firing angle α_f of the field rectifier till ω_m is equal to ω_γ. This action takes place when ω_γ is greater than the rated speed of the motor.

Problems

P.3.1   (a)   What is 4-quadrant operation and explain.

(b)   Give a simple circuit for the speed control of a dc shunt motor.

P.3.2   What are the advantages of electric braking over mechanical braking of DC motors? Explain with proper circuit diagram Speed-Torque characteristics of DC motor under dynamic braking, for the following types:

(i) separately excited dc motor,

(ii) series motor.

P3.3   Explain how the speed control of a separately excited dc motor is achieved with thyristor control. Draw a neat circuit diagram giving the triggering methods used.

P.3.4   (a)   Discuss in detail counter current braking operations of separately excited DC motor.

(b)   A 400 V, 750 rpm, 70 A dc shunt motor has an armature resistance of 0.3 Ω. When running under rated conditions, the motor is to be braked by plugging with armature current limited to 90 A. What external resistance should be connected in series with the armature? Calculate the braking torque and its value when the speed has fallen to 300 rpm.

P.3.5   Explain how a dc series motor is braked by plugging. Draw the speed toque characteristic during plugging.

P.3.6   Explain with circuit diagrams how regenerative braking is applied to :

(a)   separately excited dc motor and

(b)   series dc motor.

P.3.7   Explain how the speed of a separately excited dc motor be controlled in both the directions using a dual converter.

P.3.8   With a neat diagram, explain the operation of a dc drive in all four quadrants when fed by a single phase dual converter. Explain why circulating current mode is preferred.

P.3.9   Derive an expression for the circulating current of a 3ϕ dual converter. What is the purpose of circulating current inductor?

P.3.10   (a)   Explain the principle of closed loop control of dc drive using suitable block diagram

(b)   Draw and explain the torque-speed characteristics for dynamic braking operation of dc series motor. Why torque becomes zero at finite speed?

P.3.11   (a)   Draw the circuit diagram and explain the operation of closed-loop speed control with inner-current loop and field weakening.

(b)   A single phase fully controlled double bridge converter is operated from a 120 V, 60 Hz supply and the load resistance is 10 ohms. The circulating inductance is 40 mH. Firing delay angle for converter I and II are 60^0 and 120^0 respectively. Calculate the peak circulating current and the current through converters.

P3.12   A three phase dual converter feeds a 500 V, 50 A dc motor with separate excitation. The armature resistance $R_a = 1.2\ \Omega$. The converter is fed from a 400 V, 50 Hz supply. Assuming a voltage drop of 10 V in the converter determine the firing angle and back emf for

(a)   Motoring operation at full load current with motor terminal voltage of 450 V.

(b)   Regeneration operation at full load current with terminal voltage of 480 V.

(c)   The motor is plugged at a terminal voltage of 400 V with a current limiting resistance of 5 Ω.

(Ans : (a) 31.58^0, 390 V; (b) 150.5^0, 540 V; (c) 104.29^0, 460 V)

Objective Type Questions

1.   What are the different types of Electrical braking?

2.   While plugging a dc motor the power supply to the motor has to be cut off when the speed is almost zero. Why?

3.   Draw the equivalent circuit of dual converter driving a dc motor.

4.   A dual converter is operating in continuous current mode. The firing angle of positive converter is 120^0. What is the firing angle of negative Converter ?

5.   A dual converter consists of two converters connected in anti parallel. Both the converters should be fully controlled converters. Why?

6.   A dual converter is a .... quadrant converter.

7.   The closed loop speed control of dc motor has two loops, one inner loop and another outer loop. Which one of the two loops is current control loop?

8.   In the following figure the operating point of separately excited DC motor operating at rated speed and rated torque is A. At t = 0 dynamic braking is applied. A certain resistance R_b is connected so that braking torque equal to rated torque is developed at the instant of applying braking. Show how the operating point shifts as time t increases (Refer Fig. 3.5).

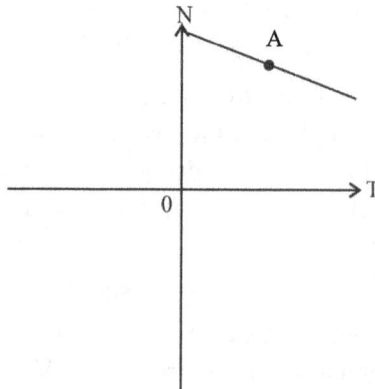

9.   In the following figure the operating point of series excited DC motor operating at rated speed and rated torque is A. At t = 0 dynamic braking is applied. A certain resistance R_b is connected so that braking torque equal to rated torque is developed at the instant of applying braking. Show how the operating point shifts as time t increases (Refer Fig. 3.9).

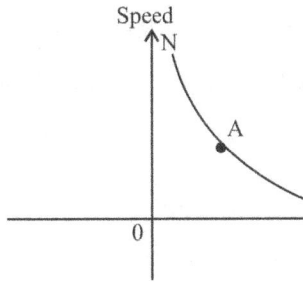

10. In the following figure the operating point of separately excited DC motor operating at rated speed and rated torque is A. At t = 0 the motor is braked by plugging. A certain resistance R_b is connected so that braking torque equal to rated torque is developed at the instant of applying braking. Show how the operating point shifts as time t increases (Fig. 3.11).

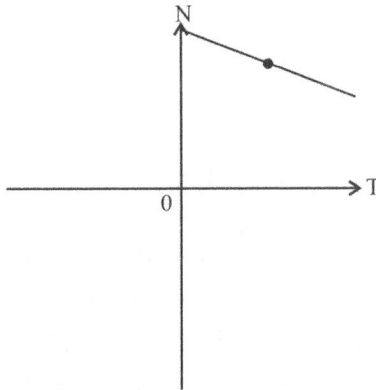

11. While applying plugging to a separately excited DC motor it is stated that the rectifier has to be disconnected from the armature at a particular speed. What is that speed ? If not disconnected what happens?

12. Is it possible for applying regenerative braking with a single half controlled converter?

13. A single phase fully controlled converter operates in Ist and IVth quadrants. When operating in Ist quadrant the range of firing angle is from .... to ....

14. The dual converter connected to 3 -ϕ, 400 V ac supply is driving a separately excited dc motor. The firing angle of positive converter is 60^0. The armature resistance is 1Ω. It draws a current of 80 A. The back emf constant $K\phi = 0.5$ V/rpm. Answer the following :

    (a) What is the average voltage applied to armature?

    (b) What is the speed at which the motor rotates?

(c)   What is the average current through negative converter?

(d)   What happens to the speed if the firing angle of negative converter is made 60°?

State whether the Following are (TRUE / FALSE)

15.   While plugging a separately excited dc motor the supply to the armature is reversed and current limiting resistance is introduced.   [   ]

16.   A dual converter is a four quadrant converter.   [   ]

17.   In a dual converter if the firing angle of positive converter is α° the firing angle of the negative converter is $(180 - \alpha)^\circ$.   [   ]

18.   A dual converter is operating in the first quadrant. The firing angle of positive converter is less than 90°.   [   ]

19.   A dual converter is operating in second quadrant. It is called farward braking.

[   ]

20.   A dual converter is operating in the second quadrant. The firing angle of positive converter is less than 90°.   [   ]

21.   A dual converter is driving a motor in third quadrant. It is called reverse motoring. The negative converter firing angle is less than 90°.   [   ]

22.   A dual converter is operating in fourth quadrant. The firing angle of positive converter is less than 90° and that of negative converter is greater than 90°.

[   ]

23.   Draw the equivalent circuit of dual converter connected to DC motor.

4

Control of DC Motors by Choppers

Control of DC motors by Choppers : Single quadrant, Two quadrant and four quadrant chopper fed dc separately excited and series excited motors – continuous current operation – output voltage and current waveforms – speed torque expressions – speed torque characteristics – problems on Chopper fed dc motors – closed loop operation (Block diagram only).

4.1   dc Chopper Drives

dc Choppers are used for obtaining variable dc voltage from fixed dc voltage source. dc voltage source may be a battery or output of diode rectifier connected to ac voltage source.

dc choppers are used for speed control of dc Motors used in industries or traction.

The following are the advantages:

1. High efficiency.
2. Flexibility in control.
3. Light weight.
4. Small size.
5. Quick response.
6. Regeneration down to very low speeds.

Comparative Advantages of Chopper Over Controlled Rectifier Chopper :

High frequency ripple in the output voltage.

Armature losses are less because of less ripple in armature current.

Lower de-rating of motor.

Region of discontinuous conduction is less.

Improved speed regulation and transient response.

With SCRs chopping frequency can be from 300 to 600 Hz.

With transistors frequency goes up to 20 KHz.

With IGBT or MOSFET frequency is greater than 20 KHz and can go up to 100 KHz.

Controlled Rectifier

With 3phase supply the ripple frequency can be 600 Hz only. This results in low utility factor and a relatively high cost.

4.2   Basic Chopper Circuit

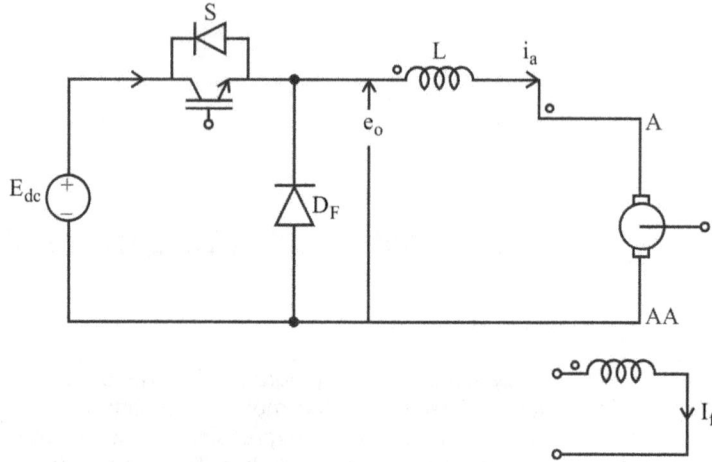

Fig. 4.1 Basic chopper circuit.

Basic chopper connected to separately excited dc motor is shown in Fig. 4.1. E_{dc} is the dc source voltage. S is switch which is operated at high speed. L is inductance connected external to dc motor to keep the armature current continuous. Continuous current operation is preferred to avoid torque pulsations.

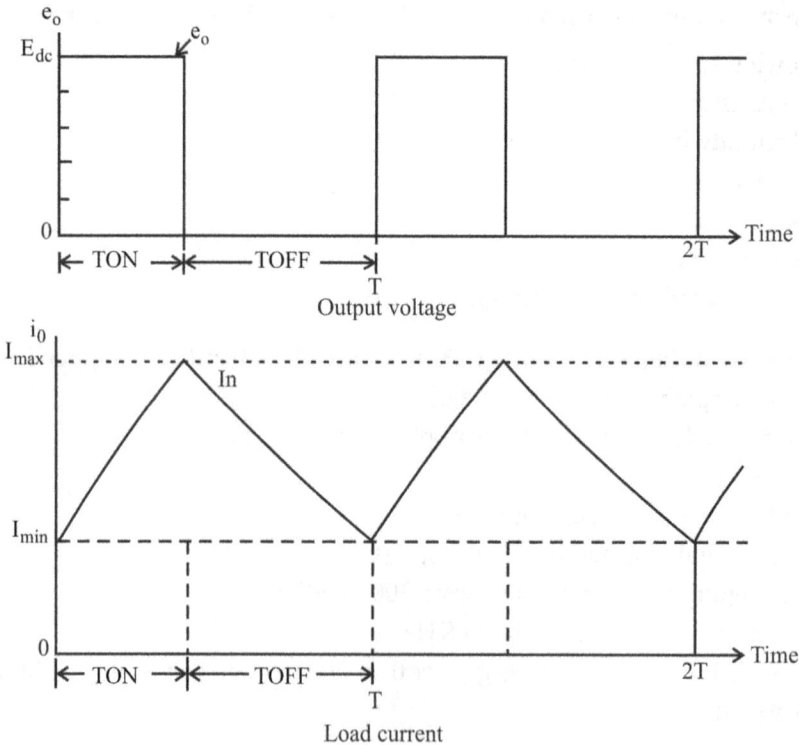

Fig. 4.1(a) Wave forms of voltage and current.

S is a solid state switch which can be turned ON or turned OFF. T_{ON} is the time for which S is ON. T_{OFF} is the time for which S is OFF.

The output voltage e_o is shown in the figure. During ON period output is equal to input voltage E_{dc}. During OFF period the output voltage is equal to zero.

$$\text{The average output voltage } E_0 = E_{dc} \frac{T_{ON}}{T_{ON} + T_{OFF}}$$

$T_{ON} + T_{OFF} = T$ the chopping period.

$$\alpha \text{ is called the duty cycle } \alpha = \frac{T_{ON}}{T}$$

Then the average output voltage $E_0 = E_{dc} \; \alpha$

The load voltage is controlled by varying the duty cycle α. There are two ways by which this can be done. They are given below.

(a) TRC or Time ratio control.

(b) Current limit control.

(a) Time ratio control

  (i) *Constant Frequency system :* T_{ON} is varied while T is held constant. It is called pulse width modulation control or PWM control.

  (ii) *Variable frequency system :* Either T_{ON} or T_{OFF} is held constant and T is varied. This system is called frequency modulation control.

(b) Current limit control : In this control the chopper is switched ON and switched OFF so that the current in the load is maintained between two limits I_{max} and I_{min}.

4.3 Types of Choppers

There are two types of choppers : 1. step down chopper and 2. step up chopper.

Step down chopper : The average output voltage E_o is less than input voltage E_{dc}.

Step up chopper : The average output voltage E_o is greater than input voltage E_{dc}.

4.3.1 Step Down Chopper

The circuit diagram is shown in Fig. 4.2. When S is ON, e_o is equal to E_{dc}. When S is OFF e_o is equal to zero.

$$\text{The average output voltage} \quad E_0 = \frac{E_{dc} \times T_{ON}}{T} = E_{dc} \, \alpha \qquad \qquad \dots\dots(4.1)$$

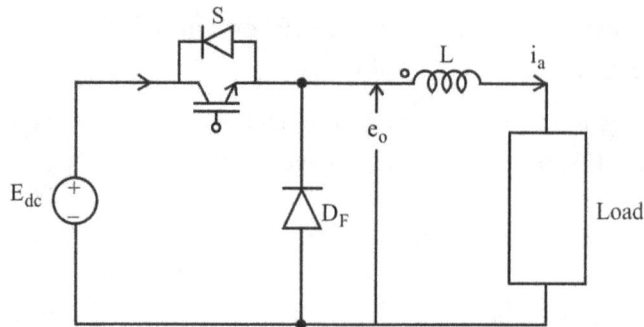

Fig. 4.2 Step down chopper.

4.3.2 Step Up Chopper

The circuit diagram is shown in Fig. 4.3.

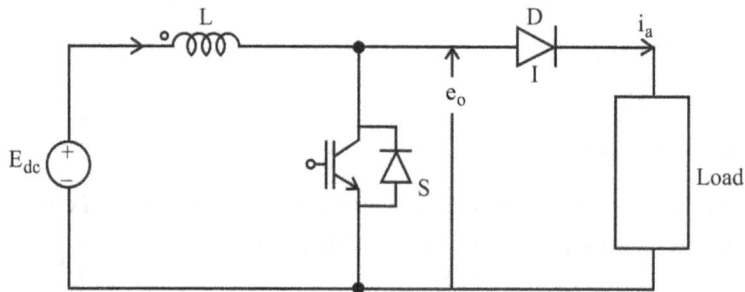

Fig. 4.3 Step up chopper.

During T_{ON} switch S is closed. The circuit diagram is shown in Fig. 4.3(a).

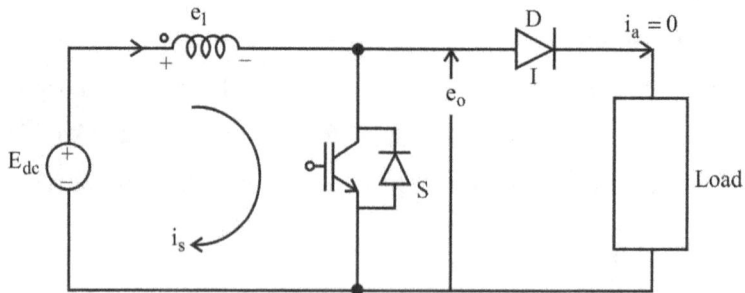

Fig. 4.3(a) Step up chopper S is ON.

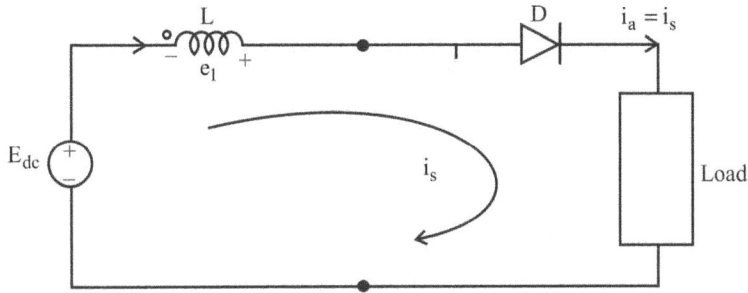

Fig. 4.3(b) Step up chopper S is OFF.

The inductor is connected across E_{dc} and the current increases and thus the stored energy in inductance L increases. Thyristor current is same as the source current.

Diode D will not allow current from the load.

During T_{OFF} the switch S is open. The circuit is as shown in Fig. 4.3(b). When the chopper is OFF, the load current is same as source current. The current decreases because the inductance L and load are in series. The voltage across L will get reversed and the load voltage is the sum of E_{dc} and e_L. This will be higher than E_{dc}

$$e_o = E_{dc} + L\frac{di_o}{dt} \qquad \text{L is very large.}$$

During T_{ON}, the energy input to the inductor $W_i = E_{dc} I_s T_{ON}$

During T_{OFF}, energy released by the inductor to load.

$$W_o = (E_0 - E_{dc}) I_s T_{OFF}$$

Under steady state these two energies are equal.

$$E_{dc} I_s T_{ON} = (E_0 - E_{dc}) I_s T_{OFF}$$

$$E_{dc} T_{ON} = E_0 T_{OFF} - E_{dc} T_{OFF}$$

$$E_0 = \frac{E_{dc}(T_{ON} + T_{OFF})}{T_{OFF}} = \frac{E_{dc}T}{T - T_{ON}} = \frac{E_{dc}}{1 - \alpha} \qquad \qquad .....(4.2)$$

4.4

The choppers are also classified based on their regions of operation in cartegion coordinates.

4.4.1 First Quadrant or Type A Chopper

The circuit diagram is shown in Fig. 4.4.

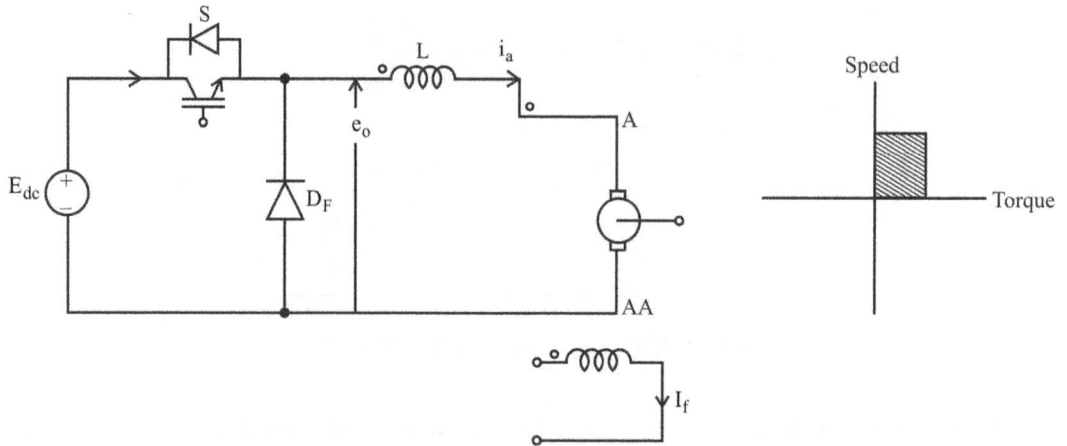

Fig. 4.4 First quadrant chopper on Type A chopper.

If the load is a dc motor or battery its equivalent circuit is as shown below in Fig. 4.4(a).

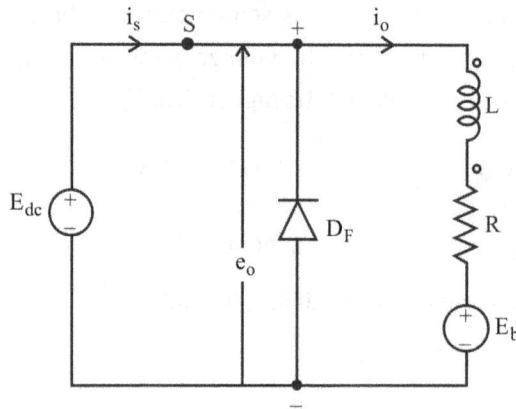

Fig. 4.4(a) Equivalent circuit during ON period.

Mode 1 : S is ON during period $0 < t < T_{ON}$

Voltages and current are related by the equation

$$E_{dc} = L\frac{di_o}{dt} + Ri_o + E_b$$ .....(4.3)

The current increases from I_{omin} to I_{omax}.

Mode 2 : S is OFF during period $T_{ON} < t < T$;

The equivalent circuit is shown in Fig. 4.4(b).

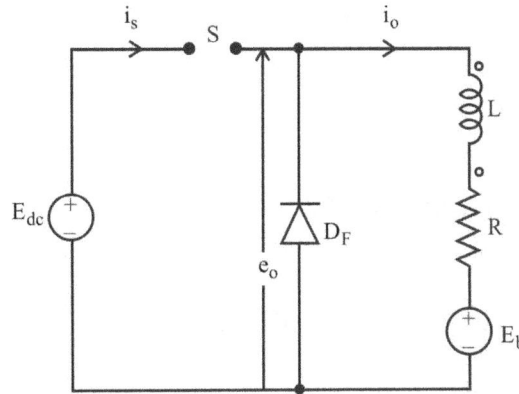

Fig. 4.4(b) Equivalent circuit during OFF period.

Current flows in the same direction. But its magnitude decreases. Voltages and current are related by the equation

$$0 = L\frac{di_o}{dt} + Ri_o + E_b \qquad \qquad .....(4.4)$$

Current decreases from $I_{omax.}$ to I_{omin}.

If $\tau = \dfrac{L}{R}$ the time constant

$$I_{0max} = \frac{E_{dc}}{R}\left[\frac{1-e^{-\frac{T_{ON}}{\tau}}}{1-e^{-\frac{T}{\tau}}}\right] - \frac{E_b}{R} \qquad \qquad .....(4.5)$$

$$I_{0min} = \frac{E_{dc}}{R}\left[\frac{e^{+\frac{T_{ON}}{\tau}}-1}{e^{\frac{T}{\tau}}-1}\right] - \frac{E_b}{R} \qquad \qquad .....(4.6)$$

The maximum and minimum currents through the motor depends on E_{dc}, E_b, L,R, T and T_{ON}

When the chopper is continuously turned on i.e., $T_{ON} = T$

Both $I_{o\ max}$ and $I_{0\ min}$ will have same value

$$I_{omax} = I_{omin} = \frac{E_{dc} - E_b}{R}$$

4.4.2 Derivation of I_{max} and I_{min} in Type A chopper

E_{dc} - Source voltage

E_o - output voltage of chopper or voltage applied to motor

I_0 - Armature current

I_s - source current

Circuit diagram of Type A chopper is shown in Fig. 4.4.

There are two modes of operation:

Mode 1: S is ON_ i.e., $0 < t < T_{ON}$

$$e_0 = E_{dc}$$

The equivalent circuit is as shown in Fig. 4.4(a).

For $\qquad\qquad 0 < t < T_{ON}$

$$e_o = E_{dc} = L\,\frac{di}{dt} + R_i + e_b$$

$$i_s = i_o$$

Mode 2 : S is OFF i.e., $T_{ON} < t < T$

The equivalent circuit is as shown in Fig. 4.4(b)

For $\qquad\qquad T_{ON} < t < T$

$$E_o = 0; \; i_s = 0$$

$$0 = L\frac{di_0}{dt} + Ri_0 + e_b$$

Wave forms of voltages and currents are shown in Fig. 4.1(a).

For $\qquad\qquad 0 < t < T_{ON}$

$$L\frac{di_0}{dt} + R\,i_o + E_b = E_{dc}$$

Taking Laplace Transform

$$(Ls + R)\,I_o - L\,I_{min} = \frac{E_{dc} - E_b}{s}$$

$\therefore \qquad\qquad$
$$I_o = \frac{E_{dc} - E_b}{s(Ls + R)} + \frac{LI_{min}}{Ls + R}$$

$$= \frac{E_{dc} - E_b}{R}\left[\frac{1}{s} - \frac{1}{s + \dfrac{R}{L}}\right] + \frac{I_{min}}{s + \dfrac{R}{L}}$$

Taking Laplace inverse

$$i_o = \frac{E_{dc} - E_b}{R}\ [1 - e^{\frac{Rt}{L}}] + I_{min}\ e^{-\frac{R}{L}t}$$

$$= \frac{E_{dc} - E_b}{R}\left[1 - e^{-\frac{t}{\tau}}\right] + I_{min}\ e^{-\frac{t}{\tau}} \qquad\qquad .....(4.7)$$

where
$$\tau = \frac{L}{R}$$

$$i_o = I_{max} \text{ at } t = T_{ON}$$

∴
$$I_{max} = \frac{E_{dc} - E_b}{R}\left[1 - e^{-\frac{T_{ON}}{\tau}}\right] + I_{min}\, e^{-\frac{T_{ON}}{\tau}} \qquad .....(4.8)$$

For
$$T_{ON} < t < T; \quad E_{dc} = 0$$

$$i_o = -\frac{E_b}{R}(1 - e^{-\frac{t_1}{\tau}}) + I_{max}\, e^{-\frac{t_1}{\tau}}$$

where
$$t_1 = 0 \text{ at } t = T_{ON}$$

At $t = T$ the current will be minimum.

i.e., at $t_1 = T\text{-}T_{ON} = T_{OFF}$, the current is equal to I_{min}

∴
$$I_{min} = -\frac{E_b}{R}(1 - e^{-\frac{T_{OFF}}{\tau}}) + I_{max}\, e^{-\frac{T_{OFF}}{\tau}} \qquad .....(4.9)$$

Substituting the value of I_{min} from (4.9) in equation (4.8) and rearranging the terms, we get

$$I_{max} = \frac{E_{dc}}{R}\left[\frac{1 - e^{-\frac{T_{ON}}{\tau}}}{1 - e^{-\frac{T}{\tau}}}\right] - \frac{E_b}{R} \qquad .....(4.10)$$

Substituting the value of I_{max} in eqn. (4.9) we get

$$I_{min} = \frac{E_{dc}}{R}\left[\frac{e^{\frac{T_{ON}}{\tau}} - 1}{e^{\frac{T}{\tau}} - 1}\right] - \frac{E_b}{R} \qquad .....(4.11)$$

Steady state ripple: The load current i_0 varies between I_{max} and I_{min}. Therefore the ripple current is $(I_{max} - I_{min})$.

From eqns. (4.10) and (4.11),

$$(I_{max} - I_{min..}) = \frac{E_{dc}}{R}\left\{\left[\frac{1 - e^{-\frac{T_{ON}}{\tau}}}{1 - e^{-\frac{T}{\tau}}}\right] - \left[\frac{e^{\frac{T_{ON}}{\tau}} - 1}{e^{\frac{T}{\tau}} - 1}\right]\right\}$$

$$= \frac{E_{dc}}{R}\left[\frac{(1 - e^{-\frac{T_{ON}}{\tau}})(1 - e^{-(T - T_{ON})/\tau})}{(1 - e^{-\frac{T}{\tau}})}\right]$$

$$\text{Ripple current} = \frac{E_{dc}}{R} \frac{(1 - e^{-\frac{\alpha T}{\tau}})(1 - e^{-\frac{(1-\alpha)T}{\tau}})}{(1 - e^{-\frac{T}{\tau}})} \qquad .....(4.12)$$

The ripple current is independent of E_b

$$\text{Per unit ripple current} = \frac{(1 - e^{-\frac{\alpha T}{\tau}})(1 - e^{-\frac{(1-\alpha)T}{\tau}})}{(1 - e^{-\frac{T}{\tau}})} \qquad .....(4.13)$$

Per unit ripple current will be maximum when $\alpha = 0.5$.

Maximum per unit ripple current is obtained by substituting $\alpha = 0.5$ in eqn. (4.13).

Output voltage: Wave form of output voltage will be rectangular as shown in Fig. 4.5.

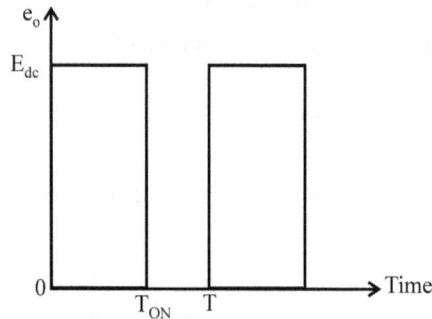

Fig. 4.5 Output voltage.

$$\text{rms value of output voltage} = \sqrt{\frac{1}{T} \int_0^{T_{ON}} E_{dc}^2 dt} = E_{dc}\sqrt{\frac{T_{ON}}{T}} = E_{dc}\sqrt{\alpha} \qquad .....(4.14)$$

$$\text{AC ripple voltage } E_r = \sqrt{E_{rms}^2 - E_0^2} = E_{dc}\sqrt{\alpha - \alpha^2} \qquad .....(4.15)$$

$$\text{Ripple factor of output voltage} = \sqrt{\left(\frac{E_{rms}^2}{E_0^2}\right) - 1} = \sqrt{\frac{E_{dc}^2 \alpha}{E_{dc}^2 \alpha^2} - 1}$$

$$= \sqrt{\frac{\alpha - \alpha^2}{\alpha^2}} = \sqrt{\frac{1 - \alpha}{\alpha}} \qquad .....(4.16)$$

4.5   Second – Quadrant or Type B chopper

The circuit diagram is shown in Fig. 4.6. This kind of arrangement is used for regenerative braking i.e., for converting kinetic energy of motor into electrical energy and feeding back to dc source. The operation is in second quadrant where speed is positive and torque is negative.

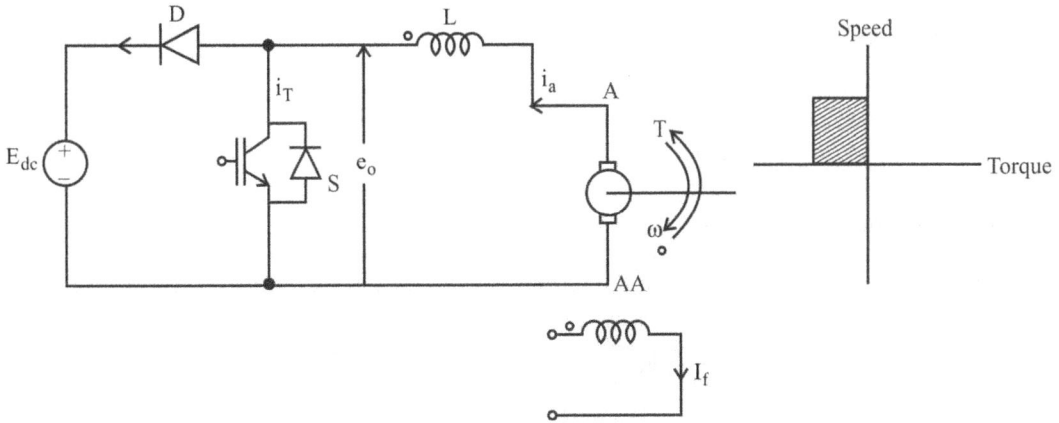

Fig. 4.6 Second quadrant chopper.

If the chopper (or switch S) is turned ON and turned OFF at regular intervals the average voltage E_0 is +ve and I_0 is −ve. The stored energy in the rotor is converted into electrical energy and fed back into the system. This is equivalent to regenerative braking.

During the period T_{ON} the switch S ON and hence $e_0 = 0$. The equivalent circuit is shown in Fig. 4.6(a)

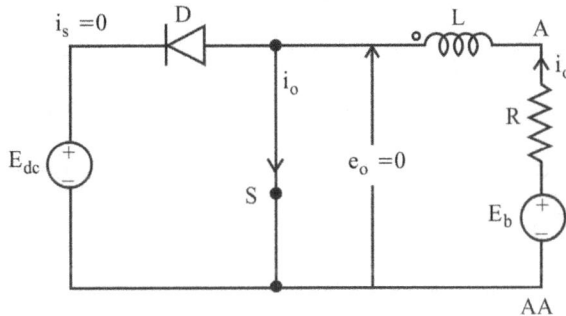

Fig. 4.6(a)

The back emf E_b forces the current to increase in opposite direction and thus increases the stored energy in the inductor. The equation showing the relation between current and voltage is given as

$$L\frac{di_0}{dt} + Ri_0 = E_b$$

Taking Laplace transfom, we get

$$I_0(s) = \frac{E_b}{s(Ls+R)} + \frac{LI_{0\min}}{(Ls+R)}$$

Taking Laplace inverse, we get

$$i_0 = \frac{E_b}{R}(1-e^{\frac{-t}{\tau}}) + I_{0\min}e^{\frac{-t}{\tau}}$$

At $t = T_{ON}$, $i_o = I_{0max}$, there fore

$$I_{0max} = \frac{E_b}{R}(1 - e^{-\frac{T_{ON}}{\tau}}) + I_{omin}e^{-\frac{T_{ON}}{\tau}} \qquad \ldots.(4.17)$$

During this period the current increases from I_{min} to I_{max}.

When switch S is OFF the equivalent circuit is as shown in Fig. 4.6(b).

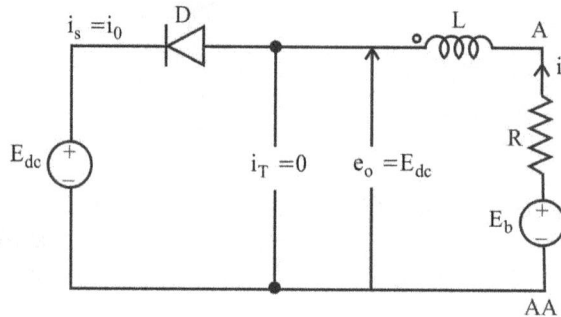

Fig. 4.6(b)

Now E_{dc} is opposite to the E_b. Hence the current decreases from I_{0max}.

The diode D is forward biased and allows the current to flow into the source.

During T_{OFF} current decreases from I_{0max} to I_{omin}. The equation for current is given as

$$L\frac{di_o}{dt} + Ri_o = E_b - E_{dc}$$

$$i_o = \frac{E_b - E_{dc}}{R}(1 - e^{-\frac{t}{\tau}}) + I_{omax}e^{-\frac{t}{\tau}}$$

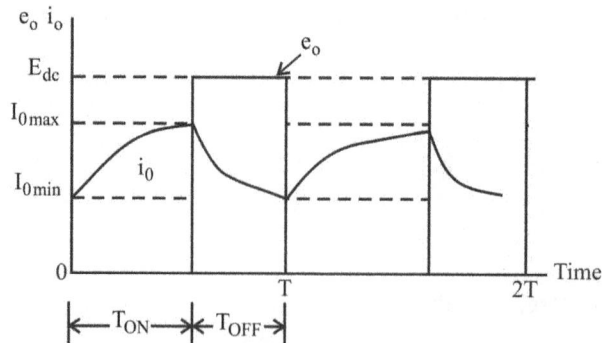

Fig. 4.6(c) Wave forms of voltage and current.

At $\qquad$ $t = T, \qquad i_o = I_{omin}$

$\therefore \qquad\qquad I_{omin} = \dfrac{E_b - E_{dc}}{R} \ (1 - e^{-\frac{T_{off}}{\tau}}) + I_{omax} \ e^{-\frac{T_{off}}{\tau}} \qquad\qquad$ .....(4.18)

4.6 Two - Quadrant or Type C Chopper

Though switching from class A to class B configuration is a satisfactory method for obtaining regenerative braking for some applications, but in applications like machine tools, driers, a very smooth transitions from driving to braking is essential.

This required drive is provided by connecting two choppers as shown in the Fig. 4.7. This is called type C or two quadrant chopper. The operating regions are in the first and second quadrants.

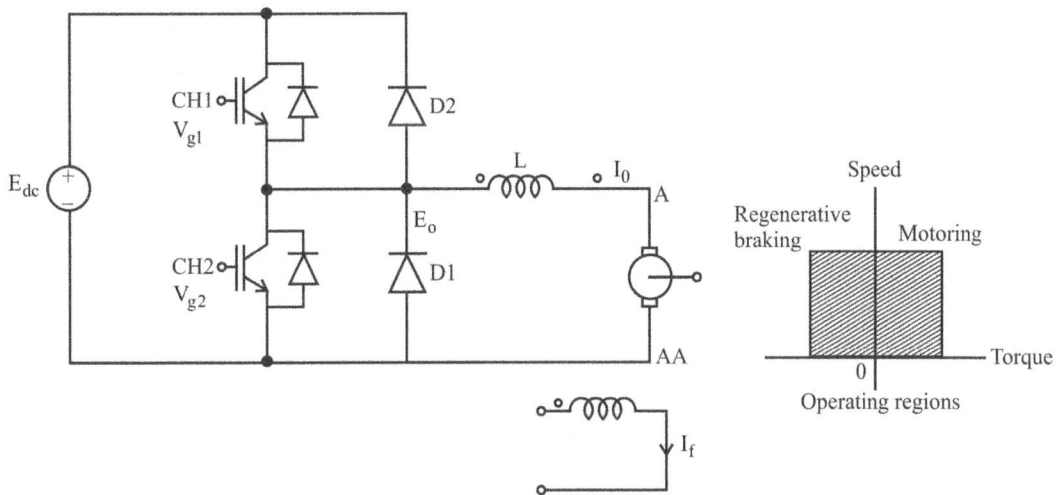

Fig. 4.7 Two quadrant chopper.

For the first quadrant operation CH1 and D1 perform the functions and if the average load current I_0 is high enough, CH2 and D2 do not conduct even though CH2 receives the gating signals because when D1 is conducting CH2 is reverse biased. For the second quadrant operation CH2 and D2 perform the functions and if the average load current I_0 has a sufficiently large negative current, CH1 and D1 do not conduct even though CH1 receives gating signal. The wave forms for two different T_{ON} and T_{OFF} times are given in the Fig. 4.8(a) and 4.8(b). V_{g1} and V_{g2} are gating signals given to CH1 & CH2 respectively.

Let us consider the operation when the duration of gating signal V_{g1} to CH1 is greater than that of V_{g2} to CH2 i.e., the gating signals are given to CH1 for duration T_{ON} greater than half periodic time $\dfrac{T}{2}$ as shown in Fig. 4.8(a).

Initially when both the choppers are OFF, the load is isolated from the supply. The wave form of the output current under steady state condition is shown. If current enters terminal A of armature it is considered as positive, if it comes out from terminal A, it is considered as negative. During interval OP the current is negative. Hence D2 is conducting CH1 is reverse biased. Hence even though the gate signal is given to CH1, it does not come to ON state. When the magnitude of current is zero at P, the direction of current changes and it becomes positive and CH1starts conducting and D2 is OFF.

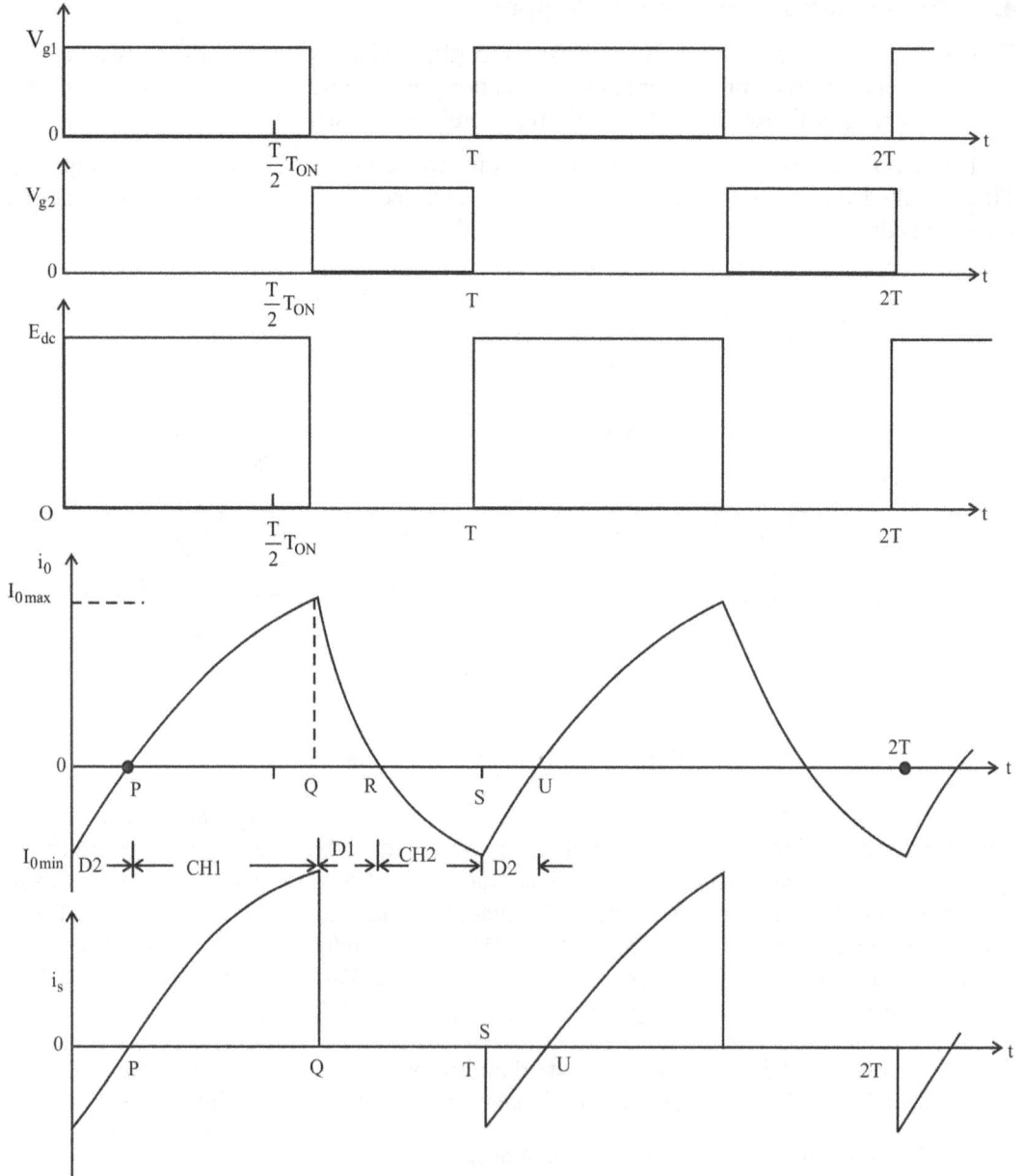

Fig. 4.8(a) Waveforms for $T_{ON} > \dfrac{T}{2}$.

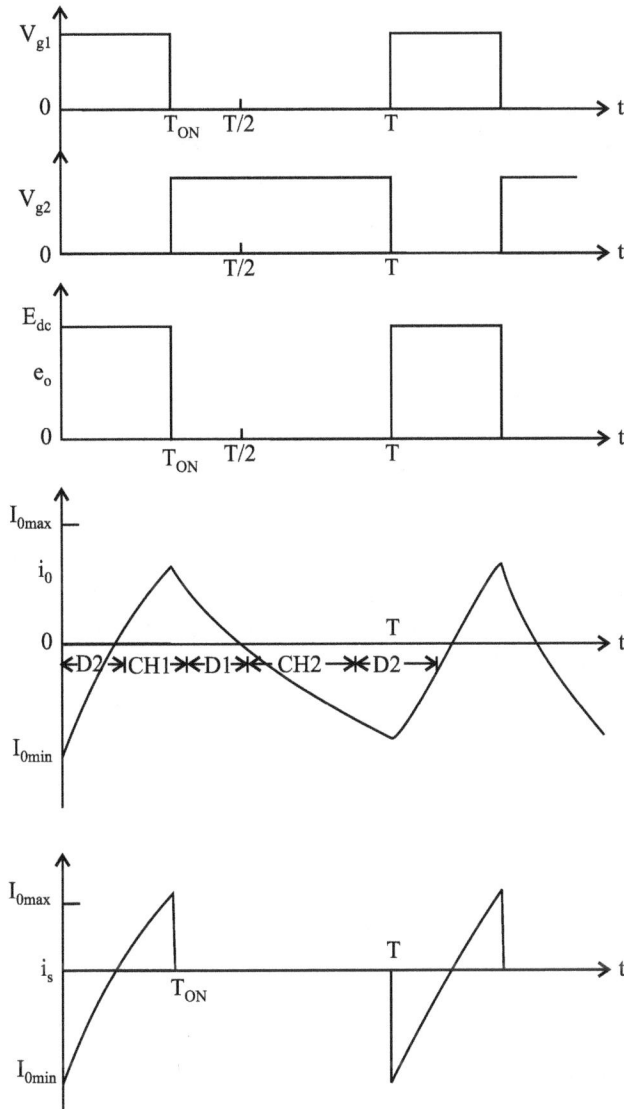

Fig. 4.8(b) Waveforms for $T_{ON} < \dfrac{T}{2}$.

The current increases to I_0 max at Q. Now CH1 is turned OFF by removing gate signal. Now the energy stored is inductance forces the current to flow through D1. Immediately after CH1 is turned OFF gate signal is given to CH2. But it does not come to ON state as D1 is still conducting. CH2 starts conducting at R when the current is zero. The current increases in opposite direction because of generated voltage in armature and the kinetic energy is converted to Electrical energy and stored in inductor. During this interval $e_0 = 0$. The current increases in opposite direction and reaches I_0 min at S. Now CH2 is turned OFF and gate signal is given to CH1. But it does not turn ON as the current is negative and diode D2 is conducting. Now the current starts decreasing as supply voltage is opposing E_b.

At U the current becomes zero, and CH1 comes to ON state and the current increases in positive direction. The average current will be positive.

By making the duration of pulses given of CH1 less than that given to CH2 as shown in Fig. 4.8(b) the average current through armature is negative.

If the average output current I_0 is positive motoring action takes places. If the average output current I_0 is negative regenerative braking occurs. By controlling T_{ON} we make I_0 either positive or negative.

$$I_0 = \frac{E_0 - E_b}{R}$$

where

$$E_0 = E_{dc} \; \alpha \; = \; E_{dc} \frac{T_{ON}}{T}$$

4.7   Two Quadrant Chopper – Type-D Chopper

The basic power circuit of this chopper is shown in Fig. 4.9. This chopper is used for operating in first and fourth quadrants. For example, a lift going up has positive torque and positive speed but while coming down speed is negative but torque is positive to limit the speed. While coming down the kinetic energy is converted to electrical energy and fed back to DC source. Thus while going up operation is in first quadrant and while coming down operation is in fourth quadrant.

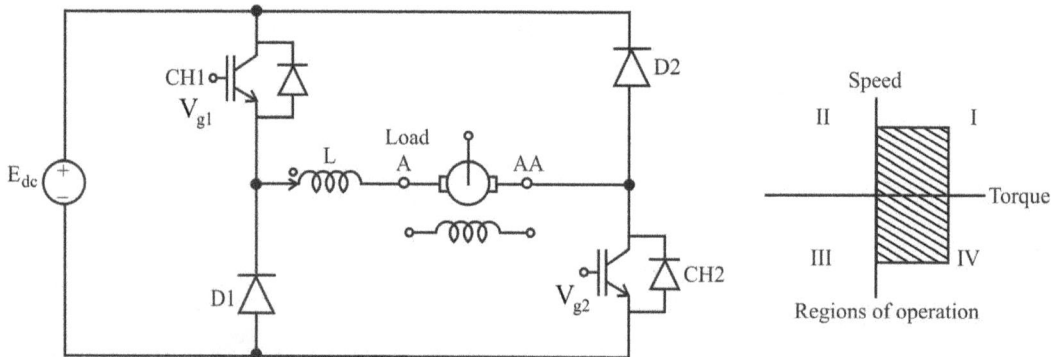

Fig. 4.9 Two quadrant chopper – Type D chopper.

There are two modes of operation.

Mode **1** *:* First quadrant operation gating signals are given to both CH1 and CH2 for duration as shown in Fig. 4.9(a).

When both CHI and CH2 are ON i.e., for $0 < t < t_1$ full voltage is applied to armature. Diodes D1 and D2 are reverse biased. Current starts rising. Energy stored in inductor increases. The equivalent circuit is as shown in Fig. 4.9(a1).

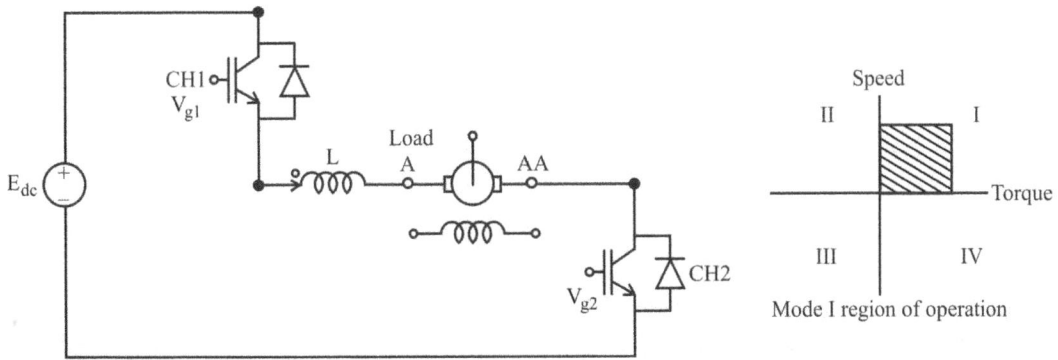

Fig. 4.9(a) CH1 and CH2 are ON.

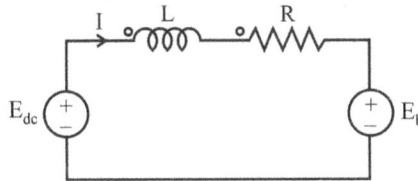

Fig. 4.9 (a1) $0 < t < t_1$.

During this interval ($0 < t < t_1$),

$$E_{dc} = L\frac{di}{dt} + R_i + E_b.$$

Current rises from I_{0min} to I_{0max}.

If any one valve is turned OFF, the current flows through other valve and one diode. The voltage across armature will be zero, current free-wheels and decreases. Energy stored in inductor decreases. As an example, if CH1 is OFF, for duration t_3 to T, the current flows through CH2 and D1 as shown in Fig. 4.9(b).

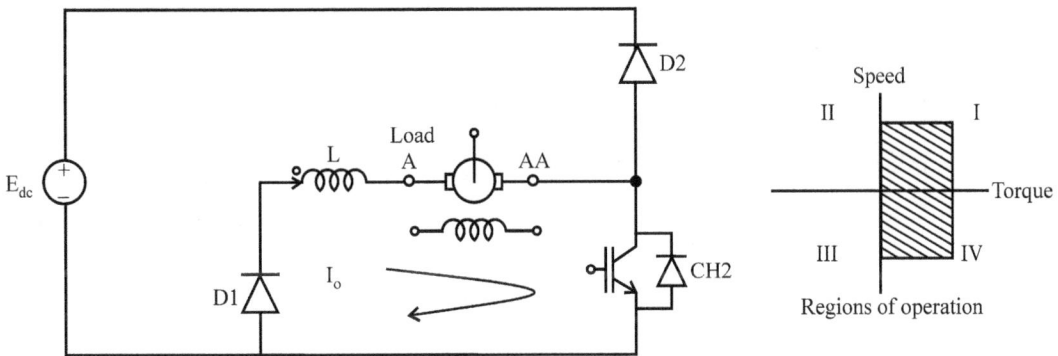

Fig. 4.9(b) CH1 is OFF.

The equivalent circuit is as shown in Fig. 4.9(b1). This interval is $t_3 < t < t_4$ in Fig. 4.9(c).

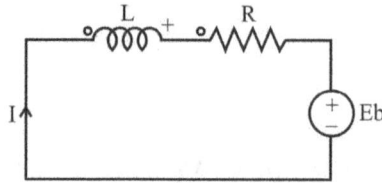

Fig. 4.9(b1)

$$0 = L\frac{di}{dt} + Ri + E_b$$

$\dfrac{di}{dt}$ is negative and current decreases from I_{0max} to I_{0min}. Same thing will happen during the interval $t_1 < t < t_2$. During the interval no gating signal to CH2 and hence if is OFF.

The current freewheels through CH1, armature and D2.

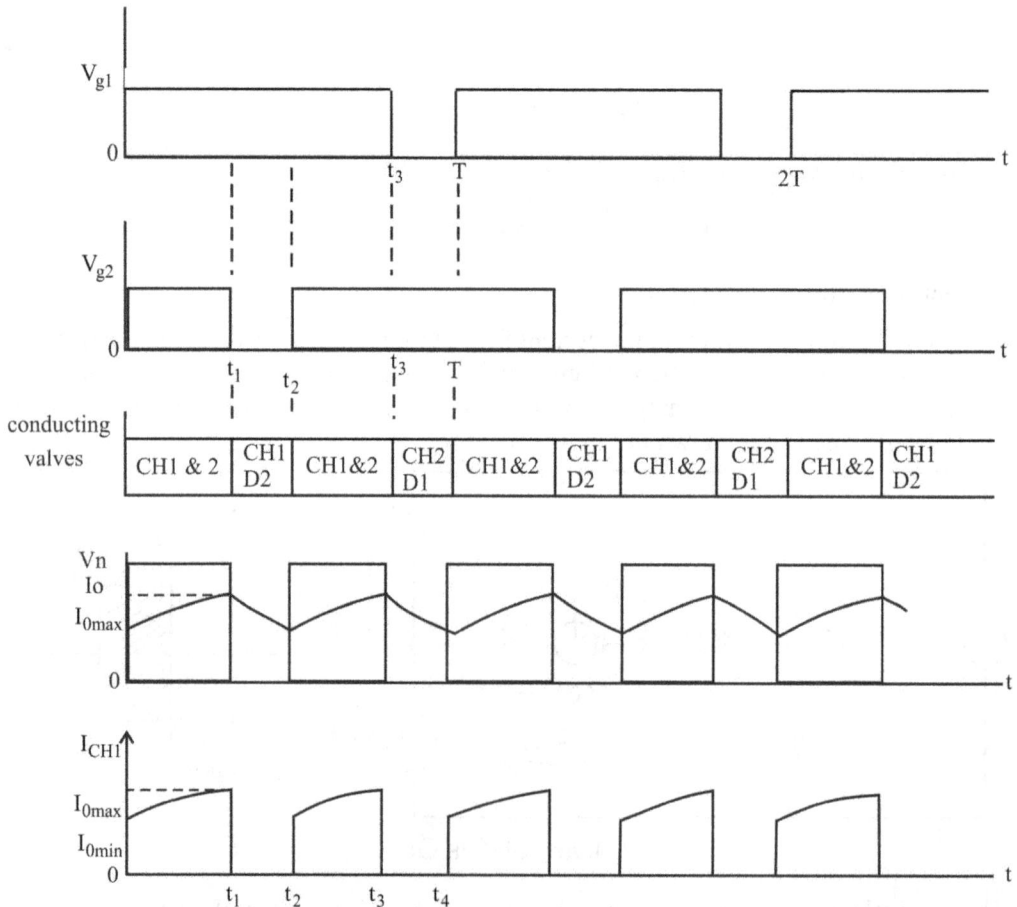

Fig. 4.9(c) Waveforms for Mode 1 operation.

Operation will still be in first quadrant as the current enters the armature and A and leaves at AA. The average current is positive. Average value of this current is varied by varying ON time and OFF time of CH1 and CH2. Thus speed and torque are controlled.

Mode 2 : Gating signals are given to CH1 and CH2 for duration less than half period i.e., $T_{ON} < T/2$. But the gating signals V_{g2} to CH2 are delayed by time t_2 as shown in Fig. 4.10(a). At any instant of time either CH1 or CH2 will be conducting. The wave forms for this mode are shown in Fig. 4.10(b).

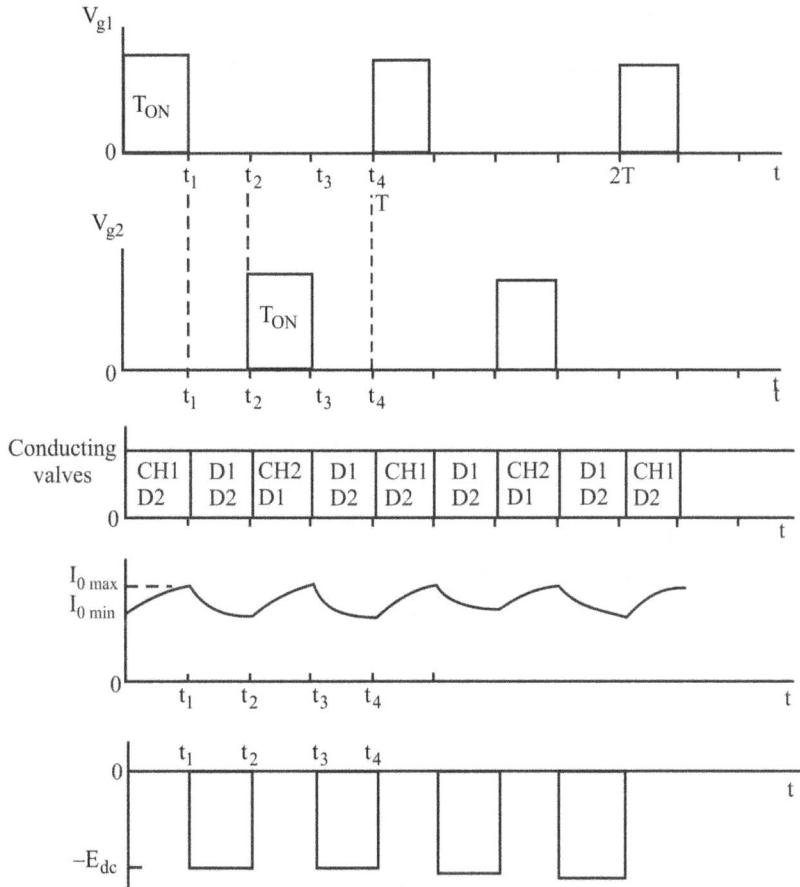

Fig. 4.10(a) Waveforms for mode 2 operation.

During the interval $0 < t < t_1$ CH1 and D2 conduct as shown in Fig. 4.10(b). The back emf forces the current to rise from I_{0min} to I_{0max}. The equivalent circuit is shown in Fig. 4.10(b1). At $t = t_1$, CH1 goes to OFF state as V_{g1} is the moved.

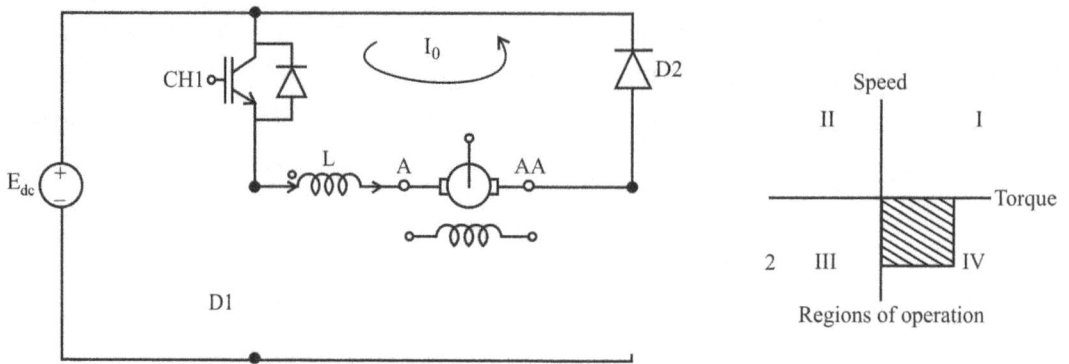

Fig. 4.10(b) CH1 is ON & CH2 is OFF.

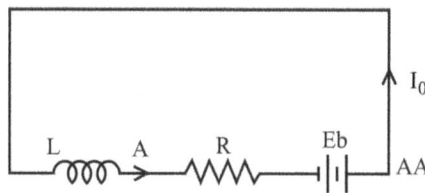

Fig. 4.10(b1) Equivalent circuit.

During the interval ($t_1 < t < t_2$) CH1 and CH2 are OFF. The current flows through D1 and D2 back into source as shown in figure 4.10(c).The equivalent circuit is as shown in Fig. 4.10(c1).

The induced emf in armature and voltage across inductance are additive and this voltage forces the current into source. This is regenerative braking.

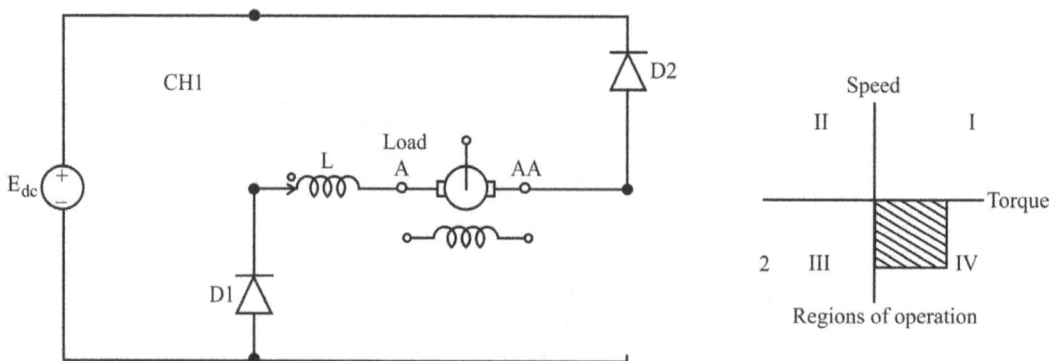

Fig. 4.10(c) CH1 & CH2 are OFF ($t_1 < t < t_2$).

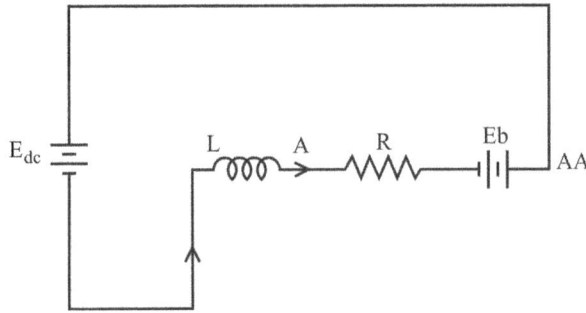

Fig. 4.10(c1) Equivalent circuit.

During interval ($t_2 < t < t_3$) CH2 is turned ON .The current flows through CH2 and diode D1 as shown in Fig. 4.10(d). The equivalent circuit is shown in Fig. 4.10(d1). The induced emf in armature E_b forces the current to increase from $I_{0\ min}$ to I_{0max}.

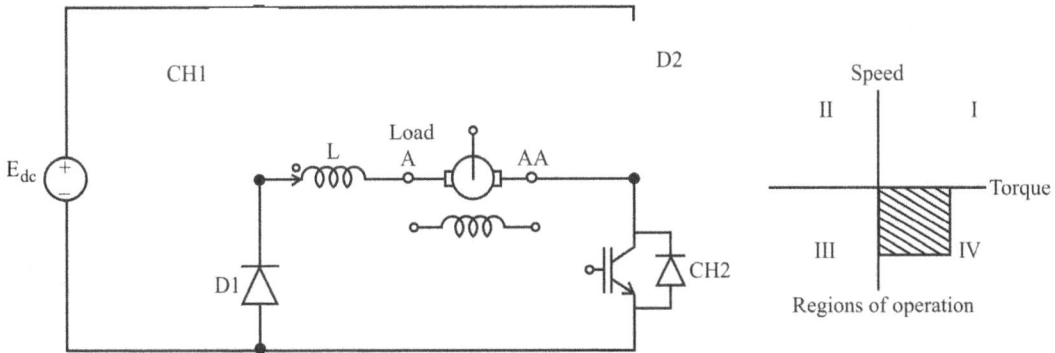

Fig. 4.10 (c) CH1 is OFF & CH2 is ON ($t_2 < t < t_3$).

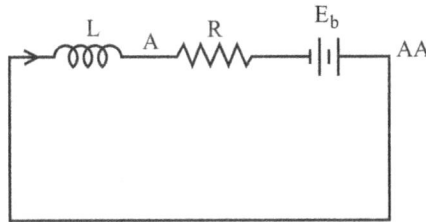

Fig. 4.10(d1) Equivalent circuit.

At t = t_3, CH2 is turned OFF. From t_3 to T regenerative action takes place as both CH1 and CH2 are OFF and the cycle repeats. The operating region is fourth quadrant. The direction of rotation has changed and hence the polarity of back emf has changed.

Average output voltage is positive if T_{ON} of CH1 and CH2 is greater than $\dfrac{T}{2}$ and becomes negative when T_{ON} of the switches CH1 & CH2 is less than $\dfrac{T}{2}$. In this configuration since average load current I_0 is positive and average voltages is reversible, the power flow is also reversible. This can be used both for motoring and regenerative braking. Type D chopper is used for operating lift or winch. When lift has to go up armature rotates in a particular direction, machine is acting as motor, back emf is opposite to supply voltage. The equivalent circuit is shown in Fig. 4.9(a1). When the lift is coming down, the direction of rotation is changed and hence the polarity of back emf has changed. The current direction has not changed. The direction of torque is not changed. Therefore braking action takes place and prevents acceleration of downward motion. The equivalent circuit is shown in Fig. 4.10(b1)(c)(d1).

4.8   Four Quadrant Chopper or Type E Chopper

Four quadrant chopper can be considered as parallel combination of two type C choppers. The power circuit diagram and its operating region is shown in Fig. 4.11. CH1 to CH4 are semi-conductor switches. They may be Transistors, SCRs or IGBTs. Important precaution is that if CH1 is ON, CH2 should be OFF and vice versa, otherwise source will be shorted. Similarly if CH3 is ON, CH4 should be OFF. The operating region of this chopper is in all four quadrants.

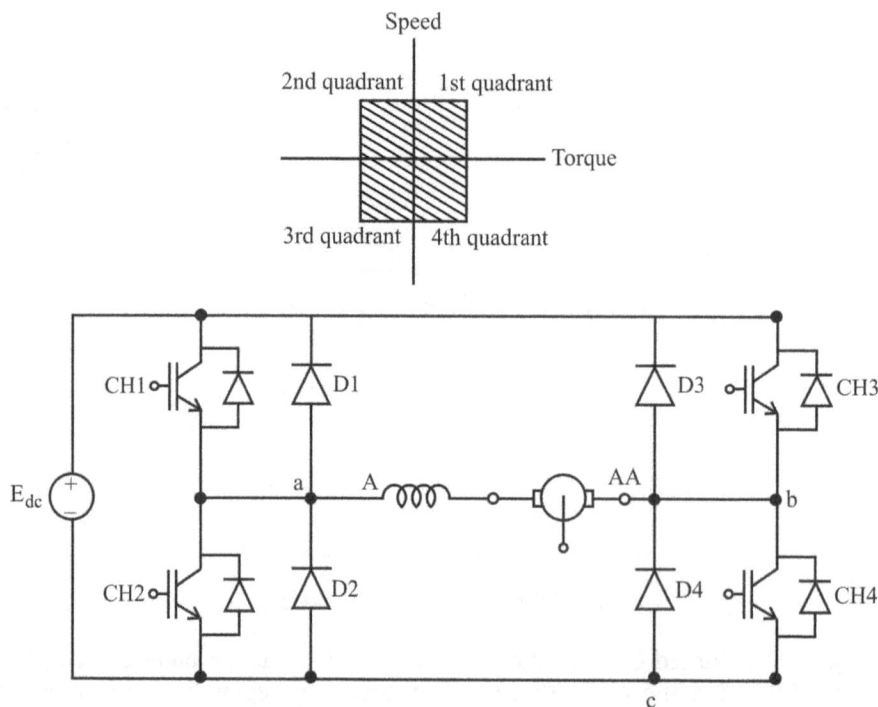

Fig. 4.11 Four quadrant chopper.

For operation in first quadrant CH4 should be continuously turned ON, so that points b and c are shorted by D4 and CH4. D3 & CH3 are reverse biased and act as open circuit. CH1 is turned ON and turned OFF at regular intervals. The circuit diagram is as shown in Fig. 4.11(a)

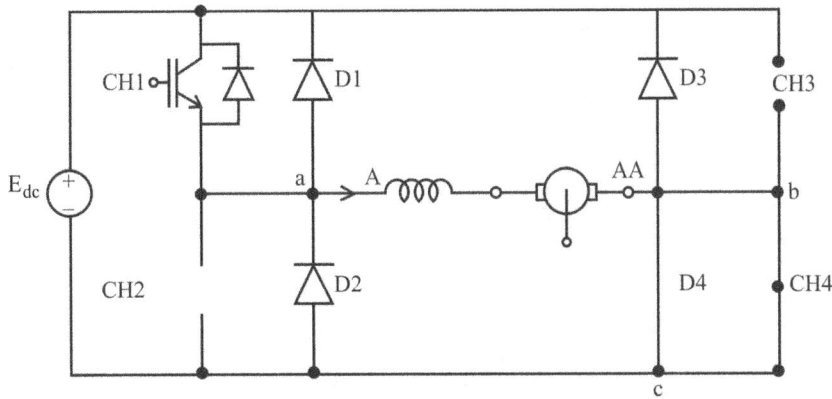

Fig. 4.11(a)

Full voltage of E_{dc} is applied across a-b. Voltage is positive, current is positive. Torque and speed are positive. The machine acts as motor and operates in Ist quadrant.

For operation in second quadrant CH4 is continuously ON. CH1 is OFF. CH2 is switched ON and OFF. The circuit is shown in Fig. 4.11(b). It acts like Type B chopper. Voltage across a-b is zero when CH2 is ON and minus E_{dc} when it is OFF. Back emf E_b forces current in opposite direction. Hence torque developed is negative and speed is positive. Operation is in 2nd quadrant. The machine operates as generator converting kinetic energy into electrical energy. It is regenerative braking.

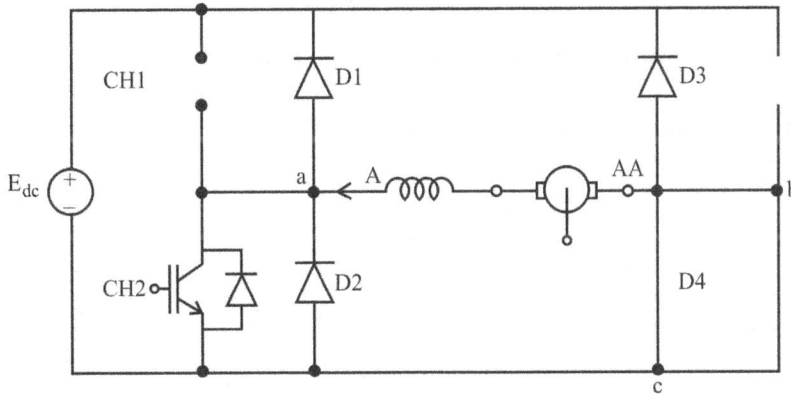

Fig. 4.11(b)

For operation in third quadrant CH2 is continuously triggered and CH1 is OFF. Points a and c are shorted. The circuit will be as shown in Fig. 4.11(c). No pulses are given to CH4 and CH3 is switched ON and OFF with certain duty cycle. The average current through armature is from AA to A. Torque developed is negative and machine operates as motor rotating in opposite direction. For operation in fourth quadrant CH3 is turned OFF and CH4 is turned ON. The back emf drives the current through CH4. The current path is AA-c-D2-a-A-AA. The current increases. If now CH4 is turned OFF, current tries to decrease, reversing the polarity of voltage across inductor. Sum of voltage across inductor and back emf will be greater than the supply voltage, which in turn forces the current to enter positive terminal of dc source. This is regenerative braking. The current path is from AA-D3-E_{dc}-c-D2-a-A-AA. This four quadrant chopper can be used for reversible regenerative dc drives.

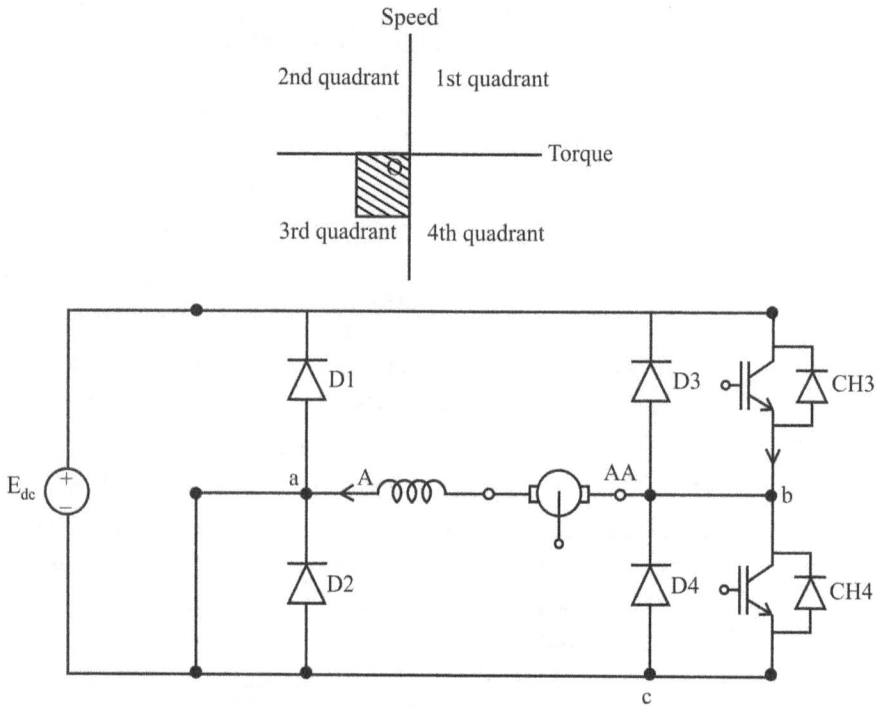

Fig. 4.11(c) Third and fourth quadrants operation.

4.9   dc Series Motor Connected to Chopper

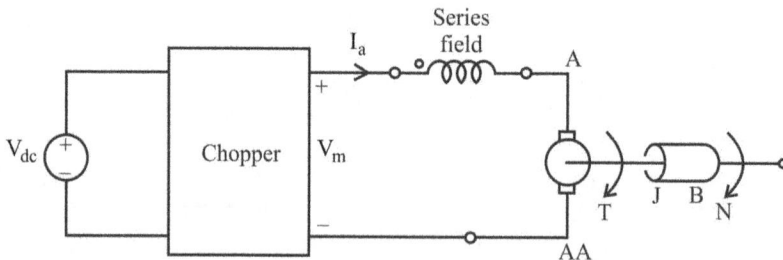

Fig. 4.12 dc series motor connected to dc chopper.

dc series motor connected to dc chopper is shown in Fig. 4.12. V_{dc} is the fixed dc source voltage. V_m is the output voltage of chopper which is applied to the motor. The applied voltage V_m to the motor can be controlled by varying the duty cycle of the chopper.

If α is the duty cycle of chopper $V_m = \alpha V_{dc}$. The equivalent circuit of series motor under steady state condition is shown in Fig. 4.13.

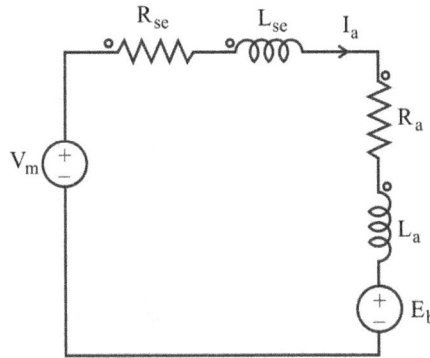

Fig. 4.13 Equivalent circuit under steady state.

The following equations hold good.

$$V_m = \alpha\, V_{dc} = I_a\,(R_{se} + R_a) + E_b \qquad \qquad .....(4.21)$$

where

V_{dc} = Source voltage in Volts.

α = Duty cycle of chopper.

V_m = Voltage applied to motor in Volts.

R_{se} = Resistance of series field in Ohms.

R_a = Resistance of armature in Ohms.

E_b = Induced voltage in armature in Volts.

$$E_b = KI_a\omega \qquad \qquad .....(4.22)$$

where ω is the speed of armature in radians per second and series field flux is assumed as proportional to field current.

T_d = Torque developed in N-m

$$T_d = KI_a^{\,2} \qquad \qquad .....(4.23)$$

T_L = Load torque in N-m

$$T_d = B\omega + T_L \qquad \qquad .....(4.24)$$

where B is frictional coefficient in N-m per rad/sec.

$$\alpha V_{dc} = I_a\,(R_{se} + R_a) + E_b = \frac{\sqrt{T_d}}{\sqrt{K}}(R_{se} + R_a) + K\omega\frac{\sqrt{T_d}}{\sqrt{K}} \qquad \qquad .....(4.25)$$

$$\omega = \frac{\alpha V_{dc}}{\sqrt{KT_d}} - \frac{(R_{se} + R_a)}{K} \qquad \qquad .....(4.26)$$

Torque speed curves for different duty cycles are shown in Fig. 4.14.

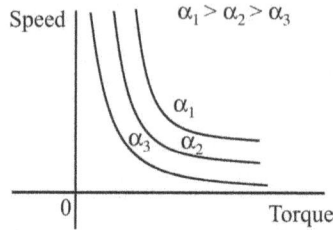

Fig. 4.14 Torque speed characteristics.

At starting ω is zero. Hence starting torque is large and is given by expression

$$T_{ds} = \frac{\alpha^2 V_{dc}^2 K}{(R_{se} + R_a)^2} \qquad .....(4.27)$$

4.10 Separately Excited DC Motor Connected Chopper

The circuit diagram is shown in Fig. 4.15. V_{dc} is the fixed DC voltage source and input to chopper. V_m is the output of chopper and is equal to αV_{dc}, where α is the duty cycle of chopper.

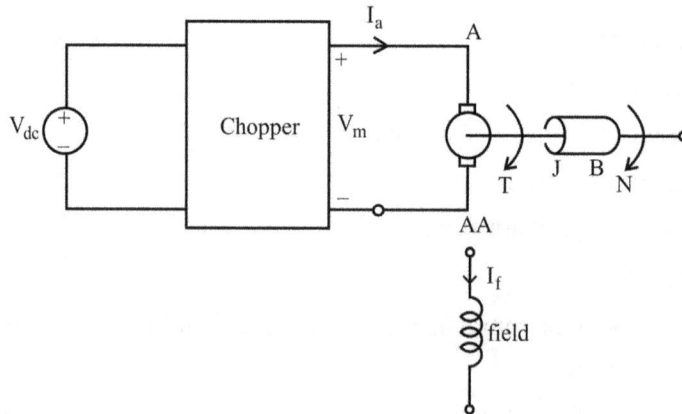

Fig. 4.15 Separately excited DC motor connected to chopper.

The equivalent circuit under steady state is shown in Fig. 4.16.

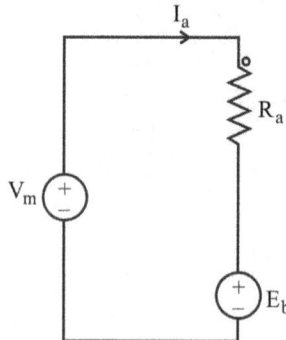

Fig. 4.16 Equivalent circuit.

The following equations hold good under steady state.

$$V_m = \alpha V_{dc} = E_b + I_a R_a \qquad \text{.....(4.28)}$$

$$E_b = K I_f \omega \qquad \text{.....(4.29)}$$

$$T_d = K I_f I_a \qquad \text{.....(4.30)}$$

$$\omega = \frac{E_b}{KI_f} = \frac{\alpha V_{dc}}{KI_f} - \frac{T_d R_a}{(KI_f)^2} \qquad \text{.....(4.31)}$$

The above equation gives relation between torque and speed. The torque speed characteristics are shown in Fig. 4.17 for different duty cycles.

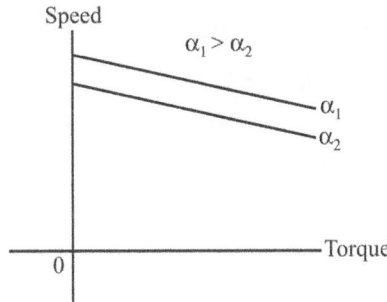

Fig. 4.17 Torque speed characteristics for different duty cycles.

Worked out Problems on dc Chopper Drive

Example **4.1**

The speed of a separately excited dc motor is controlled by a chopper as shown in the Figure 1. The dc supply voltage is 100 V, armature circuit resistance is $R_a = 0.4\,\Omega$, armature circuit inductance is $L_a = 10$ mH and the motor constant is $K_a\,\phi = 0.05$ V/rpm. The motor drives a constant torque load requiring an average armature current of 30 A. Assume that the motor current is continuous. Compute. (a) The range of speed control (b) The range of duty cycle α.

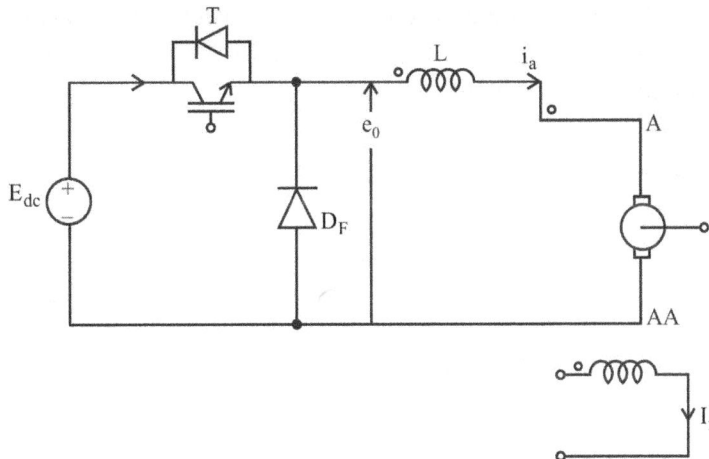

Solution:

The average value of the armature voltage E_A is given by $E_A = E_{dc} \times \alpha$ where α is the duty cycle of chopper.

$$E_A = E_b + I_a R_a; \quad E_b = K_a \phi N$$

When speed is zero $E_b = 0$

$$E_A = 0 + I_a R_a = 30 \times 0.4 = 12V$$

$$E_A = E_{dc} \times \alpha$$

Duty cycle $\alpha = 12/100 = 0.12$ for $N = 0$.

i.e., for a speed of zero, the duty cycle $= 0.12$

The maximum value of duty cycle $\alpha = 1$

$\therefore$ $E_A = E_{dc} \times 1 = 100$ V

Now back emf $= 100 - 30 \times 0.4 = 88$ V.

Speed for getting this back emf $= \dfrac{88}{0.05} = 1760$ rpm

Speed range $0 < N < 1760$

Duty cycle α range $0.12 < \alpha < 1$

Example 4.2

(a)   A dc chopper feeds power to an RLE load with $R = 2\,\Omega$, $L = 10$ mH and $E = 6$ V. If this chopper is operating at chopping frequency of 1 KHz and with duty cycle of 10% from a 220 V dc source, compute the maximum and minimum currents taken by the load.

(b)   A dc chopper is used to control the speed of separately excited dc motor. The dc supply voltage is 220 V, armature resistance $r_a = 0.2\ \Omega$ and motor constant $K_a \phi = 0.8$ V/rpm. The motor drives a constant torque load requiring an average current of 25 A. Determine (i) The range of speed control (ii) range of duty cycle. Assume the motor current to be continuous.

Solution :

$L = 10$ mH, $R = 2\ \Omega$

Time constant $\tau = \dfrac{L}{R} = \dfrac{10}{2} * 10^{-3} = 5 * 10^{-3}$ sec

$f = 1$KHz $T = 1 \times 10^{-3}$ sec

$\alpha = 10\% = 0.1$

$\therefore$ $T_{ON} = 0.1 \times 10^{-3}$ sec

$E_{dc} = 220$ V, $E_b = 6$ V

(a)

$$I_{0\,max} = \frac{E_{dc}}{R}\left[\frac{1-e^{-\frac{T_{ON}}{\tau}}}{1-e^{-\frac{T}{\tau}}}\right] - \frac{E_b}{R}$$

$$= \frac{220}{2}\left[\frac{1-e^{-\frac{0.1\times10^{-3}}{5\times10^{-3}}}}{1-e^{-\frac{1\times10^{-3}}{5\times10^{-3}}}}\right] - \frac{6}{2}$$

$$= 110\left[\frac{0.0198}{0.181}\right] - 3 = 12.033 - 3 = \textbf{9. 033 A}$$

$$I_{min} = \frac{E_{dc}}{R}\left[\frac{e^{\frac{T_{ON}}{\tau}}-1}{e^{\frac{T}{\tau}}-1}\right] - \frac{E_b}{R}$$

$$= \frac{220}{2}\left[\frac{e^{\frac{0.1}{5}}-1}{e^{\frac{1}{5}}-1}\right] - \frac{6}{2}$$

$$= 110\left(\frac{1.02-1}{1.221-1}\right) - 3 = 9.\,95 - 3 = \textbf{6. 95 A}$$

(b)   Speed will be zero when back emf is zero.

$$E_A = E_b + I_a\,R_a$$

$$= 0 + 25 \times 0.2 = 5$$

$$220 \times \alpha = 5;\ \alpha = \frac{5}{220} = \frac{1}{44}$$

Speed will be maximum when $\alpha = 1$

$$220 = E_b + I_a\,R_a = E_b + 25 \times 0.2$$

$$E_b = 215\ \text{V}$$

$$\text{Speed} = \frac{215}{0.8} = \textbf{268.7 rpm}$$

Speed range is 0 to 268.7 rpm.

Duty cycle range is 1/44 to 1

Example 4.3

(a)   A chopper fed from a 220 V dc source is working at a frequency of 50 Hz and is connected to R-L load of R = 5Ω and L = 40 mH. Determine the value of duty cycle at which the minimum load current is 20 A.

(b)   For the value of α obtained in (a) calculate the corresponding values of maximum current and the ripple factor.

Solution:

(a)

$$I_{min} = \frac{E_{dc}}{R}\left[\frac{e^{\frac{T_{ON}}{\tau}}-1}{e^{\frac{T}{\tau}}-1}\right] = \frac{220}{5}\left[\frac{e^{\frac{T_{ON}}{\tau}}-1}{e^{\frac{20\times10^{-3}}{8\times10^{-3}}}-1}\right]$$

$$\frac{(e^{\frac{T_{ON}}{\tau}}-1)45}{e^{\frac{20}{8}}-1} = 20$$

$$(e^{\frac{T_{ON}}{\tau}}-1) = \frac{20\times12.18}{44} = 5.53$$

$$e^{\frac{T_{ON}}{\tau}} = 1+4.96 = 5.96 ; \quad \frac{T_{ON}}{\tau} = \ln 5.96 = 1.78 ;$$

$$T_{ON} = 8\times10^{-3}\times1.78 = 14.28 \text{ m sec.}$$

Duty cycle $\alpha = \dfrac{T_{ON}}{T} = \dfrac{14.28}{20} = \mathbf{0.714}$

(b)   Maximum current I_{max}

$$I_{max} = \frac{E_{dc}}{R}\left[\frac{1-e^{-\frac{T_{ON}}{\tau}}}{1-e^{\frac{-T}{\tau}}}\right] = \frac{220}{5}\left[\frac{1-e^{\frac{-14.28}{8}}}{1-e^{\frac{20}{8}}}\right] = 45\frac{1-0.832}{0.9179} = \mathbf{40.78A}$$

Ripple factor of current $= \dfrac{I_{0max}-I_{0min}}{E_{dc}/R}$

$$= \frac{40.78-20}{220/5} = \frac{20.78}{45} = 0.461$$

Ripple factor of voltage $= \sqrt{\dfrac{1-\alpha}{\alpha}} = \sqrt{\dfrac{1-0.714}{0.714}} = 0.6328$

Example 4.4

A dc chopper is used in rheostatic braking of a dc separately excited motor. The armature resistance is $R_a = 0.05\,\Omega$. The braking resistor, $R_b = 5\,\Omega$. The back emf constant is $K_b = 1.52$V/A rad/sec. The average armature current is maintained constant at $I_a = 150$ A. The armature current is continuous and has negligible ripple. The field current is $I_f = 1.5$ A. If the duty cycle of the chopper is 40% determine,

(a)   The average voltage across the chopper V_{ch},

(b)   The power dissipated in the braking resistor,

(c)   The equivalent load resistance of the motor acting as generator R_{eq},

(d)   The motor speed, and

(e)   The peak chopper voltage V_p.

Solution :

$$I_a = 150 \text{ A}, K_b = 1.527 \text{ V/A rad/sec}, \alpha = 0.4, R_m = R_a = 0.05\,\Omega$$

(a)   Average voltage across the chopper:-

Average current through the resistance $I_b = 150 \times (1 - 0.4) = 90$ A

Average voltage $= R_b\, I_b = 90 \times 5 = 450$ V

(b)   Power dissipated $= I_a^2\, R_b\, (1 - \alpha) = 150 \times 150 \times 5\,(1 - 0.4) = 67.5$ KW

(c) $R_{eq} = R_b\,(1 - \alpha) + R_m = 5 \times 0.6 + 0.05 = 3.05\,\Omega$

(d)   Generated voltage $= 150\,(R_b\,(1 - \alpha) + R_m)$

$$= 150 \times 3.05 = 457.5 \text{ V}$$

$$\text{Speed } \omega_m = \frac{E_g}{K_b I_f} = \frac{457.5}{1.527 \times 1.5} = 199.73 \text{ rad/sec}$$

$$N = \frac{199.3 \times 60}{2\pi} = 1907.4 \text{ rpm}$$

(e)   The peak chopper voltage $V_p = I_a\, R_b = 150 \times 5 = 750$ V

Example 4.5

A dc chopper is used in regenerative braking of a dc series motor as shown in the Figure. The supply voltage is 600 V. The armature resistance is $R_a = 0.02\,\Omega$ and the field resistance $R_f = 0.03\,\Omega$. The back emf constant is $K_b = 15.27$ mv/A-Rad/sec. The average armature current is maintained constant at $I_a = 250$ A. The armature current is continuous and has negligible ripple. If the duty cycle of the chopper is 60%, determine

(a) The average voltage across the chopper E_0,

(b) The power regenerated to the dc supply P_g,

(c) The equivalent load resistance of the motor acting as generator R_{eq},

(d) The minimum permissible braking speed. ω_{min} ,

(e) The maximum permissible braking speed ω_{max} ,

(f) The motor speed.

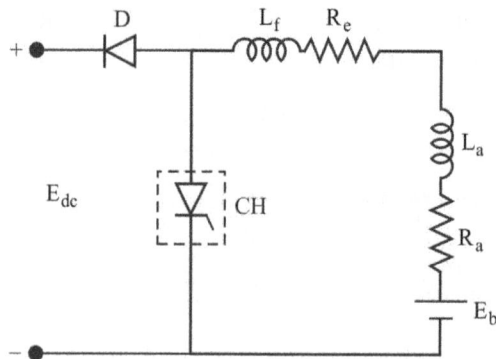

Solution :

Duty cycle $\alpha = 0.6$

$E_{dc} = 600$ V

The average voltage across the chopper $= (1 - \alpha)\, E_{dc}$ (Type B_q)

$$= (1 - 0.6)\, 600 = 240 \text{ V}$$

(a)   The power is regenerated to dc supply when the diode is conducting. In each cycle the power is fed for 0.4T sec

Average power fed back to dc source $= E_{dc}\, I_a \times 0.4$

$$= 0.4 \times 600 \times 250 = 60 \text{ KW.}$$

(b)   When the motor is acting as generator and feeding power in to mains

$$E_b = E_o + I_a\,(R_a + R_f) = 240 + 250\,(0.02 + 0.03)$$

$$= 240 + 12.5 = 252.5 \text{ V}$$

$$I_a = 250 \text{ A}$$

$\therefore$ The equivalent resistance $= \dfrac{E_b}{I_a} = \dfrac{252.5}{250} = 1.01\ \Omega$

(c)   Minimum permissible braking speed.

The braking current $= 250$ A

The minimum generated voltage required to circulate 250 A is

$$250 \times 0.05 = 12.5 \text{ V}$$

$$\omega_{min} = \frac{V}{K_a I_a} = \frac{12.5}{0.01527 \times 250} \text{ rad/sec} = 3.274 \text{ rad/sec}$$

or $$N_{min} = \frac{3.274 \times 60}{2\pi} = 31.26 \text{ rpm}$$

(d) Maximum permissible speed.

If $\qquad \alpha = 0 \qquad E_A = E_{dc} = 600 \text{ V}.$

$$E_b = 600 + 250 \times 0.05 = 612.5 \text{ V}$$

$$\omega_{max} = \frac{612.5}{0.01527 \times 250} = 160.44 \text{ rad/sec}$$

or $$N_{max} = \frac{160.44 \times 60}{2\pi} = 153.2 \text{ rpm}$$

(e) Motor speed : $\quad$ At $\alpha = 0.6 \quad E_A = 240 \text{ V}$

$\therefore \qquad E_b = 240 + 12.5 = 252.5 \text{ V}$

$$\text{Speed} = \frac{252.5}{0.01527 \times 50} \times \frac{60}{2\pi} = 631.6 \text{ rpm}$$

Example 4.6

A 230 V, 1200 rpm, 15 A separately excited dc motor has an armature resistance of 1.2 Ω. Motor is operated under dynamic braking with chopper control. Braking resistance has a value of 20 Ω.

(i) Calculate duty ratio of chopper for motor speed of 1000 rpm and braking torque equal to 1.5 times rated motor torque.

(ii) What will be the motor speed for duty ratio of 0.5 and motor torque equal to its rated torque?

Solution :

Braking resistance = 20 Ω

$$R_a = 1.2 \ \Omega$$

(i) E_b at rated speed of 1200 rpm = 230 − 15 × 1.2 = 212 V

$\therefore \qquad E_b$ at 1000 rpm $= \dfrac{212 \times 1000}{1200} = 176.67 \text{ V}$

If δ is the duty cycle, effective braking

Resistance will be $R_{be} = (1 - \delta) R_a$

The armature current at 1.5 times the rated torque is

$$I_a = 1.5 \times 15 = 22.5 \text{ A}$$

$$V_a = E_b - I_a R_a = 176.67 - 22.5 \times 1.2 = 149.67 \text{ V}$$

$$\frac{V_a}{I_a} = (1 - \delta) \, R_b$$

$$(1 - \delta) 20 = \frac{149.67}{22.5} = 6.652$$

$$1 - \delta = \frac{6.652}{20} = 0.3326$$

$$\therefore \qquad \delta = 1 - 0.3326 = 0.667$$

(ii) $\delta = 0.5$; at rated Torque $I_a = 15A$

Speed = ?

$$R_e = (1 - 0.5) \times 20 = 10 \, \Omega$$

$$E_b = 15(1.2 + 10) = 168 \text{Volts}$$

$$\text{Speed} = \frac{168}{212} \times 1200 = \mathbf{950.9 \; rpm}$$

$\mathbf{I_a = 160 \; A}$

$\delta = 0.6$

Example 4.7

A 220 V, 190 A, dc series motor has armature and field resistance of 0.03 and 0.02 ohms respectively. Running on no load as a generator with field winding connected to a separated source it gave the following magnetization characteristics at 500 rpm.

Field Current, A	40	80	120	160	200
Back emf, V	52	108	148	176	189

Motor is controlled by a chopper in dynamic braking with a braking resistance of 2 ohms.

(i)   Calculate motor speed for a duty ratio of 0.6 and motor current of 160 A.

(ii)   What will be the motor speed for a duty ratio of 0.75 and motor torque equal to half of rated torque?

Solution :

Effective Braking

(i) R_{eb} Resistance $= R (1 - \delta)$

$$= 2(1 - 0.6) = 0.8 \, \Omega$$

Motor current = 160 A

$$E_b = 160 (0.02 + 0.03 + 0.8) = 136 \text{ V}$$

At field current of 160 A the E_b is 176 V at speed of 500 rpm

$\therefore$ Speed for E_b of 136 V will be $\dfrac{136}{176} \times 500 = 386.36$ rpm

(ii)   Duty ratio $\delta = 0.75$, $R_{eb} = 2(1 - 0.75) = 0.5\ \Omega$

we know $T = K\phi\ I_a$ and $E_b = K\omega\phi$. $N = 500$ rpm or $\omega = 52.36$ rad/sec

At constant speed of 500 rpm relation between E_b and I_a is given. $(k\phi)$ at different values of E_b are calculated using the relation $k\phi = \dfrac{E_b}{\omega} = \dfrac{E_b}{52.36}$.

The values are tabulated

I_a	40	80	120	160	200
E_b	52	108	148	176	189
$K\phi = E_b/\omega$	0.993	2.06	2.826	3.36	3.61
$T = k\phi I_a$	39.7	164.8	339.1	537.66	721.9

From the tabulated values a graph is drawn between I_a and torque as shown in the figure.

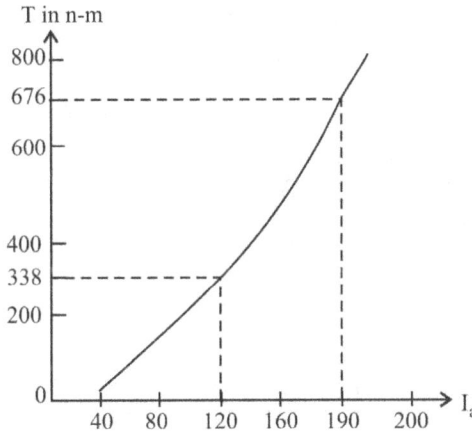

From the graph the torque at rated current of 190 A is obtained as 676 N-m

Torque equal to half the rated torque $= \dfrac{676}{2} = 338$ N-m

At this torque $I_a = 120$ A, from the graph

$\therefore$ $E_b = I_a (R_a + R_{se} + R_{bc}) = 120 (0.02 + 00.3 + 0.5) = 66$

$E_b = 148$ V and speed $= 500$ rpm at $I_a = 120$ A

For the same I_a of 120A and E_b of 66 speed $= \dfrac{66}{148} \times 500 = 222.9$ rpm.

Example **4.8**

A 230 V 1000 rpm, 30 A separately excited dc motor has armature resistance and inductance of 0.7 Ω and 50 mH respectively. Motor is controlled in regenerative braking by a chopper operating at a frequency of 800 Hz from a dc source of 230 V. Assuming continuous conduction,

(i)   calculate duty ratio of chopper for rated torque and speed of 800 rpm,

(ii)   what will be the motor speed for duty ratio of 0.6 and rated motor torque?

(iii)   what will be the maximum allowable speed of the motor, if the chopper has maximum duty ratio of 0.9 and maximum allowable motor current is twice the rated current,

(iv)   calculate power fed to the source for operating condition in (iii),

(v)   motor field is also controlled along with armature voltage. Rated field current is 0.5 A. Calculate the field current for duty ratio of 0.9 and motor speed of 1500 rpm and armature current of 30 A

Solution :

(i)   Rated current $I_a = 30$ A

Speed = 800 rpm, E_b at 800 rpm = $\dfrac{(230 - 0.7 \times 30) \times 800}{1000}$ = 167.2 Volts

Under regenerative braking $E_{dc} \, \delta = E_b - I_a R_a$

$\therefore$ $230 \, \delta = E_b - I_a \, R_a$

$= 167.2 - 30 \times 0.7 = \mathbf{146.2}$

$$\delta = \frac{146.2}{230} = \mathbf{0.6356}$$

(ii) $\delta = 0.6 \therefore$ Voltage across A-AA = $230 \times 0.6 = 138$ V

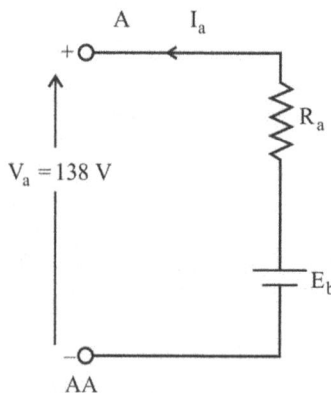

$\mathbf{I_a = 30 \ A;}$

E_b at rated speed of 1000 rpm   is equal to $230 - 30 \times 0.7 = 209$ V

Under regeneration when the voltage across armature is 128 V,

Back emf $E_b = V_a + I_a R_a$

∴ $E_b = 138 + 30 \times 0.7 = 159$ V

$$\text{Speed} = \frac{159}{209} \times 1000 = 760.76 \text{ rpm}$$

(iii) δ max = 0.9. Maximum voltage across A-AA = 0.9 × 230 = 207 V

Maximum allowable current = 2 × 30 = 60 A

Maximum value of E_b = 207 + 60 × 0.7 = 249 V

$$\text{Maximum speed} = \frac{249}{209} \times 1000 = 1191.4 \text{ rpm}$$

(iv)   Power fed back to source

$$= 207 \times 60 = 12.42 \text{ KW}$$

(v) δ = 0.9 ∴ Voltage across the A-AA = 0.9 × 230 = 207 V

Armature current = 30 A

∴ E_b should be = 207 + 30 × 0.7 = 218 V

$$\frac{E_{b1}}{E_{b2}} = \frac{I_{f1}N_1}{I_{f2}N_2}$$

$$\frac{209}{218} = \frac{0.5 \times 1000}{I_{f2} \times 1500}$$

∴ $$I_{f2} = \frac{0.5 \times 1000}{1500} \times \frac{218}{209} = 0.3636 \text{ A}$$

Problems

P4.1 (a)   Draw a neat sketch of a four quadrant chopper for variable speed reversible drive of a dc series motor. Discuss its operation.

(b)   A dc chopper has a resistive load of 10 ohms and input voltage of 220 V. When the chopper switch remains ON, its voltage drop is 2 V and the chopping frequency is 1 KHz. If the duty cycle is 50%, determine

    (i)   The average output voltage and

    (ii)   The output power

P4.2 (a)   Explain the principle of speed control of a dc motor and show how it can be achieved by a chopper.

(b)   A 230 V, 1200 rpm, 15 A separately excited motor has an armature resistance of 1.2 Ω. Motor is operated under dynamic braking with chopper control. Braking resistance has a value of 20 Ω.

    (i)   Calculate duty ratio of chopper for motor speed of 1000 rpm and braking torque equal to 1.5 times rated motor torque.

    (ii)   What will be the motor speed for duty ratio of 0.5 and motor torque equal to its rated torque.

P4.3 (a)   Derive the expressions for average motor current, RMS motor current, torque and average motor voltage, for chopper-fed dc series motor.

    (b)   A dc chopper controls the speed of dc series motor. The armature resistance $R_a = 0.04\,\Omega$, field circuit resistance $R_f = 0.06\,\Omega$, and back emf constant $K_v = 35$ mV/rad/s, the dc input voltage of the chopper $V_s = 600$ V. If it is required to maintain a constant developed torque of $T_d = 547$ N-m, plot the motor speed against the duty cycle K of the chopper.

P4.4 (a)   Derive the expressions for average motor current, currents I_{max} and I_{min} and average torques for chopper –fed dc separately excited motor.

    (b)   A dc chopper controls the speed of a separately excited motor. The armature resistance is $R_a = 0.05\ \Omega$. The back emf constant is $K_v = 1.527$ V/A-rad/s. The rated field current is $I_f = 2.5$ A. The dc input voltage to the chopper is $Vs = 600$ V. If it is required to maintain a constant developed torque of $T_d = 547$ N-m, plot the motor speed against the duty cycle k of the chopper.

P4.5 (a)   With neat circuit diagram and wave form, explain dynamic braking of separately excited motor by chopper control.

    (b)   A dc shunt motor has the armature resistance of $0.04\,\Omega$ and the field winding resistance of $10\,\Omega$. Motor is coupled to an over hauling load with a torque of 400 N-m following magnetization curve was measured at 600 rpm:

Field Current A	2.5	5	7.5	10	12.5	15	17.5	20	22.5	25
Back emf V	25	50	73.5	90	102.5	110	116	121	125	129

Calculate the value of R_B when the motor is required to hold overhauling load at 1200 rpm.

P4.6 (a)   Draw and explain the Torque-Speed characteristics for dynamic braking operation of dc series motor. Why torque becomes zero at finite speed?

    (b)   A 220 V, 24 A, 1000 rpm separately excited dc motor having an armature resistance of $2\,\Omega$ is controlled by a chopper. The chopping frequency is 500 Hz and the input voltage is 230 V. Calculate the duty ratio for a motor torque of 1.2 times rated torque at 50 rpm.

P4.7 A 230 V separately excited dc motor takes 50 A at a speed of 800 rpm. It has armature resistance of 0.4 ohms. This motor is controlled by a chopper with an input voltage of 230 V and frequency of 500 Hz. Assuming continuous conduction throughout, calculate and plot speed – torque characteristics for:

(i) Motoring operation at duty ratios of 0.3 and 0.6.

(ii) Regenerative braking operation at duty ratios of 0.7 and 0.4.

P4.8 (a) Discuss with the suitable diagrams First quadrant and Second quadrant choppers.

(b) A constant frequency TRC system is used for the speed control of dc series traction motor from 220 V dc supply. The motor is having armature and series field resistance of 0.025 Ω and 0.015 Ω respectively. The average current in the circuit is 125 A and the chopper frequency is 200 Hz. Calculate the pulse width if the average value of back emf is 60 volts.

P4.9 (a) A 230 V, 960 rpm, 200 A separately excited dc motor has an armature resistance of 0.02 Ω. The motor is fed from a chopper, which is capable of providing both motoring and braking operations. The source has a voltage of 230 V. Assuming continuous conduction:

(i) Calculate the time ratio of chopper for the motoring action at rated torque and 350 rpm.

(ii) Determine the maximum possible speed if maximum value of time ratio is 0.95 and maximum permissible motor current is twice the rated value.

(iii) Draw the necessary wave forms for the above problem.

P4.10 A class-A chopper, operating in time-ratio control, is supplying the armature of the separately excited dc motor. Show that the motor speed-torque relationship is

$$\omega_m = \frac{\partial.V}{K} - \frac{R_a}{K^2}T_a,$$ Where V-chopper input voltage, R_a – Armature resistance,

T_a-motor torque, K-torque constant. δ-duty cycle of chopper.

P4.11 A dc series motor is powered by a simple (first quadrant) chopper from a 600 V dc source. Armature resistance = 0.03 Ω, field circuit resistance = 0.05 Ω. The back emf constant of the motor K_v = 15.27 mV/A-rad/s. The average armature current I_a = 450 A. The armature current is continuous and has negligible ripple. If the duty cycle of the chopper is 75%, determine

(a) the input power from the source,

(b) the equivalent input resistance of the chopper drive,

(c) speed and developed torque of the motor.

P4.12 (a) Explain with neat circuit diagram the basic principle of operation of a class A type of chopper. The chopper is connected to R-L- E load. Analyze the same for continuous current mode of operation.

(b)   A dc supply of 200 V supplied power to a separately excited dc motor via a class A thyristor chopper. The motor has an armature circuit resistance of 0.33 ohms and inductance of 11 mH. The chopper is fully ON at the rated motor speed of 1200 rpm when the armature current is 20 A. If the speed is to be reduced to 800 rpm with the load torque constant, calculate the necessary duty cycle. If the chopper frequency is 500 Hz, is the current continuous?

Objective Type Questions

Choose correct answer for the following :

1. Type B Chopper is a   [   ]
   (a)   Single quadrant chopper and operates in Ist quadrant.
   (b)   Single quadrant chopper and operates in IInd quadrant.
   (c)   Two quadrant chopper and operates in Ist & IInd quadrants.
   (d)   Four quadrant chopper.

2. Type C chopper is a   [   ]
   (a)   Single quadrant chopper and operates in Ist quadrant.
   (b)   Single quadrant chopper and operates in IInd quadrant.
   (c)   Two quadrant chopper and operates in Ist & IInd quadrants.
   (d)   Four quadrant chopper.

3. Type D chopper is a   [   ]
   (a)   Single quadrant chopper and operates in Ist quadrant.
   (b)   Single quadrant chopper and operates in IInd quadrant.
   (c)   Two quadrant chopper and operates in Ist & IVth quadrants.
   (d)   Four quadrant chopper.

4. Type E chopper is a   [   ]
   (a)   Single quadrant chopper and operates in Ist quadrant.
   (b)   Single quadrant chopper and operates in IInd quadrant.
   (c)   Two quadrant chopper and operates in Ist & IInd quadrants.
   (d)   Four quadrant chopper.

5. A four quadrant chopper is considered as parallel combination of two type C Choppers (TRUE / FLASE)

6. The steady state ripple current of dc chopper is independent of back emf of the motor (TRUE / FALSE)

7. Per unit ripple current is maximum when α = ...

8. A type A chopper connected to 200 V DC source is supplying continuous current to a load. It is operating at chopping frequency of 500 Hz. Duty cycle is 0.4. The ON period T_{ON} = …………………………..

9. Type A Chopper connected to 100 Volts DC source is operating with $\alpha = 0.4$. What is the average output voltage?

10. Give three advantages of Chopper drives compared to Controlled rectifier drives.

11. What are the two control strategies of dc Chopper?

12. A dc Chopper is operating with duty cycle $\alpha = 0.4$. What is the value of ripple factor of output voltage?

13. A dc Chopper is operating with duty cycle $\alpha = 0.3$. Chopping frequency = 1 KHz; R = 1 Ohm, L = 1 mH. What is the value of ripple factor of current?

14. If type A chopper connected to 100 V DC source is operating with duty cycle of 0.4 and supplying continuous current, what is the value of AC ripple voltage E_r?

15. A separately excited DC motor has R_a = 0.5 Ω and L_a = 10 mH. It draws continuous current from DC chopper connected to 200 V DC source. Duty cycle of chopper is 0.5. Back emf of the motor is 20 V. What is the average armature current?

16. A DC chopper is used to control the speed of separately excited DC motor. The DC supply voltage is 220 V, armature resistance r_a = 0.2 Ω and motor constant $K_a \phi$ = 0.8 V /rpm. The motor drives a constant torque load requiring an average current of 25 A. Determine,

    1. The range of speed control

    2. Range of duty cycle. Assume motor current as continuous.

17. Draw the power circuit diagram of Type A chopper connected to separately excited DC motor.

18. Draw the power circuit diagram of Type B chopper connected to separately excited DC motor.

19. Draw the power circuit diagram of Type C chopper connected to separately excited DC motor.

20. Draw the power circuit diagram of Type E (four quadrant) chopper connected to separately excited DC motor.

21. The circuit diagram of four quadrant Chopper is given in Fig.1. State clearly the operation you perform for the Chopper to work in first quadrant.

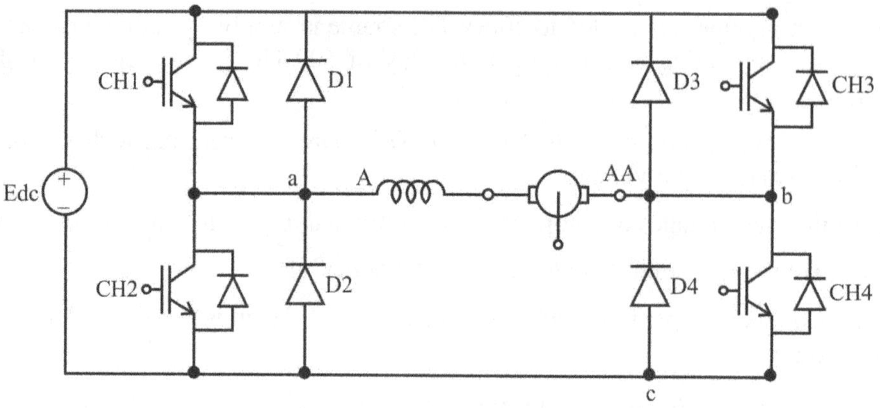

Fig. 1

22. The circuit diagram of four quadrant Chopper is given in Fig. 1. State clearly the operation you perform for the Chopper to work in second quadrant.

23. The circuit diagram of four quadrant Chopper is given in Fig.1 State clearly the operation you perform for the Chopper to work in third quadrant.

24. The circuit diagram of four quadrant Chopper is given in fig. State clearly the operation you perform for the Chopper to work in fourth quadrant.

25. Draw torque-speed characteristics of separately excited dc motor connected to type A chopper for two different values of duty cycles α_1 and α_2 where α_1 is greater than α_2.

26. Draw torque-speed characteristics of series excited dc motor connected type A chopper for two different values of duty cycles α_1 and α_2 where α, is greater than α_2.

5

Control of Induction Motor

Control of induction motor by AC voltage controllers – Waveforms – speed torque characteristics.

5.1 Speed Control of Induction Motors

Various methods of controlling the speed of induction motors are :
1. Stator voltage control
2. Variable frequency control
3. Rotor resistance control
4. Slip- energy recovery scheme

Methods 1 & 2 are applicable to both squirrel cage and slip ring induction motors.

Methods 3 & 4 are applicable to slip ring induction motors only.

5.2 Equivalent Circuit of Induction Motor

Per phase equivalent circuit of 3-phase induction motor is shown in Fig. 5.1.

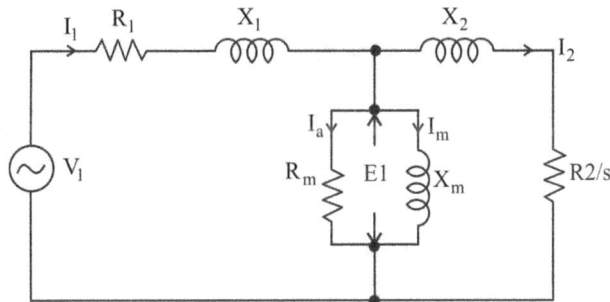

Fig. 5.1 Equivalent circuit.

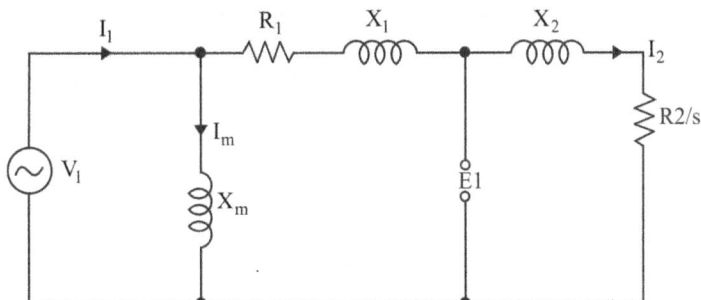

Fig. 5.2 Approximate equivalent circuit.

151

Without introducing much error the voltage drop in R1 and X1 due to I_a and I_m can be neglected. Then the equivalent circuit in Fig. 5.1 can be approximated as shown in Fig. 5.2.

1. Power input to motor $= 3V_1 I_1 \cos\phi_1$ .....(5.1)

   where,

   V_1 - Phase Voltage,

   I_1 - Phase Current,

   $\cos \phi_1$- Input power factor.

The different losses that occur in induction motor are:

     (a) Stator winding losses $= 3I_1^2 R_1$

     (b) Core loss P_{core} $= I_a^2 R_m$

     (c) Rotor winding losses $= 3 I_2^2 R_2$

     (d) Friction & winding losses $= P_{fw}$

     (e) Stray losses

The sum of core losses, friction, winding and stray losses i.e., (b + d + e) is called rotational losses.

2. Air gap power P_g = Power input – stator winding losses

$$= P_{in} - 3I_1^2 R_1$$

$$= 3V_1 I_1 \cos\phi_1 - 3I_1^2 R_1 \qquad \text{.....(5.2)}$$

3. Mechanical power developed $P_m = P_g - 3I_2^2 R_2 = P_g (1 - s)$ .....(5.3)

4. Useful power output or shaft power $P_{shaft} = P_m -$ rotational losses

$$= P_m - P_{rot} \qquad \text{.....(5.4)}$$

5. Shaft torque $= \dfrac{P_{shaft}}{\omega_r} = \dfrac{P_m - P_{rot}}{\omega_r} = \dfrac{P_m - P_{rot}}{\omega_s (1-s)}$ .....(5.5)

   where ω_r – rotor speed in rad/sec & ω_s – synchronous speed in rad/sec

6. Electromagnetic torque developed in the air gap $= \dfrac{\text{Power input to rotor}}{\omega_s}$

$$= \dfrac{P_g}{\omega_s} = \dfrac{P_m}{\omega_r} = \dfrac{P_m}{\omega_s (1-s)} \qquad \text{.....(5.6)}$$

7. Rotor winding losses or rotor copper losses $= s P_g$ watts .....(5.7)

Power flow diagram of induction motor is shown below

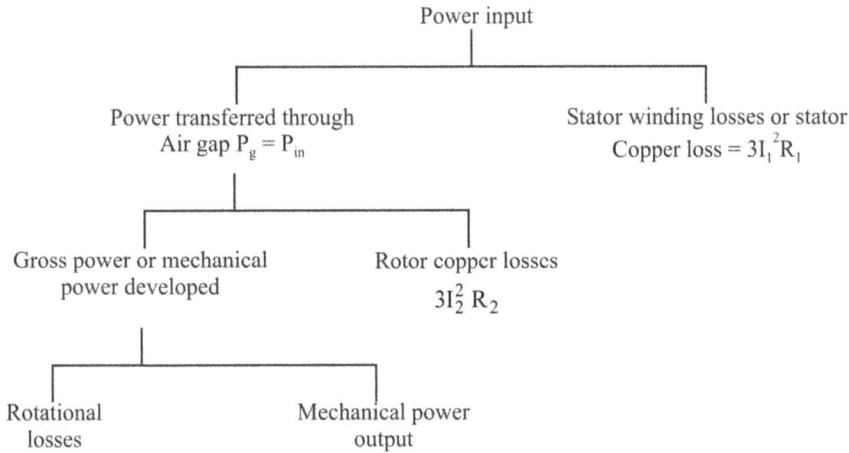

5.2.1 Expressions for Torque and Slip

From approximate equivalent circuit shown in Fig.5.3 we can determine the maximum torque and slip at maximum torque

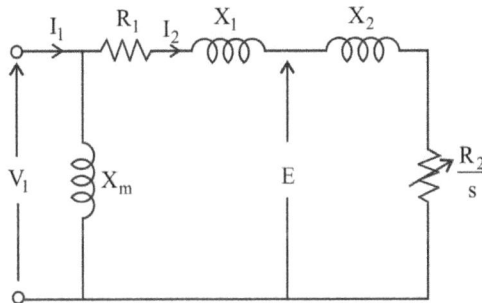

Fig. 5.3 Approximate equivalent circuit

Rotor current $I_2 = \dfrac{V_1}{(R_1 + \dfrac{R_2}{s}) + j(X_1 + X_2)} = \dfrac{V_1}{\sqrt{(R_1 + \dfrac{R_2}{S})^2 + (X_1 + X_2)^2}}$ .....(5.8)

Rotor copper losses $= 3\, I_2^2\, R_2 = 3\, R_2 \dfrac{V_1^2}{(R_1 + \dfrac{R_2}{s})^2 + (X_1 + X_2)^2}$

Air gap power $= P_g = \omega_s T_d$

Mechanical power developed $= P_m = \omega_r T_d$

Electromagnetic torque developed in the air gap T_d is given by the following expression.

$$T_d = \frac{P_g}{\omega_s} = \frac{3I_2^2 R_2}{s\omega_s}$$

$$= 3 \frac{R_2 V_1^2}{\left[(R_1 + \frac{R_2}{s})^2 + (X_1 + X_2)^2\right] s\omega_s}$$

$$= \frac{3V_1^2 R_2}{s\omega_s \left[(R_1 + \frac{R_2}{s})^2 + (X_1 + X_2)^2\right]} \qquad .....(5.9)$$

Torque developed is a function of slip s.

The maximum torque is obtained by taking the derivative of T_d with respect to slip and then equating it zero.

$$\frac{dT_d}{ds} = \frac{3V_1^2 R_2}{\omega_s} \frac{-\left[(R_1 + \frac{R_2}{s})^2 + (X_1 + X_2)^2\right] + s2(R_1 + \frac{R_2}{s})(-\frac{R_2}{s^2})}{(.....)^2} = 0$$

$$(R_1 + \frac{R_2}{s})^2 + (X_1 + X_2)^2 = \frac{2}{s}(R_1 + \frac{R_2}{s}) R_2$$

$$\frac{R_2^2}{s^2} = R_1^2 + (X_1 + X_2)^2$$

$$s_{max} = \pm \frac{R_2}{\sqrt{R_1^2 + (X_1 + X_2)^2}} \qquad .....(5.10)$$

The maximum torque $T_{d\,max}$ is obtained by substituting s_{max} in eq. 5.9.

$$T_d = \frac{3V_1^2 R_2}{\omega_s s \left[(R_1 + \frac{R_2}{s})^2 + (X_1 + X_2)^2\right]}$$

$$S_m = \frac{R_2}{\sqrt{R_1^2 + (X_1 + X_2)^2}}$$

$$T_{d\,max} = \frac{3V_1^2 R_2}{\omega_s} \left[\cfrac{1}{s_m \left[R_1^2 + \cfrac{R_2^2}{s_m^2} + \cfrac{2R_1 R_2}{s_m} + (X_1 + X_2)^2 \right]} \right]$$

$$= \frac{3V_1^2 R_2}{\omega_s} \left[\cfrac{1}{s_m (R_1^2 + (X_1 + X_2)^2) + \cfrac{R_2^2}{s_m} + 2R_1 R_2} \right]$$

$$= \frac{3V_1^2 R_2}{\omega_s} \left[\cfrac{1}{\pm 2R_2 \left[R_1^2 + (X_1 + X_2)^2 \right]^{\frac{1}{2}} + 2R_1 R_2} \right]$$

$$= \frac{3V_1^2}{2\omega_s (R_1 \pm \sqrt{R_1^2 + (X_1 + X_2)^2})} \qquad \text{.....(5.11)}$$

Note that the maximum torque is independent of rotor resistance and Slip at maximum torque S_{mt} is directly proportional to rotor resistance.

In the expressions for slip & torque the plus sign represents motor action and minus sign represents generator action.

5.2.2 Maximum Gross Power Output or
Maximum Mechanical Power Developed P_{mm}

The approximate equivalent circuit is redrawn in Fig. 5.4. The mechanical power developed P_m is equivalent to the power loss in the resistance $R_2(1-s)/s$.

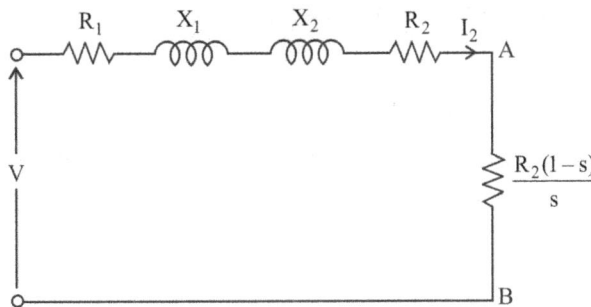

Fig. 5.4 Approximate equivalent circuit.

Mechanical power developed $P_m = \dfrac{3I_2^2 R_2}{s}(1-s)$

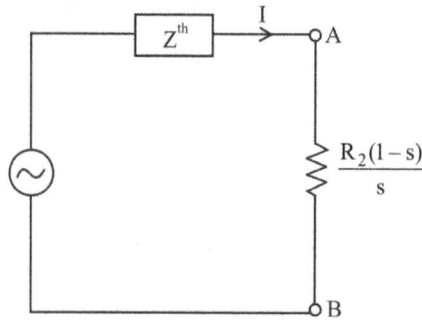

Fig 5.4(a) Equivalent of fig 5.4.

P_m will be maximum, when $\dfrac{R_2(1-s)}{s} = Z_{th}$ according to maximum power transfer theorem.

$$\therefore \qquad \dfrac{R_2(1-s)}{s} = Z_{th} = \sqrt{(R_1 + R_2)^2 + (X_1 + X_2)^2}$$

$$\therefore \qquad s = \dfrac{R_2}{R_2 + Z_{th}}$$

Let slip at maximum mechanical power developed $= s_{mm}$

Then $\qquad s_{mm} = \dfrac{R_2}{R_2 + \sqrt{(R_1 + R_2)^2 + (X_1 + X_2)^2}}$ .....(5.12)

Maximum mechanical power developed

$$P_{mm} = 3I_2^2 R_2(1 - S_{mm})/S_{mm} \ \text{ W}$$

$$= 3I_2^2 \sqrt{(R_1 + R_2)^2 + (X_1 + X_2)^2}$$ .....(5.13)

5.3   Torque–speed Characteristic

From eqn. 5.9 it can be seen that torque developed by the motor depends on voltage applied to stator and slip. The torque speed (torque slip) characteristic of induction motor at rated voltage is shown in Fig. 5.5.

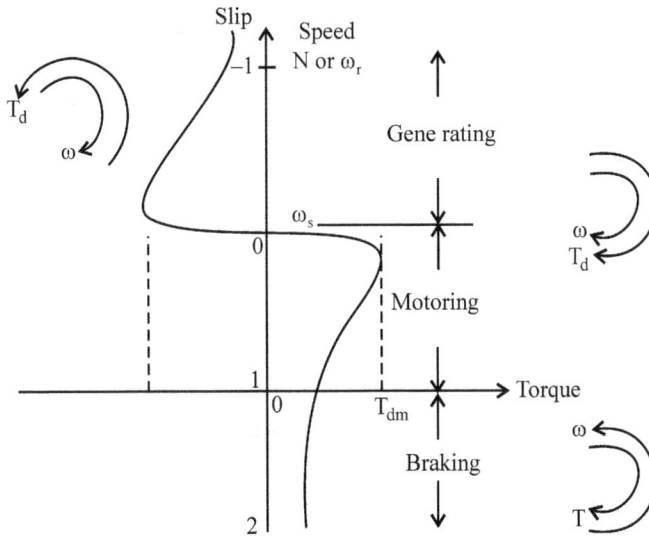

Fig. 5.5 Torque –speed characteristic of induction motor.

$$\text{Torque developed } T_d = \frac{3V_1^2 R_2 s}{\omega_s \left[\left(R_1 s + R_2 \right)^2 + s^2 (X_1 + X_2)^2 \right]}.$$

1. Motoring action $0 < \omega_r < \omega_s$

   Slip $s = \dfrac{\omega_s - \omega_r}{\omega_s}$ $0 < s < 1$ I^{st} Quadrant operation

2. Generating action $\omega_r > \omega_s$

   $s = \dfrac{\omega_s - \omega_r}{\omega_s}$ $s < 0$ IInd Quadrant operation

3. Plugging action ω_r is –ve

   $s = \dfrac{\omega_s - (-\omega_r)}{\omega_s}$ $s > 1$ IV th Quadrant operation

As torque developed is proportional to square of applied voltage, the speed of motor can be controlled by varying the voltage. Variable ac voltage can be obtained from fixed ac voltage by solid state ac voltage controller.

5.4   Single Phase ac Voltage Controller

The circuit diagram of single phase ac voltage controller with inductive load is shown in Fig. 5.6

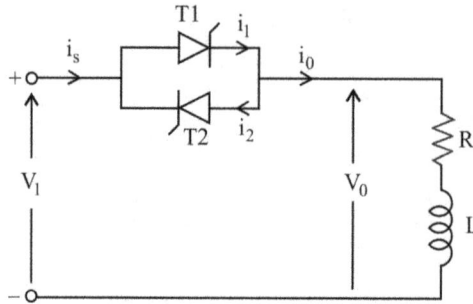

Fig. 5.6 Single phase ac voltage controller

Let us assume that thyristor T1 is fired at $\omega t = \alpha$, in the positive half cycle. T_1 comes to ON state and the load current increases. Due to inductance in the circuit, the current of thyristor T1 would not fall to zero at the instant when the input voltage starts to be negative. Thyristar T1 will continue to conduct until its current i_1 falls to zero at $\omega t = \beta$. The conduction angle of Thyristor T1 is $\delta = \beta - \alpha$ and depends on the delay angle α and the power factor angle ϕ of the load. Thyrister T2 is triggered in negative half cycle. Since it is forward biased it comes to ON state and allows current to flow in opposite direction. The wave forms of the input voltage v_i, load voltage v_0, load current i_0 and gating pulses are shown in the Fig. 5.7. By changing firing angle rms value of voltage can be changed.

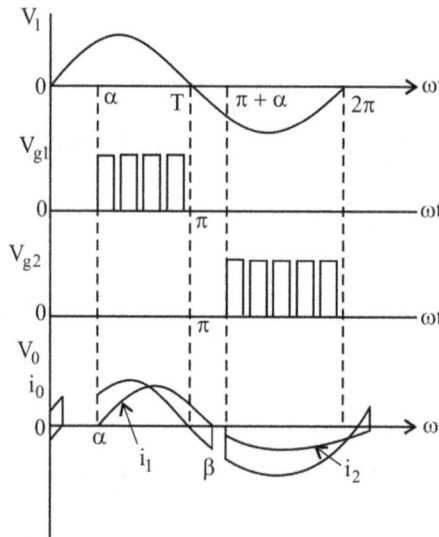

Fig. 5.7 Wave forms of voltage and current

5.4.1 Derivation of RMS Value of Output Voltage

If $v_I = \sqrt{2} \, V \sin \omega t$, and T1 is fired at delay angle α, then the Thyristor current i_1 can be found from

$$L\frac{di_1}{dt} + Ri_1 = \sqrt{2} \, V \sin \omega t$$

The solution of this equation is of the form

$$I_1 = \frac{\sqrt{2}V}{Z}\sin(\omega t-\phi) + A\,e^{-\frac{Rt}{L}}$$

where load impedance $Z = \sqrt{R^2 + (\omega L)^2}$ and Load angle $\phi = \tan^{-1}\dfrac{\omega L}{R}$

The constant A can be determined from initial conditions.

At $\qquad\qquad \omega t = \alpha,\ i_1 = 0$

$$\frac{\sqrt{2}V}{Z}\sin(\alpha - \phi) + A\,e^{-\frac{R\alpha}{\omega L}} = 0$$

$\therefore \qquad\qquad A = -\dfrac{\sqrt{2}V}{Z}\sin(\alpha - \phi)\,e^{\frac{R\alpha}{\omega L}}$

$\therefore \qquad\qquad i_1 = \dfrac{\sqrt{2}V}{Z}\left[\sin(wt - \phi) - \sin(\alpha - \phi)e^{\frac{R}{L}\left(\frac{\alpha}{\omega}-t\right)}\right] \qquad\qquad .....(5.14)$

The angle β where the current i_1 falls to zero and thyristor turns off is determined from eqn. 5.14

$$\text{Sin}(\beta - \phi) = \sin(\alpha - \phi)e^{\frac{R}{\omega L}(\alpha-\beta)} \qquad\qquad .....(5.15)$$

The value of β can be determined from this transcendental equation.

The rms value V_0 of output voltage :-

$$V_0 = \left[\frac{1}{\pi}\int_\alpha^\beta 2V_i^2\,\omega t d\,(\omega t)\right]^{\frac{1}{2}}$$

$$= V_i\left[\frac{1}{\pi}(\beta - \alpha + \frac{\sin 2\alpha}{2} - \frac{\sin 2\beta}{2})\right]^{\frac{1}{2}} \qquad\qquad .....(5.16)$$

1. The rms value of current in Thyristor

$$I_{IR} = \left[\frac{1}{2\pi}\int_\alpha^\beta i_1^2 d(\omega(\omega)\right]^{\frac{1}{2}}$$

$$= \frac{V}{Z}\left[\frac{1}{\pi}\int_\alpha^\beta \left\{\sin(\omega t - \phi) - \sin(\alpha - \phi)e^{\frac{R}{L}\left(\frac{\alpha}{\omega}-t\right)}\right\}^2 d(\omega t)\right]^{\frac{1}{2}} \qquad .....(5.17)$$

The wave form of load current is identical in both half cycles. Hence the rms value I_{OR} of load current is given as $I_{OR} = \left(I_{1R}^2 + I_{2R}^2 \right)^{\frac{1}{2}} = \sqrt{2} . I_1$ $\because I_{1R} = I_{2R.}$

Note : If firing angle α is equal to impedance angle ϕ, from eqn. 5.15

$$\sin (\beta - \alpha) = \sin (\alpha - \alpha) \, e^{\frac{R}{\omega L}(\alpha - \beta)} = 0$$

$\therefore$ $\beta - \alpha = \pi$

$\therefore$ Each SCR conducts for 180^0 and the current becomes continuous and pure sinusoidal.

Since the conduction angle δ cannot exceed π and the load current must pass through zero, for getting ac, the delay angle α should not be less than ϕ. Therefore for effective control of output voltage, the range of delay angle α should be such that

$$\phi \leq \alpha \leq \pi$$

For three phase loads three such controllers are required. The circuit diagram for controlling voltage in three phase load is given in Fig. 5.8.

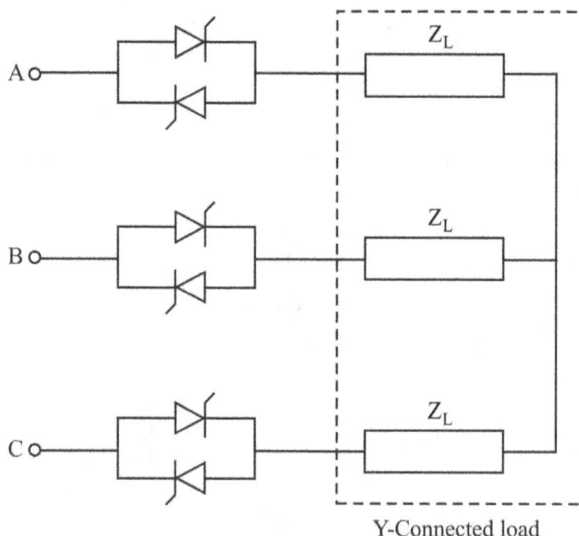

Y-Connected load

Fig. 5.8 Three phase ac voltage controller.

If we adopt phase control as explained above for controlling output voltage, the power factor is poor for low values of load current and appreciable amount of harmonics is present. For improvement of system power factor and reduction of harmonics in the source current and load voltage, sequence control of ac regulators is employed. Sequence control of ac regulators means, the use of two or more stages of voltage controllers in parallel for the regulation of output voltage. The term sequence control means, that the stages of voltage controllers in parallel are triggered in proper sequence one after the other so as to obtain variable output with low harmonic content. Sequence controlled ac regulators can be used for speed control of induction motors.

5.5 Two Stage Sequence Control

The circuit diagram of single phase two stage voltage controller is given in Fig. 5.9(a) and wave forms in Fig. 5.9(b)

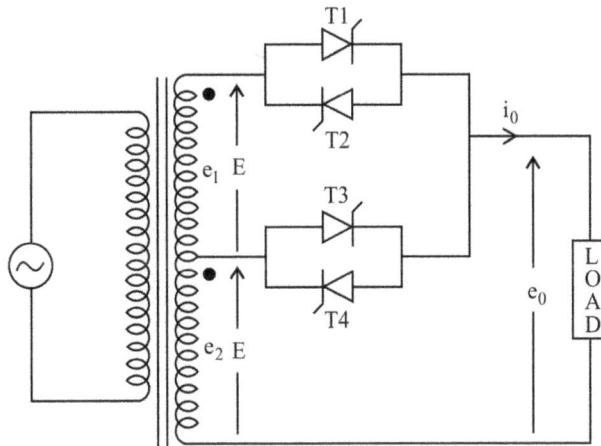

Fig. 5.9(a) Two stage sequence control of single phase ac regulator circuit.

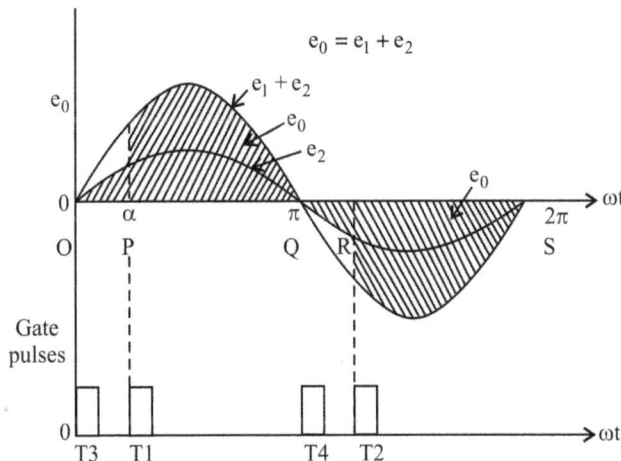

Fig. 5.9(b) Wave form of output voltage.

The output can be varied from E to 2E. Firing angles for T3 & T4 are zero degrees. Firing angles for T1 & T2 can be varied from 0 to 180^0. If the firing angle of T3 is zero and firing angle of T1 is 180^0 the rms value of output voltage is E. If the firing angles of T3 & T1 are zero the output voltage will be 2E.

During the period OP – T3 conducts

During the period PQ – T1 conducts – T3 goes OFF by line commutation.

During the period QR – T4 conducts

During the period RS – T2 conduct – T4 goes OFF

5.6   Stator Voltage Control of Three Phase Induction Motor

Circuit diagram of symmetrical 3-Φ voltage controller for driving a three phase induction motor is shown in Fig. 5.10

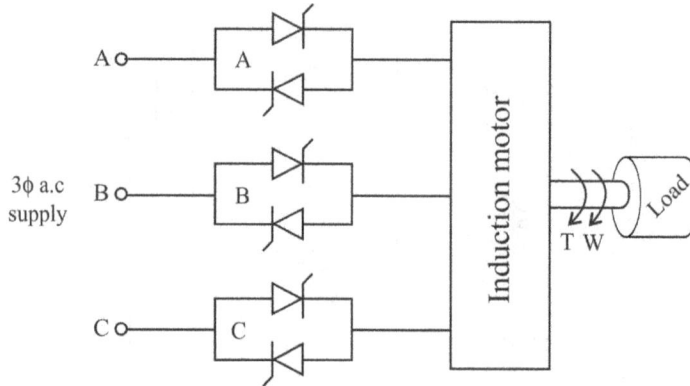

Fig. 5.10 Three phase ac voltage controller for driving induction motor.

Circuit diagram of reversible symmetrical 3Φ voltage controller for driving reversible three phase induction motor is shown in Fig. 5.11.

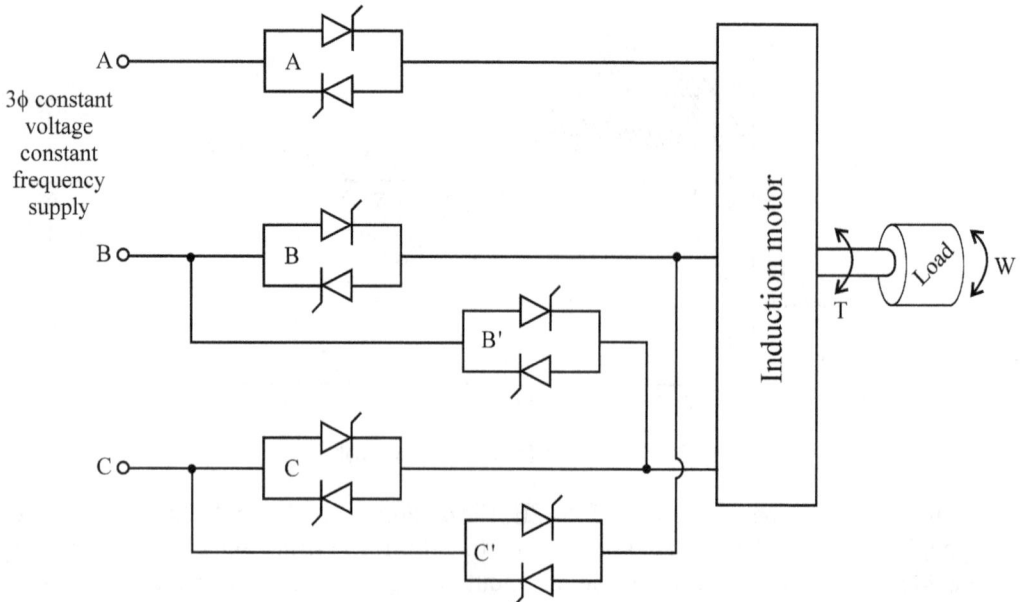

Fig. 5.11 Three phase ac voltage controller for reversible induction motor drive.

Thyristor pairs A, B, C provide operation in first quadrant for motor operation and plugging in fourth quadrant, while the Thyristor pairs A B^1 C^1 provide operation in third quadrant for motor operation and plugging in second quadrant. Torque-speed curves for reversible drive are shown in Fig. 5.12. Solid line curve for operation of thyristor pairs ABC. Dashed line for operation of thyristor pairs A B^1 C^1

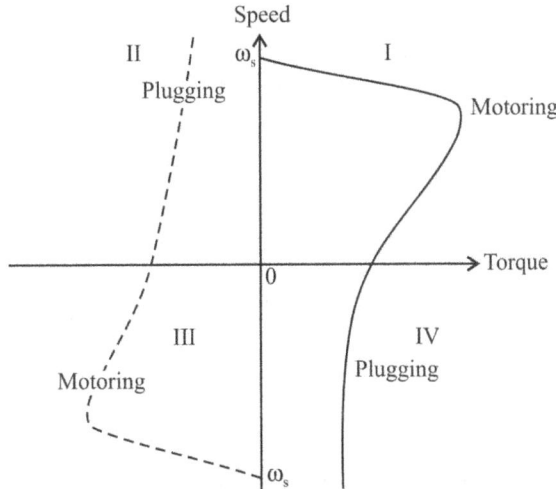

Fig. 5.12 Torque speed curves for reversible drive.

5.7   Speed Control by Stator Voltage

Torque developed is proportional to square of applied voltage i.e., $T_d \propto V_1^2$. This method of speed control is suitable for loads whose torque is proportional to square of speed like centrifugal pumps.

The torque speed characteristics for different voltages are shown in Fig. 5.13 for normal squirrel cage induction motor. Usually rotor resistance of induction motor will be very small. For such motors the speed can be changed over a small range only by stator voltage control. This method of speed control is not suitable for constant torque loads like lifts. N1 is the speed for stator voltage V1 and N4 is the speed for stator voltage V4. The change from N1 to N4 is small.

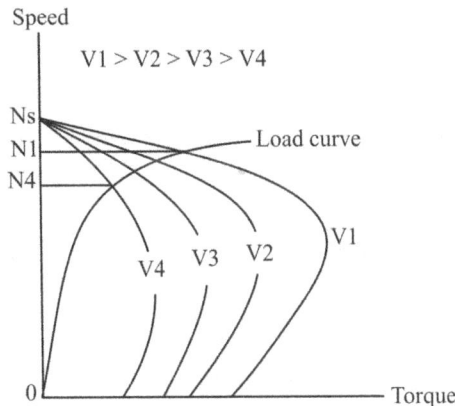

Fig. 5.13 Torque-Speed characteristic of Induction Motor for different stator voltages.

The torque speed characteristics for different stator voltages of induction motors designed for large rotor resistance are shown in Fig. 5.14. Higher speed range is possible in this case as evident from the figure. Now the difference between N1 and N4 is more than that shown in Fig. 5.14.

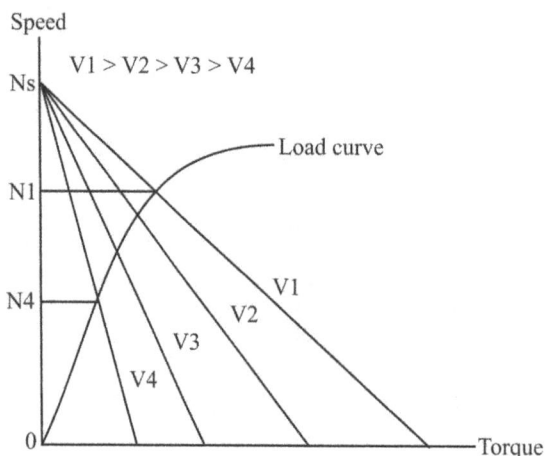

Fig. 5.14 Torque –Speed characteristic of Induction Motor with large rotor resistance for different stator voltages

5.8   Closed Loop Speed Control

5.8.1 Single Quadrant Closed Loop Speed Control

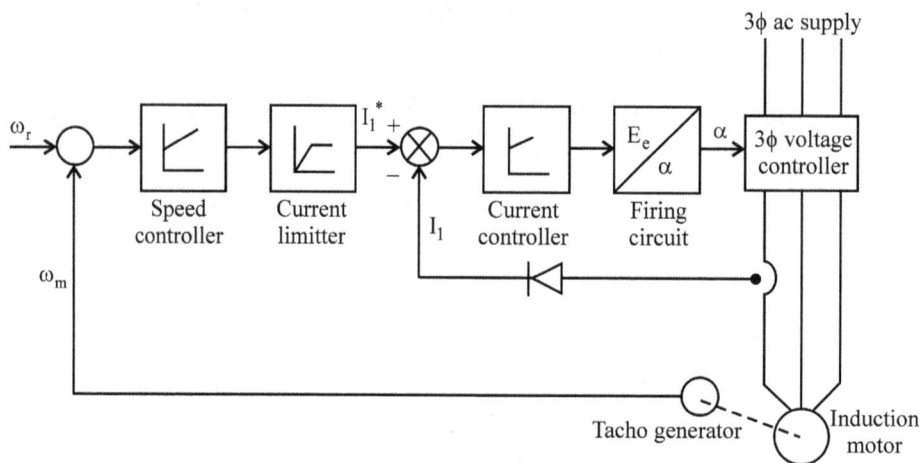

Fig. 5.15 Block diagram of single quadrant speed control.

A block diagram for single quadrant speed control of three phase induction motor with stator voltage control is shown in Fig. 5.15. Three phase ac voltage controller shown in Fig. 5.10 is used. The set speed ω_r is compared with actual speed ω_m and the difference signal is given to speed controller whose output is given to current controller with saturation. The output signal is limited to maximum allowable current through stator winding. The actual current of stator is sensed and compared with set value and the difference is used for changing stator current by changing the firing angle of controller. As usual the inner loop is for current and outer loop is for speed control.

5.8.2 Four Quadrant Closed Loop Speed Control

A block diagram for four quadrant speed control of three phase induction motor with stator voltage control is shown in Fig. 5.16. In this diagram voltage controller shown in Fig 5.11 is used. In this scheme apart from usual voltage and current controllers an absolute value circuit block and a master controller block are added.

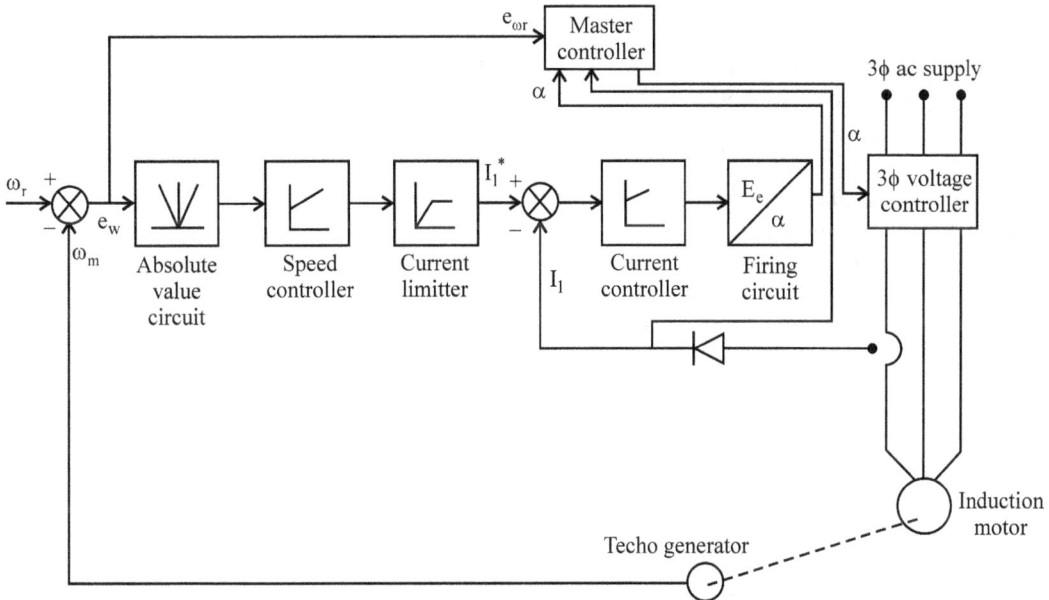

Fig. 5.16 Block diagram of four quadrant speed control.

The absolute value circuit gives positive voltage as the output whether the input signal is positive or negative depending on the direction of rotation, so that necessary voltage is developed. For master controller there are three input signals and one output signal. The information regarding the required direction of rotation is given by speed error signal e_ω. The output of firing circuit and actual stator current signals are also given as inputs. After processing this information the master controller sends firing pulses to thyrister sets ABC if positive direction of rotation is required and sends firing pulses to thyrister sets AB'C' if negative direction of rotation is required. Let us assume that motor is rotating in positive direction.

When the speed command is set for the reverse direction the speed error e_ω reverses and exceeds a prescribed limit. The speed error signal is given by equation

$$e_\omega = \omega_r - \omega_m$$

When the speed command signal ω_r is +ve, e_ω is +ve and equal to the difference between ω_r and ω_m.

If the speed command signal ω_r is negative

$e_\omega = -\omega_r - \omega_m$

Therefore e_ω is – ve and very large initially.

The master controller senses this large value, withdraws the gate pulses to thyrister sets ABC forcing the current to zero. The master controller provides a delay of 5 to 10 milliseconds after the zero current is sensed, for ensuring that the outgoing thyristors are turned OFF. The gate pulses are now released to other set of thyristors (A B^1 C^1). The drive first decelerates and accelerates in the other direction at constant maximum allowable current and finally settles at the desired speed.

Example 5.1

A 440 V, 3ϕ, 50 Hz, 6 pole, 945 rpm delta connected induction motor has the following parameters referred to stator.

$R_S = 2.0\ \Omega$, $R_r = 2.0\ \Omega$, $X_S = 3\ \Omega$, $X_r = 4\ \Omega$.

When driving a fan load at rated voltage it runs at rated speed. The motor speed is controlled by stator voltage control. Determine the motor terminal voltage, current and torque at 800 rpm.

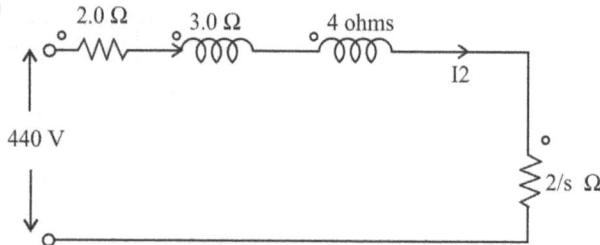

Rated voltage = 440 V;   Rated speed = 945 rpm

$$\text{Slip} = \frac{1000 - 945}{1000} = 0.055$$

$$\text{Rotor phase current at rated slip} = \frac{440}{(2 + \frac{2}{0.055}) + j7} = \frac{440}{38.36 + j7} = \frac{440}{38.99} = 11.28\ A$$

$I_2 = 11.28\ A$

Rotor copper losses $= 3I_2^2 R_2 = 3 \times 11.28^2 \times 2 = 763\ W$

$$\text{Torque developed} = \frac{3I_2^2 R_2}{s.\omega_s} = \frac{763 \times 60}{0.055 \times 1000 \times 2\pi} = 132.9\ \text{N-m}$$

Load is fan. Hence T_L is proportional to ω^2

$$\text{Torque developed at 800 rpm} = 132.9 \left(\frac{800}{945}\right)^2 = 95.06\ \text{N-m}$$

$$\text{Slip at } 800 = \frac{1000 - 800}{1000} = 0.2$$

$$\frac{3I_2^2 R_2}{s\omega_s} = \text{Torque developed}$$

$$I_2^2 = \frac{T_d s w_s}{3R_2} = \frac{95.06 \times 0.2}{3 \times 2} \cdot \frac{1000 \times 2\pi}{60} = 331.8$$

$$I_2 = 18.21 \text{ A}$$

$$\text{Impedance } \bar{z} = 2 + \frac{2}{0.2} + j7 = 12 + j\,7$$

$$z = 13.89\,\Omega$$

∴   Supply voltage $= IZ = 18.21 \times 13.89 = 253.02$ V

Thus :

Terminal voltage $= 253.02$ V

Phase current $= 18.21$ A

Torque $= 95.06$ N-m

Example 5.2

A 3 KW, 400 V, 50 Hz, 4 pole, 1370 rpm, Y connected induction motor has the following parameters.

$R_s = 2\,\Omega$, $R_r^1 = 5\,\Omega$, $X_s = X_r^1 = 3\,\Omega$. Load characteristics are matched with motor, such that motor runs at 1370 rpm with full voltage across its terminals. The motor is controlled by terminal voltage control and load torque is proportional to speed. Calculate the motor terminal voltage and current at half the rated speed.

Solution :

The approximate equivalent circuit is shown below.

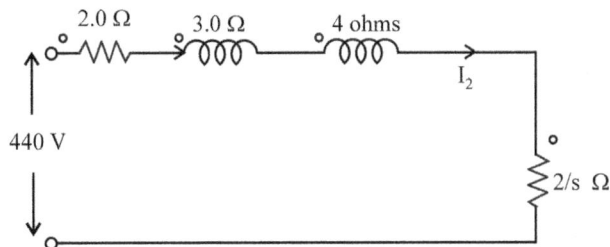

Speed = 1370; $V_p = \dfrac{400}{\sqrt{3}} = 230.94$

Synchronous. Speed = 1500 rpm = 157.08 rad /sec

$$s = \dfrac{1500 - 1370}{1500} = 0.0866$$

Rotor current at s = 0.0866 :

$$I = \dfrac{230.94}{(2 + \dfrac{5}{0.0866}) + j6} = \dfrac{230.94}{(59.69 + j6)} = \dfrac{230.94}{59.99} = 3.849 \text{ A}$$

Torque developed $= \dfrac{3I_2{}^2 R_2}{s.\omega_s} = \dfrac{3 \times 3.849^2 \times 5}{0.0866 \times 157.08} = 16.33 \text{ N-m}$

Torque required at $\dfrac{1}{2}$ rated speed (Tα ω)

$$T_d = \dfrac{16.33}{2} = 8.165 \text{ N-m}$$

Slip at half rated speed $= \dfrac{1500 - 685}{1500} = 0.543$

Per phase impedance $= (2 + \dfrac{5}{0.543}) + j6 = 11.20 + j6 = 12.70 \text{ Ohms}$

$$\dfrac{3I_2{}^2 R_2}{s\omega_s} = T_d$$

$$I_2{}^2 = \dfrac{T_d s \omega_s}{3 R_2} = \dfrac{8.165 \times 0.543 \times 157.08}{3 \times 5} = 46.42$$

$$I_2 = \sqrt{46.42} = 6.81 \text{ A}$$

Supply voltage $V_p = 6.81 \times 12.70 = 86.53$ V

Motor terminal voltage $= \sqrt{3} \times 86.53 = 149.88$ V

Line current = phase current = 6.81 A

Example 5.3

A pump has a torque-speed curve given by $T_L = (1.4 \times 10^{-3} \times \omega^2)$ N-m. It is proposed to use a 240 V, 50 Hz, four pole, star connected induction motor with the equivalent circuit parameters (referred to stator turns)

$R_1 = 0.2\,\Omega$, $X_1 = 0.36\ \Omega$, $R_2 = 0.65\,\Omega$, $X_2 = 0.36\,\Omega$, $X_m = 17.3\,\Omega$. The pump speed N is to vary from full speed 1250 rpm to 750 rpm by voltage control using pairs of inverse – parallel-connected SCRs in the lines. Calculate the range of firing angles required.

Solution :

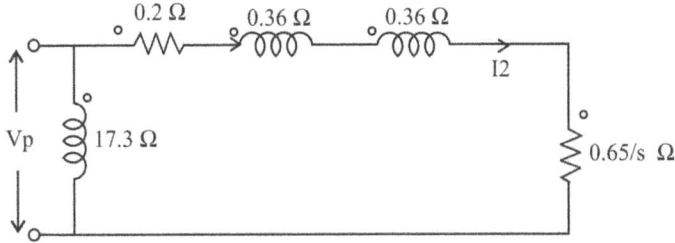

$N_1 = 1250$ rpm; $N_s = 1500$ rpm; $S_1 = \dfrac{1500 - 1250}{1500} = 0.167$

$N_2 = 750$ rpm; $S_2 = \dfrac{1500 - 750}{1500} = 0.5$

At 1250 rpm the impedance $Z_1 = (R_1 + \dfrac{R_2}{s}) + j\,(X_1 + X_2)$

$$= (0.2 + \dfrac{0.65}{0.167}) + j(0.36 + 0.36) = 4.15 + j\,0.72$$

$$Z_{in\ 1} = j17.3 // (4.15 + j\,0.72)$$

$$= 3.65 + j1.54 = 3.96\ \angle 22.9^0$$

Therefore the range of firing angle is $22.9^0 < \alpha < \pi$

At 750 rpm $Zin_2 = j\,17.3 // \left[(0.2 + \dfrac{0.65}{0.5}) + j(0.36 + 0.36) \right]$

$$= j\,17.3 // (1.55 + j\,0.72)$$

$$= 1.42 + j\,0.815 = 1.64\ \angle 29.9^0$$

Therefore the range of firing angle is $29.9^0 < \alpha < \pi$

Load torque at speed 1250 rpm $T_L = 1.4 \times 10^{-3}\ (\dfrac{1250}{60} \times 2\pi)^2 = 24$ N-m

T_L at 750 rpm $= 1.4 \times 10^{-3} \times (\dfrac{750}{60} \times 2\pi)^2 = 8.635$ N-m

Torque developed by the motor at 1250 rpm

$$= \dfrac{3 I_2^2 R_2}{s\omega_s} = 24;\ \omega_s = \dfrac{1500 \times 2\pi}{60} = 157.07 \text{ rad/s};$$

$$\frac{3I_2^2 \times 0.65}{0.167 \times 157.07} = 24$$

$$\therefore \qquad 0.0743\,I_2^2 = 24; \qquad I_2 = \sqrt{\frac{24}{0.0743}} = 17.96 \text{ A}$$

Similarly I_2 at 750 rpm :

$$T_d = \frac{3I_2^2 R_2}{sW_s} = \frac{3 \times 0.65 I_2^2}{0.5 \times 157.07} = 8.635$$

$$I_2 = \sqrt{\frac{8.635 \times 0.5 \times 157.07}{3 \times 0.65}} = 18.64 \text{ A}$$

Voltage per phase at 1250 rpm = $\overline{Z_2}.\overline{I_2}$

$$= |(4.15 + j0.72)|\,17.96 = 4.21 \times 17.96 = 75.64 \text{ V}$$

Power factor angle $= \tan^{-1}\dfrac{1.54}{3.65} = 22.9^0$

Voltage per phase at 750 rpm $= |Z_2.I_2| = |1.55 + j0.72|.|18.64| = 31.85 \text{ V}$

Impedance angle at 1250 rpm $= 22.9^0$

Impedance angle at 750 rpm $= 29.9^0$

Range of α is $22.9 < 29.9^0 < \pi$

Example 5.4

A 2.8 KW, 400 V, 50 Hz, 4 pole, 1370 rpm delta connected squirrel cage induction motor has the following parameters referred to stator.

$R_s = 2\Omega$, $R_r^1 = 5\ \Omega$, $X_s = X_r^1 = 5\Omega$, $X_m = 80\ \Omega$.

Motor speed is controlled by stator voltage control. When driving a fan load it runs at rated speed at rated voltage. Calculate

(i) Motor terminal voltage, current and torque at 1200 rpm

(ii) Motor speed, current and torque for a terminal voltage of 300 V

Solution :

$$T_d = \frac{3}{\omega_s}\frac{V_1^2 R_2}{s\left[(R_1 + \dfrac{R_2}{s})^2 + (X_1 + X_2)^2\right]}$$

Synchronous speed $\quad N_s = \dfrac{120f}{p} = \dfrac{120 \times 50}{4} = 1500 \text{ rpm};$

$$\omega_s = \frac{1500}{60} \times 2\pi = 50\pi \text{ rad/sec}$$

At full load slip $s = \dfrac{1500-1370}{1500} = 0.0867$

Torque developed at full load

$$T_d = \frac{3}{50\pi} \frac{400^2 \times 5}{0.0867\left[(2+\frac{5}{0.0867})^2 + (5+5)^2\right]} = 48.13 \text{ N-m}$$

For a fan load torque is proportional to square of speed.

$$\omega_m = \omega_s(1-s) \ \& \ \omega_s \text{ constant.}$$

∴ $T_d = K(1-s)^2$

At full load T_d is 48.13 N-m & s = 0.0867

∴ $K = \dfrac{48.13}{(1-0.0867)^2} = 57.7$

(i) At 1200 rpm $s = \dfrac{1500-1200}{1500} = 0.2$

Load torque $T_L = 57.7(1-0.2)^2$

$$= 36.9 \text{ N-m}$$

$$\frac{3}{50\pi} \frac{V^2 \times 5}{0.02\left[(2+\frac{5}{0.2})^2 + (5+5)^2\right]} = 36.9;$$

∴ $V = 253.2$ V

Rotor current I_2 referred to stator $= \dfrac{V}{Z} = \dfrac{253.2}{(2+\frac{5}{0.2})+j10} = 8.246 - j3.054$

$$I_m = \frac{V}{jX_m} = \frac{253.2}{j80} = -j\,3.165$$

∴ $I_1 = 8.246 - j\,3.054 - j\,3.166 = 10.328 \ \angle -37^0$

Line current $= \sqrt{3} \times 10.328 = 17.89$ A

(ii) At 300 V

$$T = \frac{3}{50\pi} \frac{300^2 \times 5}{s\left[(2+\frac{5}{s})^2 + 10^2\right]} = \frac{27 \times 10^4 s}{10\pi\left[104 s^2 + 20 s + 25\right]}$$

In steady state $T = T_L$ But $T_L = K(1-s)^2 = 57\text{-}7(1-s)^2$

$$\therefore \qquad \frac{27 \times 10^4 s}{10\pi \left[104s^2 + 20s + 25 \right]} = 57\text{-}7(1-s)^2$$

or $\qquad 104\,s^4 - 188s^3 + 89s^2 - 179s + 25 = 0$

which gives $\qquad s = 0.147$

Torque produced by the motor $= 57.7(1- 0.147)^2$

$$= 41.94 \text{ N-m}$$

Speed $= N_s (1-s) = 1500 (1-0.147) = 1279$ rpm

Current : $\dfrac{300}{j80} + \dfrac{300}{(2 + \dfrac{5}{0.147}) + j10} = 9.75 \quad \angle -37.3^0$

Line current $= \sqrt{3} \times 9.75 = 16.88$ A

Example 5.5

A Three phase 460 V, 60 Hz four pole Y- connected induction motor has the following parameters.

$R_s = 1.01\ \Omega$; $R_r = 0.69\ \Omega$; $Xs = 1.3\ \Omega$; $X_r = 1.94\ \Omega$ and $X_m = 43.5\ \Omega$.

The no load loss, $P_{no\text{-}load}$, is negligible. The load torque, which is proportional to the speed squared, is 41 N-m at 1740 rpm. If the motor speed is 1550 rpm determine

(a) The load torque T_L; (b) The rotor current I_r; (c) The stator supply voltage V_a; (d) The motor input current I_i; (e) The motor input power P_i; (f) The slip for the max^m current s_a; (g) The maximum rotor current $I_{r(max)}$ (h) The speed at maximum rotor current ω_a; and (i)The torque at the maximum current T_a.

Solution :   The load toque at 1740 rpm $= 41$ N-m

(a)   The load toque at 1550 rpm $= 41 \times (\dfrac{1550}{1740})^2 = $ **32. 5 N-m**

(b) $N_s = \dfrac{120 \times 60}{4} = 1800$ rpm; $\omega_s = \dfrac{2\pi \times 1800}{60} = 60\pi = 188.5$ rad/sec ;

$N = 1550$ rpm; Slip $s = \dfrac{1800-1550}{1800} = \dfrac{250}{1800} = 0.139$

Rotor copper losses $\times \dfrac{(1-s)}{s} =$ Mech. Power developed $= T_L\, \omega_m + P_{no\ load}$

$$= 32.5 \times \dfrac{2\pi \times 1550}{60}$$

Rotor copper losses $= \dfrac{32.5 \times 2\pi \times 1550}{60} \times \dfrac{0.139}{(1-0.139)}$

If I_r is the rotor current, total rotor copper loss $= 3I_r^2 R_r$

$$\therefore \qquad I_r = \sqrt{\dfrac{1}{3 \times 0.69} \times \dfrac{32.5 \times 2\pi \times 1550}{60} \times \dfrac{0.139}{(1-0.139)}} = \mathbf{20.\,28\ A}$$

(c)   The stator supply voltage

$$V_a = I_r \times \sqrt{(R_s + \dfrac{R_r}{s})^2 + (X_s + X_r)^2}$$

$$= 20.28 \times \sqrt{(1.01 + \dfrac{0.69}{.139})^2 + (1.3 + 1.94)^2} = \mathbf{137.82\ V}$$

(d) $I_i = 137.82 \left[\dfrac{1}{j43.5} + \dfrac{1}{(1.01 + \dfrac{0.69}{.139}) + j(1.3 + 1.94)} \right] = \mathbf{22} \angle -35.75^0$

(e)   The motor input pf $= \cos(-35.75^0) = 0.812$ lagging

Power input $= 3V_i I_i \cos\phi_1 = 3 \times 137.82 \times 22 \times 0.812 = \mathbf{7386\ W}$

(f) $T_L = K\,\omega_m^2;\ \omega_m = \omega_s(1-s)$

The mechanical power output $= T_L . \omega_m = 3I_r^2 R_r \dfrac{(1-s)}{s}$

$$I_r^2 = \dfrac{T_L \omega_m S}{3R_r(1-S)} = \dfrac{K\omega_m^2 \omega_m S}{3R_r(1-S)}$$

$$= \dfrac{K\omega_s^3 (1-s)^3 s}{3R_r(1-s)} = \dfrac{K}{3R_r} \times \omega_S^3 (1-s)^2 s$$

To find the value of slips S_a for maximum rotor current we take the derivative of I_r with respect to s and equate to zero and solve for S.

$$I_r = (\dfrac{K\omega_s^3}{3R})^{\frac{1}{2}} (1-s)\, s^{\frac{1}{2}}; \quad \dfrac{dI_r}{ds} = \dfrac{1}{2}(1-s)s^{-\frac{1}{2}} + s^{\frac{1}{2}}.-1 = 0;$$

$$\dfrac{1-s}{s^{\frac{1}{2}}} = 2s^{\frac{1}{2}} \text{ which gives } s_a = 1/3$$

The rotor current is maximum at $s_a = \dfrac{1}{3}$

(g)   Maximum rotor current I_{rm} is obtained by substituting $s = \dfrac{1}{3}$

$$I_r = \sqrt{\frac{1.235 \times 10^{-3}}{3 \times 0.69}} \times 188.5^3 (1 - \frac{1}{3})^2 \times \frac{1}{3} = \sqrt{592} = \textbf{24. 3 A}$$

The speed at maximum current $N = N_s (1-s_n) = 1800\ (1\text{-}1/3)$

$$= 1200\ \text{rpm}$$

$$= 1200 \times 2 \times \pi/60 = \textbf{125.66 rad/sec}$$

(h)   The torque T_a at maximum current

$$T_a = 41 \times (\frac{1200}{1740})^2 = \textbf{19. 45 N-m}$$

Problems

P.1 (a)   Using 3 phase solid state AC voltage controllers explain clearly how it is possible to achieve 4 quadrant operation of 3 phase Induction motors.

(b)   Draw a closed loop block schematic diagram for the above speed control technique. Mention the merits of the above method of speed control α

P.2 (a)   Draw the speed torque characteristics which are obtained by stator voltage variation of 3 phase Induction motor.

(b)   Draw the circuit diagrams of AC voltage controller for delta connected controller and star connected controller. How it is possible to change the direction of rotation of 3 phase Induction motor using AC voltage controllers?

P.3 (a)   A 3 phase, 4 pole, 50 Hz squirrel cage induction motor has the following circuit parameters.

$R_1 = 0.05$ ohm, $R_2{}' = 0.09$ ohm, $X_1 + X_2{}' = 0.55$ ohm. The motor is star connected and rated voltage is 400 V. It drives a load whose torque is proportional to the speed and is given as $T = 0.05 \times \omega$ Nw-m. Determine the speed and torque of the motor for a firing angle of 45^0 of the AC voltage controller on a 400 V, 50 Hz supply.

P.4 (a)   What is an AC voltage controller?

(b)   Explain with suitable diagrams the various types of solid state 3 phase ac voltage controllers which can be used for speed control of 3 phase induction motor from stator side. Mention the advantages of the ac voltage controllers over the other methods of solid state speed control techniques of 3 phase induction motor.

P.5   Explain why stator voltage control is suitable for speed control of induction motors in fan and pump drives. Draw a neat circuit diagram for speed control scheme of 3 phase induction motor using ac voltage controller.

P.6.(a) For stator voltage control scheme of a 3 phase induction motor discuss about speed range, regeneration, harmonics, torque pulsating, power factor, cost, efficiency and applications.

(b)   Draw a block schematic diagram for automatic speed control of 3 phase cage induction motor using solid state ac voltage controller on stator side.

P.7.(a) Discuss how the soft start scheme for the 3 phase induction motor drive can be implemented using ac voltage controllers. Mention the restrictions of this scheme.

(b)   A 440 V, 3 phase, 50 Hz, 6 pole, 945 rpm delta connected induction motor has the following parameters referred to stator side:

$R_s = 2.0\Omega$, $R_r = 2.0\ \Omega$, $X_s = 3\ \Omega$, $X_r = 4\ \Omega$. Rated voltage is impressed at the terminals for driving a fan load at rated speed. Stator voltage control employed to get variable speeds. (a) If the induction motor is running at 800 rpm, determine the motor terminal voltage, current flow into the stator and developed torque, (b) what is the motor speed, current and torque if the terminal voltage is 280 V.

P.8   A three phase AC voltage controller is used to start and control the speed of a three phase, 100 Hp, 460 V, 4 pole induction motor driving a centrifugal pump. At full load output the power factor of the motor is 0.85 and the efficiency is 80 percent. The motor current is sinusoidal. The controller and motor are connected in delta. Draw relevant circuit diagram.

(a)   Determine the rms current rating of the thyristors

(b)   Determine the peak voltage rating of the thyristors

(c)   Determine the control range of the firing angle α

P.9   A 3 phase, 8 pole, 50 Hz induction motor has the following parameters:

$R_2 = 0.15$ ohm, $X_2 = 0.7$ ohm. The motor speed is controlled by varying the applied voltage by an AC voltage controller, which operates from a 380 V, 50 Hz supply. Determine the applied voltage per phase of the motor to have a slip of 0.15. The motor drives a load with a characteristic of $T_1 = 0.014\omega^2$ Nw-m. Determine the firing angle of the converter.

P.10   A three phase, 6 pole star connected squirrel cage induction motor is to be controlled by terminal voltage variation using pairs of inverse parallel-connected thyristors in each supply line. Sketch a diagram of this arrangement and list the advantages and disadvantages of thyristor control using symmetrical phase – angle triggering compared with sinusoidal voltage variation.

Objective Type Questions

1. A three phase ac voltage controller feeding a three phase induction motor has an output`t of:

    (a)   constant voltage of variable frequency

    (b)   variable voltage of variable frequency

    (c)   variable voltage of constant frequency

    (d)   constant voltage of constant frequency   [   ]

2. The method of speed control using a three phase voltage controller is suitable for loads,

    (a)   whose torque is proportional to speed

    (b)   having constant torque

    (c)   whose torque is proportional to $\sqrt{N}$

    (d)   having torque proportional to N^2 [   ]

3. A three phase ac voltage controller feeds an induction motor. The motor has,

    (a)   very good efficiency and power factor at all speeds

    (b)   very good efficiency but poor power factor at all speeds

    (c)   poor efficiency but good power factor at all speeds

    (d)   poor efficiency and poor power factor at low speeds   [   ]

4. Full load slip of 3 ϕ squirrel cage induction motor will be about

    (a)   5 to 7%   (b)   7 to 10%

    (c)   0.5%   (d)   20%   [   ]

5. From fixed AC voltage, variable voltage can be obtained by connecting two SCRs in anti-parallel and controlling the firing angle α. If θ is the impedance angle of the load we will be able to change the load voltage if the firing angle α is

    (a)   Less than θ (b)   Greater than θ

    (c)   Less than 90^0 (d)   Lies between θ and 90^0 [   ]

6. Slip for maximum torque is directly proportional to

    (a)   Stator resistance   (b)   Stator reactance

    (c)   Rotor reactance   (d)   Rotor resistance   [   ]

7. Complete torque speed characteristics of 3-ϕ induction motor lies in

    (a)   I & II quadrants   (b)   I, II & III quadrants

    (c)   I , II , & IV quadrants   (d)   I ,II, III & IV quadrants   [   ]

    where X-axis is torque.

8.   The slip at maximum torque of 3 φ induction motor varies with the applied voltage

[True / False]   [   ]

9.   In single phase AC controller if the firing angle α is equal to impedance angle, the voltage and current in the load will be sinusoidal   (True / False)   [   ]

10.   The pf of AC controller operating on light loads is very low (True /False)   [   ]

11.   Induction motor operating with AC controller draws currents having lot of harmonics. (True /False)   [   ]

12.   For controlling the speed of 3 φ induction motor over wide range by voltage control the rotor resistance should be ……………………… (Large / Small)

13.   If the slip s lies between 0 and 1 of induction motor, it is called motoring. If the slip is greater than 1, the operation is called …………………………………..

14.   If the slip is negative the induction motor is working as ……………………………

      (Generator/Motor)

15.   The torque speed characteristics of 3 φ induction motor is given below. Draw on the same diagram how the characteristics will change, if the applied voltage is reduced. (Refer Fig. 5.13)

16.   The torque speed characteristic of 3 φ induction motor is given below. Draw on the same diagram the characteristics if the rotor resistance is increased and keeping the applied voltage constant. (Refer Fig. 5.14)

17.   For controlling the speed of 3 φ induction motor coupled to centrifugal pump by AC controller how many SCRs are required?

18.   For controlling the speed of 3 φ induction motor in both directions by AC controller how many SCRs are required.?

19.   What are the two methods of controlling the speed of 3 φ squirrel cage induction motor?

20.   Draw the circuit diagram of the 1 φ AC voltage controller and draw the wave form of load voltage and current if the load contains R & L.

21.   Draw the complete speed torque characteristics of 3 φ induction motor and indicate the mode of operation in different regions.

Control of Induction Motor Through Stator Frequency

Control of induction motor through stator frequency : Variable frequency characteristics – variable frequency control of induction motor by voltage source and current source inverter and cyclo-converters – PWM control – comparison of VSI and CSI operations – speed torque characteristics – numerical problems on induction motor drives – closed loop operations of induction motor drives (Block diagram only)

6.1   Control of Induction Motor Through Stator Frequency

We know that the synchronous speed of induction motor $N_s = \dfrac{120f}{p}$ rpm

By changing frequency of ac supply to the motor, the synchronous speed changes.

By changing the supply frequency the speed of the motor can be controlled below and above the normal full load speed.

Per phase equivalent circuit of induction motor is given in Fig.6.1

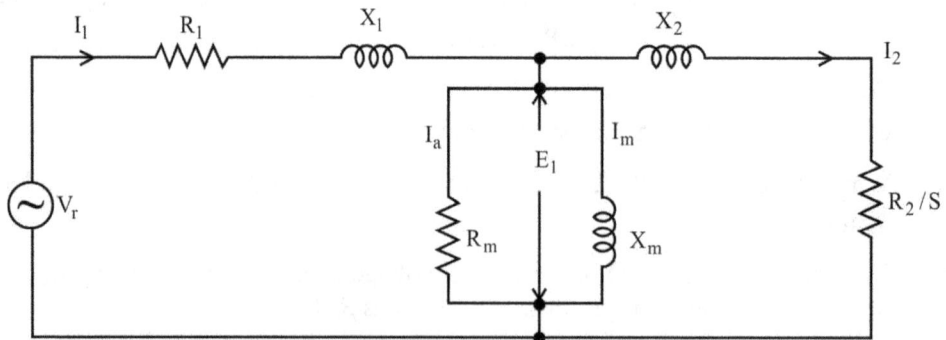

Fig. 6.1 Per phase equivalent circuit of 3 phase induction motor.

The induced emf E is proportional to the product of flux and frequency.

$$E \, \alpha \, \phi \, f$$

If voltage drop in R_1 and X_1 is neglected, the supply voltage V is proportional to the product of flux and frequency. The approximate equivalent circuit is shown in Fig.6.2

178

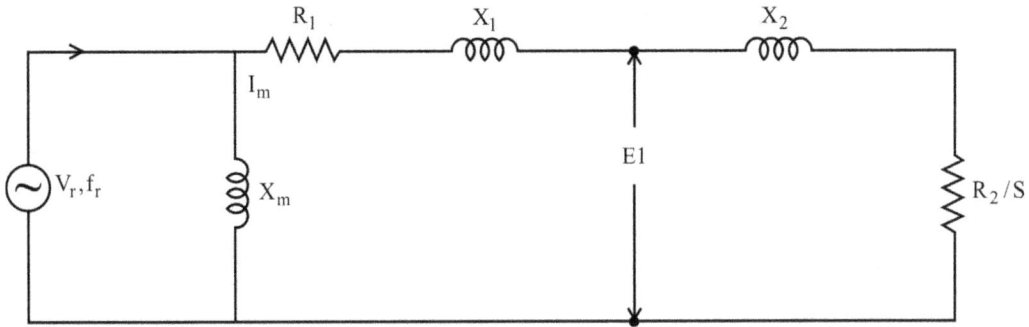

Fig. 6.2 Approximate equivalent circuit.

Keeping the applied voltage constant if the frequency is reduced the flux density will increase. If the flux is above the normal flux, the iron may get saturated resulting in the distortion of line current and voltage. The core loss and stator copper loss increases and produce noise.

If the frequency is increased flux density decreases thus the torque produced decreases.

If the induction motor is to be run at speed less than its rated speed, supply frequency is to be reduced. If the frequency is reduced the supply voltage is also to be reduced to avoid magnetic saturation. To keep the flux constant, the ratio $\dfrac{V}{f}$ should be kept constant.

Let V_r be the rated voltage and f_r be the rated frequency of the motor.

Let V be the operating voltage and f be the operating frequency. Then let

$$K = \frac{f}{f_r} = \frac{V}{V_r} \qquad \text{K is known as per unit frequency.}$$

At rated voltage V_r and rated frequency f_r the rotor current I_2 from the equivalent circuit is given by the equation,

$$I_2 = \frac{V_r}{\sqrt{(R_1 + \frac{R_2}{s})^2 + (X_1 + X_2)^2}} \qquad .....(6.1)$$

Torque developed at slip s is T_d and

$$T_d = \frac{3}{s\omega_{sr}} \cdot \frac{V_r^2 R_2}{(R_1 + \frac{R_2}{s})^2 + (X_1 + X_2)^2} \qquad .....(6.2)$$

where ω_{sr} is the rated synchronous speed in rad/sec.

$$T_{d\,max} = \frac{3}{2\omega_{sr}} \cdot \frac{V_r^2}{(R_1 \pm \sqrt{R_1^2 + (X_1 + X_2)^2})} \qquad .....(6.3)$$

6.2   Operation of Induction Motor at Frequency Less Than Rated Frequency (K < 1)

At stator voltage $V = K V_r$ & $f = K f_r$. Parameters in the equivalent circuit change and the modified equivalent circuit is shown in Fig. 6.3.

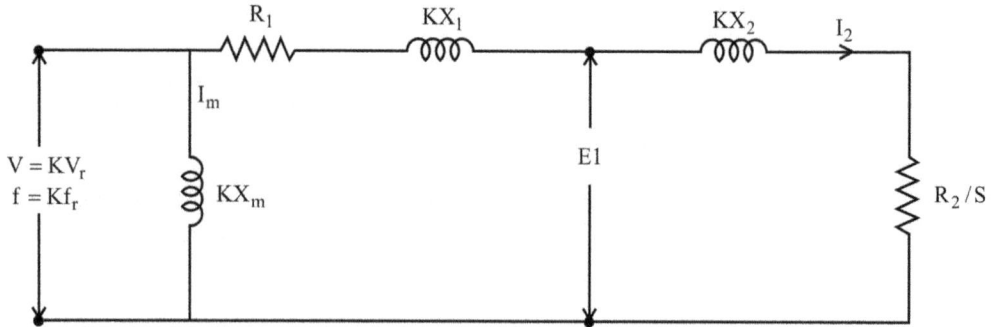

Fig. 6.3 Approximate equivalent circuit at frequency $f = Kf_r$.

From the equivalent circuit   rotor current I_2 referred to stator is given as

$$I_2 = \frac{KV_r}{\sqrt{(R_1 + \frac{R_2}{s})^2 + K^2(X_1 + X_2)^2}} \qquad .....(6.4)$$

Torque developed, $$T = \frac{3}{K\omega_{sr}} \cdot \frac{K^2 V_r^2 R_2/s}{(R_1 + \frac{R_2}{s})^2 + K^2(X_1 + X_2)^2}$$

$$= \frac{3}{\omega_{sr}} \frac{R_2}{sK} \frac{V_r^2}{(\frac{R_1}{K} + \frac{R_2}{sK})^2 + (X_1 + X_2)^2} \qquad .....(6.5)$$

$$T_{max} = \frac{3}{2K\omega_{sr}} \frac{K^2 V_r^2}{R_1 \pm \sqrt{R_1^2 + K^2 + (X_1 + X_2)^2}}$$

$$= \frac{3}{2\omega_{sr}} \left[\frac{V_r^2}{\frac{R_1}{K} \pm \sqrt{\frac{R_1^2}{K^2} + (X_1 + X_2)^2}} \right] \quad \text{for K<1} \ .....(6.6)$$

When f is large $\frac{R_1}{K} \ll (X_1 + X_2)$ giving a constant value of T_{max}

$$T_{max} \cong \frac{3}{2\omega_{sr}} \cdot \frac{V_r{}^2}{(X_1 + X_2)} \text{ for range of frequency near } f_r.$$

For low values of f, the maximum torque capability is altered. It decreases for motoring action and increases for braking action. Refer Fig. 6.4

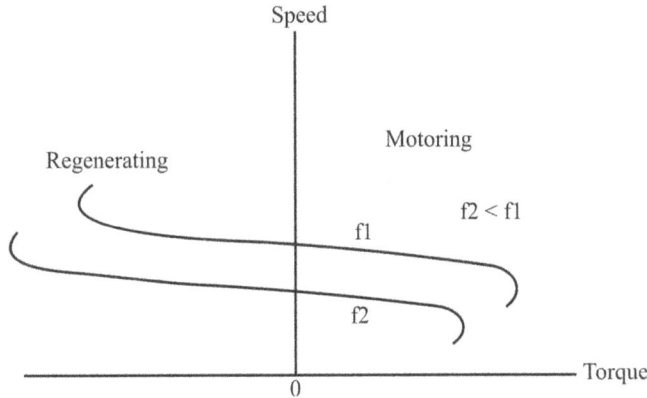

Fig. 6.4 Torque-speed characteristics at low speeds

At low frequencies the applied voltage and all reactances decrease, but the stator resistance is fixed. The resistance drop, which is negligible for high values of f, becomes appreciable in comparison with terminal voltage at low values of f. As a result, the $\dfrac{E}{f}$ ratio decreases, decreasing flux and the motor torque capability.

When working in the regenerative mode, the flux and braking torque will have higher values at low frequencies.

The above Fig. 6.4 shows the speed torque characteristics for constant V/f and for $f < f_r$.

To make full use of the motor's torque capability at the start and for low speed the V/f ratio is increased to compensate for the stator resistance drop at low frequencies.

The starting and low speed performance of a variable frequency drive is superior compared to constant frequency drive.

6.3 Operation Above the Rated Frequency (K > 1)

The operation at a frequency higher than the rated frequency takes place at constant voltage V_r or at the maximum voltage that can be obtained from the variable frequency source. Thus the flux decreases and hence torque also deceases as the frequency is increased with constant voltage. The equivalent circuit for K > 1 is shown in Fig. 6.5.

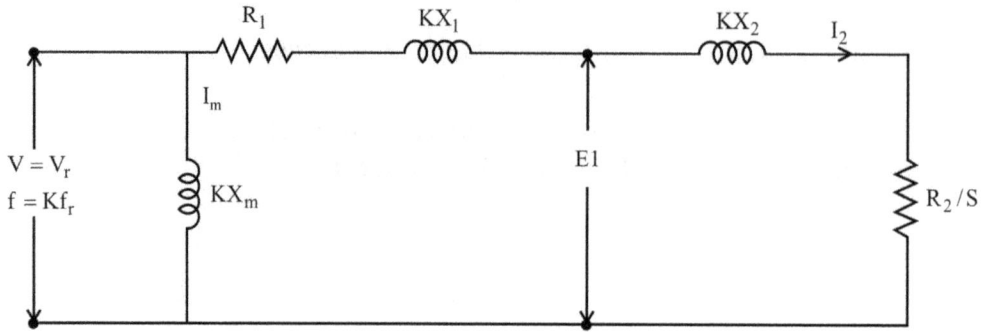

Fig. 6.5 Equivalent circuit for K > 1.

$$I_2 = \frac{V_r}{\left[(R_1 + \frac{R_2}{s})^2 + K^2(X_1 + X_2)^2\right]^{\frac{1}{2}}} \quad\quad .....(6.7)$$

$$T = \frac{3}{K\omega_{sr}} \cdot \frac{V_r^2 R_2/s}{(R_1 + \frac{R_2}{s})^2 + K^2(X_1 + X_2)^2}, \quad K>1 \quad .....(6.8)$$

$$T_{max} = \frac{3}{2\omega_{sr}K}\left[\frac{V_r^2}{R_1 \pm \sqrt{R_1^2 + K^2(X_1 + X_2)^2}}\right], \quad K>1 \quad .....(6.9)$$

Since K > 1, the breakdown torque decreases with the increase in frequency and speed.

The speed- torque curves for operation with K > 1 and K < 1 are shown in the Fig. 6.6

Performance curves of 3-phase induction motor driven by variable voltage, variable frequency are shown in Fig. 6.6(a). $\frac{V}{f}$ is held constant for frequencies less than rated frequency i.e., for K<1. In this region torque T, stator current I_s, slip speed ω_{sl} remain constant and power developed increases. In the region of frequencies greater than rated frequency and upto twice the rated frequency i.e., for 1 < K < 2 stator current I_s, slip speed ω_{sl} and power developed p_m remain constant but torque decreases. Above twice the rated frequency i.e., for K > 2 parameters T, P_m, I_s decrease but ω_{sl} may remain constant. These variations are shown in Fig. 6.6(a).

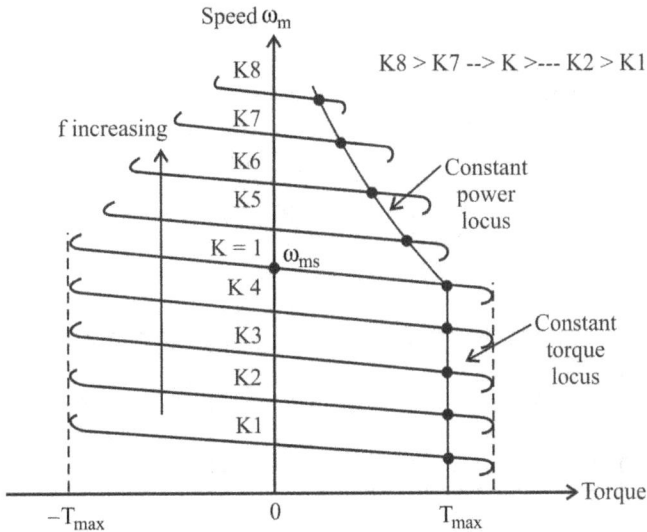

Fig. 6.6 Torque speed curves for different frequencies.

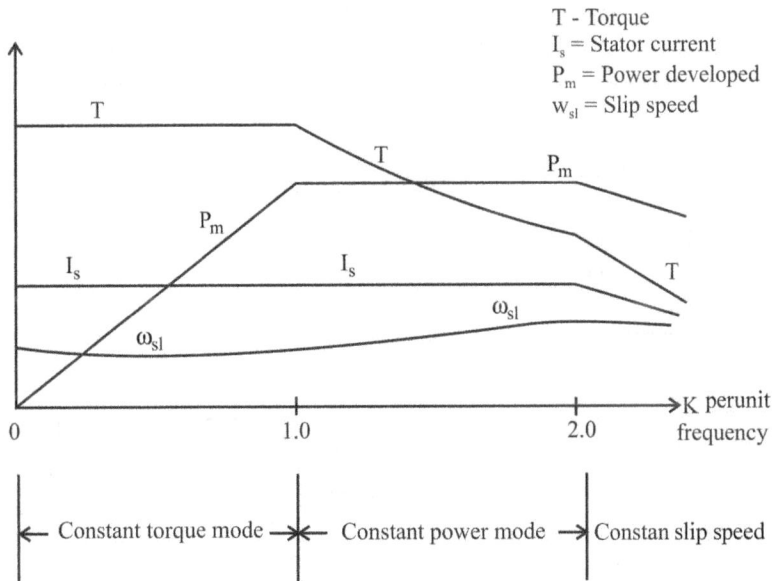

Fig. 6.6(a) Modes of operations and variation of I_s, ω_{sl}, T and P_m with per unit frequency K.

6.4 Variable Voltage Variable Frequency Drive of Three Phase Induction Motor

The variable frequency supply to an induction motor for speed control can be made available using:

    1. Voltage source inverter 2. Current source inverter 3. Cyclo- converter

6.4.1 Control of Induction Motor by Voltage Source Inverter (VSI)

VSI has low internal impedance, voltage remains substantially constant with variations in load. Any short circuit across its terminals cause current to rise very fast. The fault current cannot be regulated by current control and must be cleared by fast acting fuse links.

Variable frequency, variable voltage required for speed control of induction motor can be obtained from inverters. The power supply available may be either ac or dc. The different schemes available for driving the induction motor using inverters are shown in Fig. 6.7.

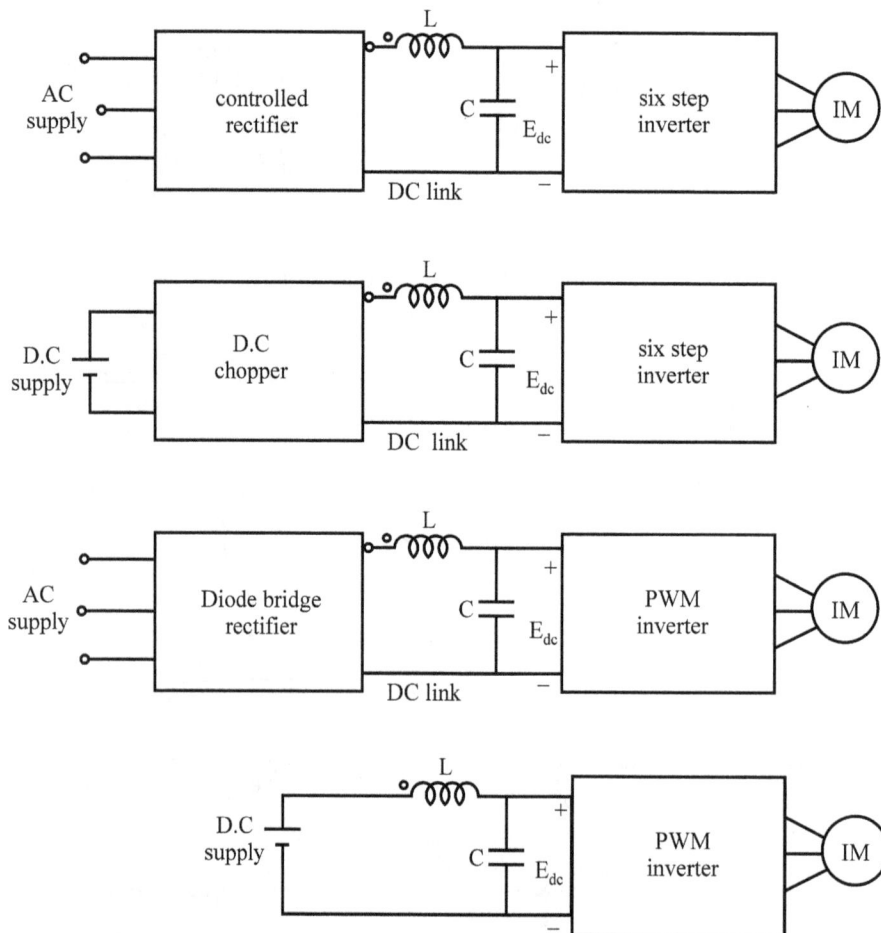

Fig. 6.7 Different schemes for getting variable voltage variable frequency supply.

6.4.2 Three Phase Voltage Source Inverter with 180⁰ Conduction Mode

The circuit diagram of three phase voltage source inverter connected to star connected induction motor and wave form of line voltage are shown in Fig. 6.8.

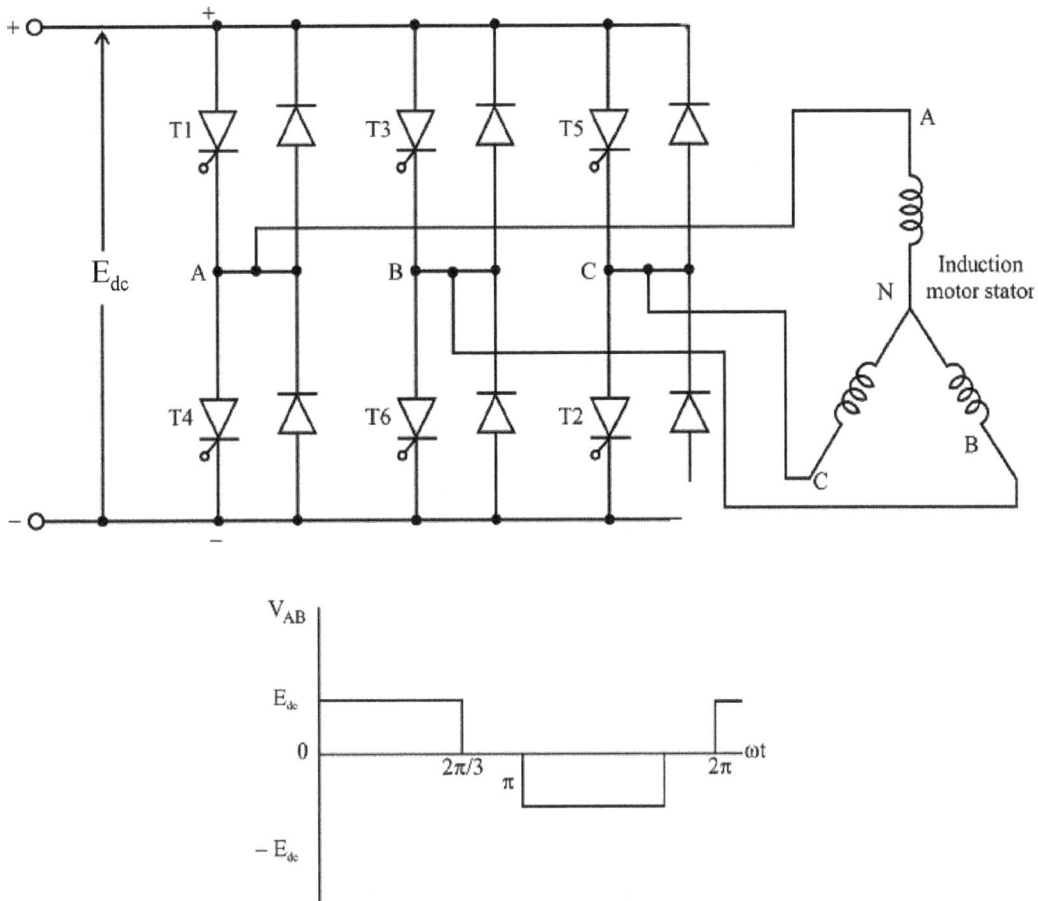

Fig. 6.8 Voltage source inverter with 180^0 conduction mode and voltage wave form.

The instantaneous line voltage for star connected induction motor is shown in the Fig. 6.8. It is rectangular wave. It can be expressed by Fourier series as

$$e_{AB}(t) = \sum_{n=1,3,5}^{\infty} \frac{4E_{dc}}{n\pi} \cos\frac{n\pi}{6} \sin n(\omega t + \frac{\pi}{6}) \qquad .....(6.10)$$

$$e_{BC}(t) = \sum_{n=1,3,5}^{\infty} \frac{4E_{dc}}{n\pi} \cos\frac{n\pi}{6} \sin n(\omega t - \frac{\pi}{2})$$

$$e_{CA}(t) = \sum_{n=1,3,5}^{\infty} \frac{4E_{dc}}{n\pi} \cos\frac{n\pi}{6} \sin n(\omega t - \frac{7\pi}{6})$$

T1 to T6 are slid state switches. They may be either SCRs, IGBTs, MOSFETS or transistors. These switches are closed in serial order from T1 to T6 at 60^0 intervals in a cycle. Or at one sixth the periodic time T. By changing the periodic time the frequency of the output voltage is changed. In the Fig. 6.8(a) waveforms of the triggering signals I_{g1} to I_b are shown. At any given instant three values will be conducting as shown. The waveforms of phase and line voltage applied to three phase star connected induction motor are shown.

Waveform of line voltage V_{AB} is shown in Fig. 6.8 also for analysis. Phase voltage is a six stopped wave.

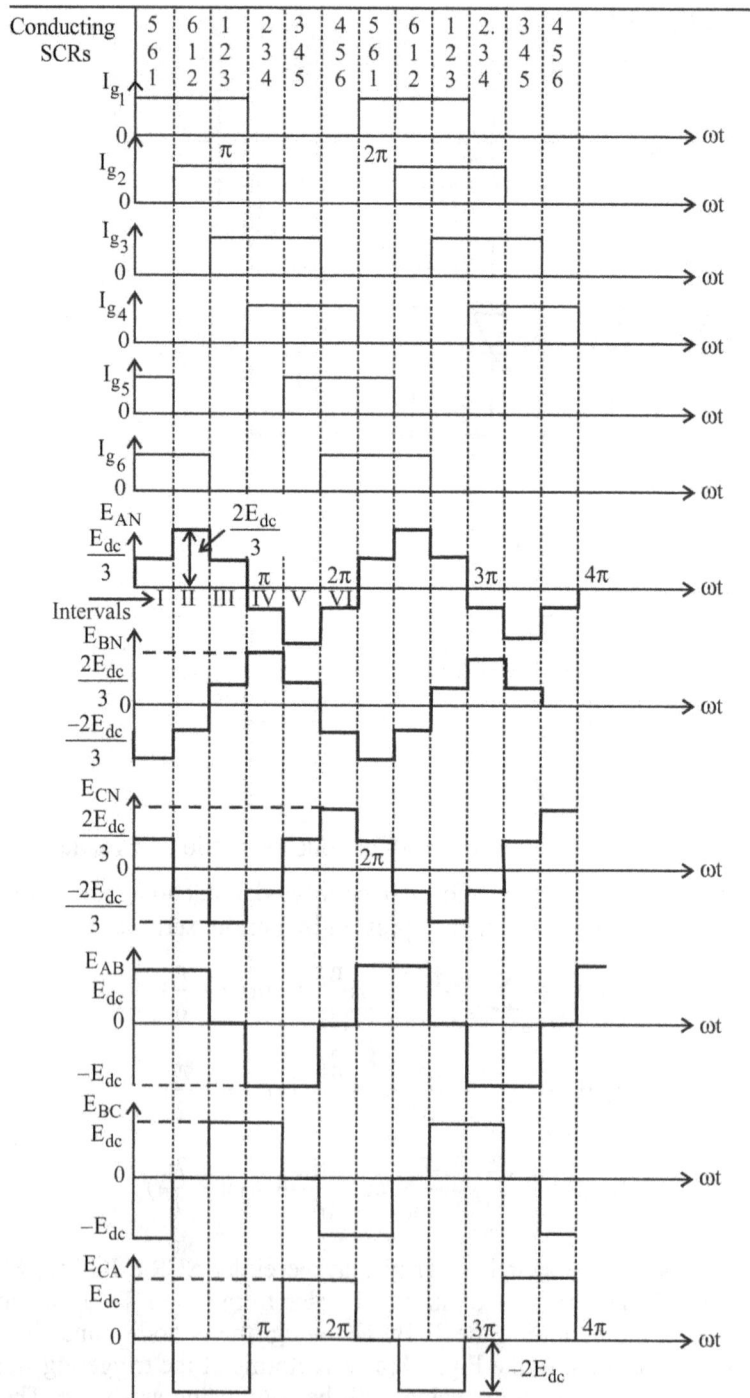

Fig. 6.8(a) Voltage waveforms for 180° conduction mode 3 SCRs conduct at any instant.

All triplen harmonics are zero because $\cos\dfrac{3k\pi}{6} = 0$ for k integer.

Fundamental component of line voltage

$$e_{AB1}(t) = \frac{4E_{dc}}{\pi}\frac{\cos\pi}{6}\sin(\omega t + \frac{\pi}{6})$$

$$= \frac{4E_{dc}}{\pi}\frac{\sqrt{3}}{2}\sin(\omega t + \frac{\pi}{6}) = \frac{2\sqrt{3}}{\pi}E_{dc}\sin(\omega t + \frac{\pi}{6}) \quad .....(6.11)$$

rms value of the fundamental component of the line voltage

$$E_{L1} = \frac{\sqrt{6}}{\pi}E_{dc} = 0.7797 E_{dc} \qquad\qquad .....(6.12)$$

rms value of the fundamental component of the phase voltage

$$E_{p1} = \frac{E_{L1}}{\sqrt{3}} = \frac{0.7797}{\sqrt{3}}E_{dc} = 0.45 E_{dc} \qquad\qquad .....(6.13)$$

R.M.S Value

Total rms value of Line to Line voltage

$$= \sqrt{\frac{1}{\pi}\int_{0}^{\frac{2\pi}{3}} E_{dc}^{2}d(\omega t)} = E_{dc}\sqrt{\frac{2\pi}{3\pi}} = E_{dc}\sqrt{\frac{2}{3}} = 0.8165\ E_{dc} \qquad ......(6.14)$$

Total rms value of phase voltage $= \dfrac{0.8165 E_{dc}}{\sqrt{3}} = 0.471\ E_{dc}$ $\qquad .....(6.15)$

The wave form of the phase voltage for Y connected induction motor driven by three phase VSI operating with 180^{0} conduction mode is shown in the Fig. 6.9.

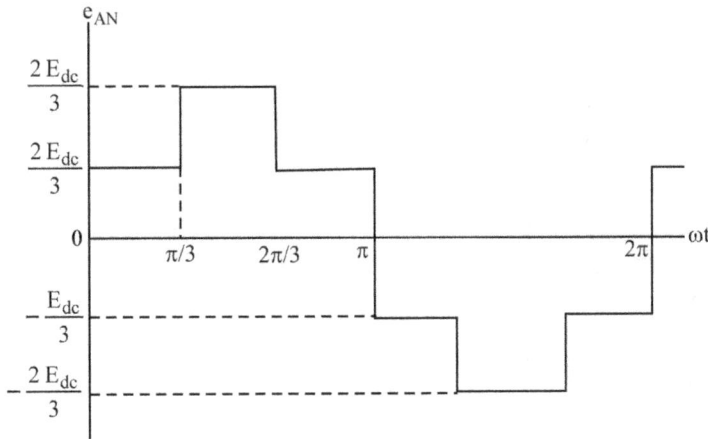

Fig. 6.9 Wave form of phase voltage.

$$\text{rms value phase voltage} = \sqrt{\frac{1}{\pi}[\int_0^{\frac{\pi}{3}} (\frac{E_{dc}}{3})^2 \omega t + \int_{\frac{\pi}{3}}^{\frac{2\pi}{3}} (\frac{2}{3}E_{dc})^2 d\omega t + \int_{\frac{2\pi}{3}}^{\pi} (\frac{E_{dc}}{3})^2 d\omega t]}$$

$$= E_{dc}\sqrt{\frac{1}{\pi}\left[\frac{1}{9}(\frac{\pi}{3} - 0) + \frac{4}{9}(\frac{2\pi}{3} - \frac{\pi}{3}) + \frac{1}{9}(\pi - \frac{2\pi}{3})\right]}$$

$$= E_{dc}\sqrt{\frac{\pi}{\pi}\left[\frac{1}{27} + \frac{4}{27} + \frac{1}{27}\right]}$$

$$= E_{dc}\sqrt{\frac{6}{27}} = \frac{\sqrt{2}}{3}E_{dc} = 0.471\ E_{dc} \qquad \text{.....(6.16)}$$

Same as the previous expression.

6.4.3 Three-phase Voltage Source Inverter with 120^0 Conduction Mode – star Connected Load

In this mode two SCRs will be conducting at any instant as shown in Fig. 6.10.Wave forms of phase and line voltages are shown for star connected load.

The switches T1 to T6 are closed in serial order from T1 to T6 at 60° intervals or at one sixth the periodic time of T seconds. The frequency of the output voltage can be changed by changing the periodic time T. The wave form of each phase and line voltahes are shown. The phase voltahe is rectangular wave and the line voltage is six stepped wave.

Fourrier analysis of the wave form shows that

Line to neutral voltage.

$$e_{AN} = \sum_{n=1,3,5}^{\infty} \frac{2E_{dc}}{n\pi}\cos\frac{n\pi}{6}\sin n(\omega t + \frac{\pi}{6}) \qquad \text{.....(6.16)}$$

$$e_{BN} = \sum_{n=1,3,5}^{\infty} \frac{2E_{dc}}{n\pi}\cos\frac{n\pi}{6}\sin n(\omega t - \frac{\pi}{2}) \qquad \text{.....(6.17)}$$

$$e_{CN} = \sum_{n=1,3,5}^{\infty} \frac{2E_{dc}}{n\pi}\cos\frac{n\pi}{6}\sin n(\omega t - \frac{7\pi}{6}) \qquad \text{.....(6.18)}$$

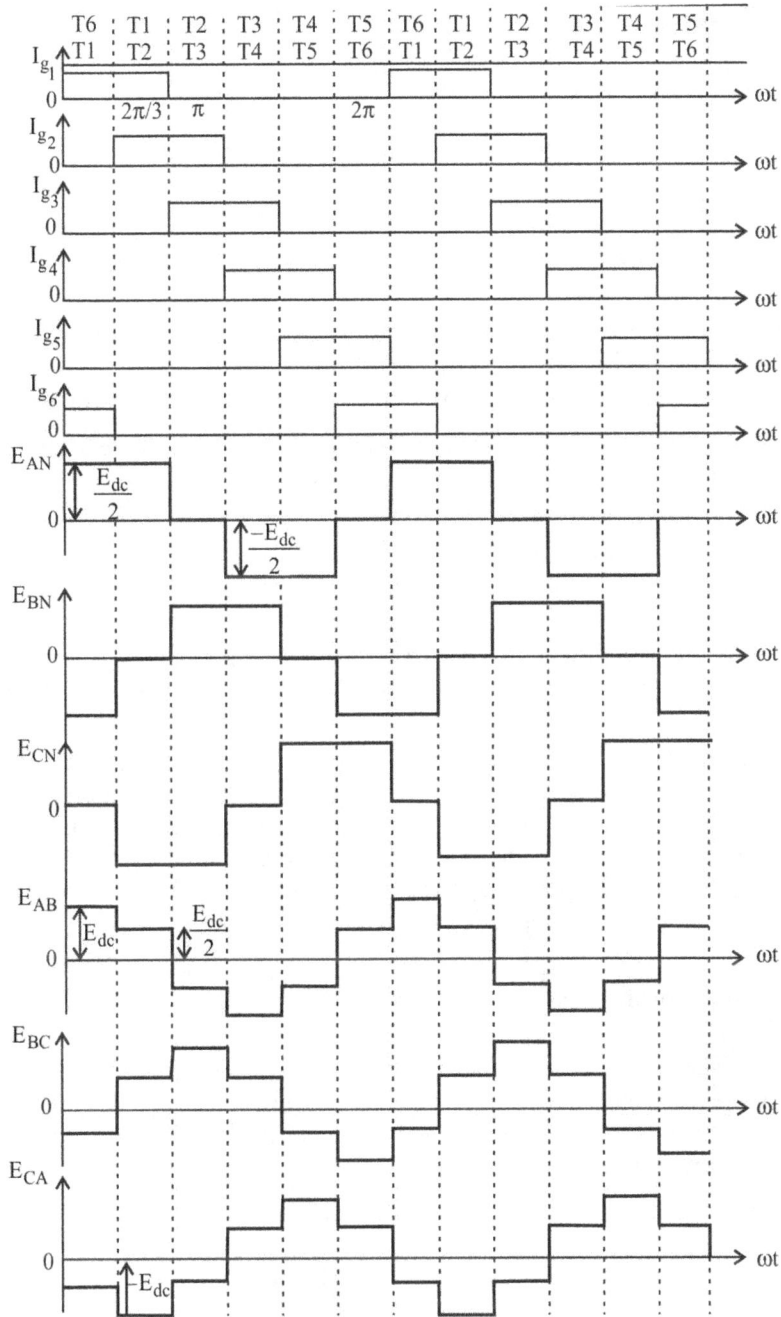

Fig. 6.10 Gain signals and voltage waveform for 120° conduction two SCRs conduct at any instant.

The fundamental component of phase voltage

$$e_{ANI} = \frac{2E_{dc}}{\pi} \cos\frac{\pi}{6} \sin(\omega t + \frac{\pi}{6})$$ .....(6.19)

$$\text{rms value of } E_{ANI} = \frac{2E_{dc}}{\sqrt{2}\pi} \frac{\sqrt{3}}{2} = \frac{E_{dc}}{\pi}\sqrt{\frac{3}{2}} = 0.389 \ E_{dc}$$ .....(6.20)

rms value of line voltage $E_{ABI} = \sqrt{3} \times 0.389 \ E_{dc} = 0.675 \ E_{dc}$ .....(6.21)

6.4.4 Operation of Three Phase Induction Motor with Periodic Non-sinusoidal Supply Voltages

We have seen that the output voltage of voltage source inverter is nonsinusoidal.

The application of symmetrical, non-sinusoidal voltages of constant periodicity to the motor terminals results in symmetrical, non-sinusoidal three phase motor currents. These currents may be thought to consist of a fundamental component plus higher time harmonics.

Current time harmonics of order $n = 3h + 1$ for h integer result in forward rotating mmf waves while time harmonics of order $n = 3h + 2$ produce backward rotating mmf waves. Triplen (i.e., multiples of three) order harmonics are usually suppressed by the circuit connections in three phase supply systems. But triplen order harmonics are present in delta connected windings as zero sequence components and they do not produce rotating mmf.

For steady state motoring operation the physical nonlinear ties of an induction motor such as magnetic saturation and the resistance and inductance variation with current can be neglected. If the motor can properly be regarded as a linear system a periodical non-sinusoidal supply voltage can be resolved in to Fourier components and each sinusoidal higher order harmonic component applied separately to an appropriate equivalent circuit, the total effect being obtained by applying the principle of super position.

Compared with the fundamental frequency equivalent circuit the following criteria apply to the n th time harmonic equivalent circuit.
(i)   All reactances have a value n times the fundamental frequency value.
(ii) The operating slip is the harmonic slip s_n.
(iii) Skin effect should be taken into account in calculating the primary and the secondary resistance and reactance of cage rotors for high frequency harmonics.

The feature of an induction motor fed from a square wave inverter can be summarized as follows:
1.   The inverter has impressed dc voltage of variable amplitude.
2.   Advantageous for multi motor drive.
3.   Commutation is load independent. Converter and load need not be matched. The converter represents a source to which the motor can be just plugged on.
4.   At present converter output frequencies up to 1500 Hz are possible. This drive is very much suitable for (motors of) high speed operation. The drives are available up to rating of 200 KVA.

5. The lowest operating frequency as limited by commutation is about 5 Hz. Speed range is 1 : 20.
6. Not suitable for acceleration on load and sudden load changes.
7. Dynamic behavior is fairly good at high speeds.
8. Dynamic braking is possible. Regeneration (Four quadrant operation) requires an additional converter connected in anti-parallel to the line side one. Speed reversal is obtained by changing the phase sequence.
9. The input voltage to the motor is non-sinusoidal. This results in additional losses, heating and torque pulsations.
10. Motor should have sufficiently large leakage inductance to limit the peak currents and decrease the harmonic content.
11. Open loop control of the motor is possible, but may have stability problem at low speeds.
12. Line power factor is poor due to phase control.
13. It can be operated as a slip controlled drive.
14. It finds application as a general purpose industrial drive for low to medium power.

The specific features of PWM inverter induction motor drive can be summarized as follows:

1. The inverter has constant dc link voltage and employs PWM principle for both voltage control and harmonic elimination.
2. The output voltage wave form is improved, with reduced harmonic content. The amplitude of torque pulsations is minimal even at low speeds.
3. Parallel operation of many inverters on the same dc bus.
4. Uninterrupted operation is possible when a buffer battery is used.
5. The power factor of the system is good as a diode rectifier can be employed on the line scale.
6. The control is complicated.
7. Four quadrant operation is possible. During regeneration a battery or another converter with phase control may be used. Dynamic braking can also be employed.
8. Single and multi motor operations are possible. Smooth change over of voltage and frequency values at zero crossing for speed reversal with full torque capability at standstill.
9. Due to switching losses, the highest operating frequency is 150 Hz.
10. The inverter and load need not be matched. The converter operates as a source to which the motor can be plugged.
11. The leakage reactance of the motor further smoothens the motor current, effectively decreasing the harmonic content and peak values of stator current.
12. The size of the filter also decreases.
13. The efficiency may be affected by switching losses.
14. Open loop operation is possible. The drive has a very good dynamic and transient response.
15. Selected harmonic elimination techniques may be employed to eliminate lower order harmonics.

16. The drive finds application in low to medium power. Commercial drives are available up to 450 KVA.

17. Transistors may be used in place of thyristors.

6.4.5 Closed Loop Operation of Induction Motor Driven by Voltage Source Inverter

The block diagram of the control scheme that can be used for controlling the speed of three phase induction motor driven by voltage source inverter is given in Fig. 6.11.

The three phase controlled rectifier gives variable dc voltage. This dc voltage is filtered and given as input to VSI. The VSI gives variable frequency. This scheme can be used for driving the induction motor at any desired speed by keeping ω_r corresponding to the desired speed. The actual speed ω_m is sensed and compared with the reference speed ω_r and suitable action is taken by controllers till ω_m is equal to ω_r. The output of current controller K2, generates the reference frequency ω_s and reference voltage V_r such that V_r/ω_s is held constant. Voltage controller and firing circuit changes the firing angle α till the output voltage of rectifier is equal to reference voltage V_r. The block K4 changes the periodic time of the VSI till the actual frequency of 3ϕ voltage applied to induction motor is equal to reference frequency. This change in frequency has to be done slowly so that the stator current of induction motor should not exceed its rated value. For this purpose current control loop with current limiter is used. Thus the speed of the motor is changed to the desired speed without exceeding the rated value of the current and also keeping V/f constant at fixed value.

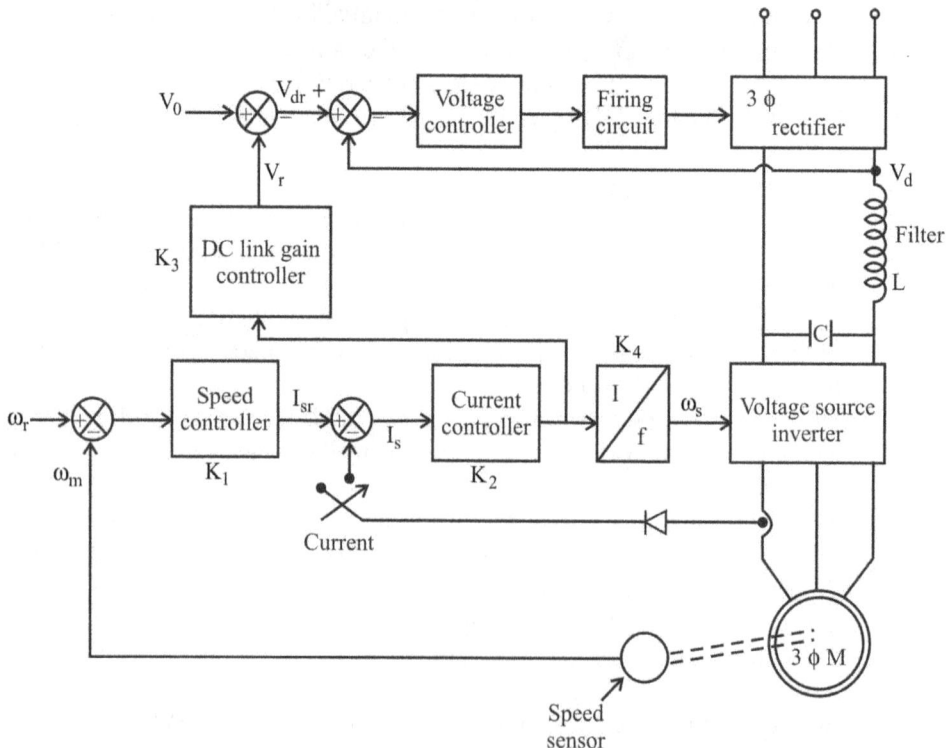

Fig. 6.11 Closed loop control of induction motor fed from VSI V/f constant.

6.5   Speed Control of 3-phase Induction Motor with Current Source Inverter

The speed of 3 phase induction motor can be controlled below and above its rated speed using variable frequency current source inverter. As its name indicates the input current is to be kept constant by putting a large inductance in series with the source. The circuit diagram of CSI is shown in Fig. 6.12.

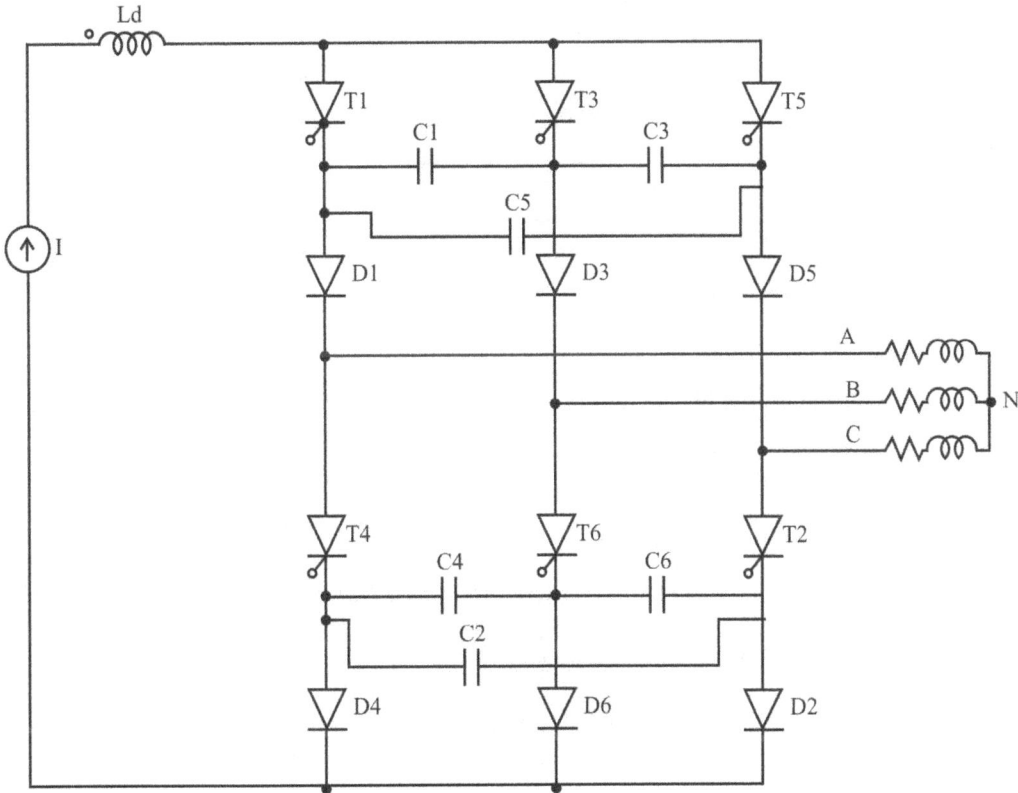

Fig. 6.12 3-phase current source inverter.

The Thyristors T1 to T6 are triggered in serial order at 60^0 interval. When one thyristor is triggered one thyristor will go to OFF state as indicated. Two thyristors will be in the ON state at any instant of time. Six pairs of thyristors conduct in a periodic time, each pair for one sixth period as shown in the figure. The wave form of current in each phase of star connected induction motor are shown in Fig. 6.13 for one period. The current in each phase is of rectangular wave shape. Fourier analysis of this rectangular wave will gave

$$i_A = \frac{2\sqrt{3}\,I}{\pi}\,[\sin wt - \frac{1}{5}\sin 5\,wt - \frac{1}{7}\sin 7\,\omega t\;\frac{1}{11}\sin 11\,\omega t + ...]$$

Fundamental components $i_{A1} = \dfrac{2\sqrt{3}\,I}{\pi}\sin \omega t.$

ON	T1	T2	T3	T4	T5	T6	T1
OFF	T5	T6	T1	T2	T3	T4	T5
SCRs	T6,T1	T1, T2	T2, T3	T3, T4	T4, T5	T5, T6	T6, T1

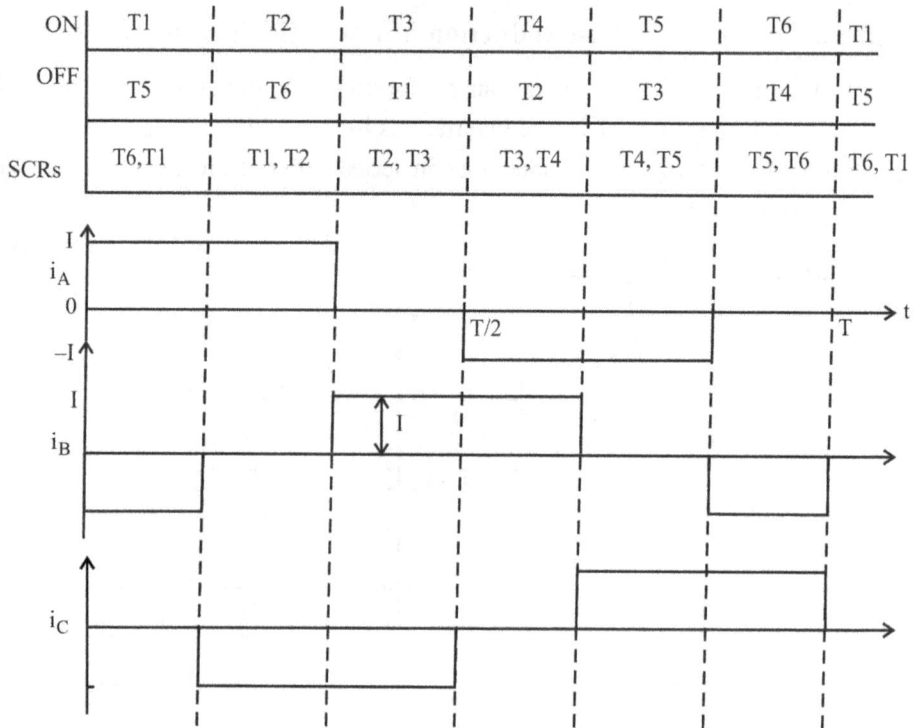

Fig. 6.13 Idealized current waveform of 3φ CSI.

R.M.S value of fundamental components of line current

$$I_{A1} = \frac{\sqrt{6}\,I}{\pi} = 0.78\ I$$

Per phase equivalent circuit of induction motor fed from current source inventor at rated frequency f_r shown in Fig. 6.14. All parameters are referred to stator.

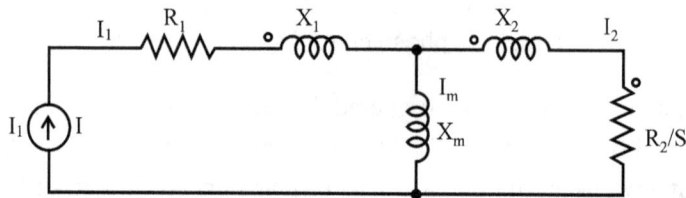

Fig. 6.14 Per phase equivalent circuit of induction motor fed from CSI at rated frequency f_r.

From the equivalent circuit drawn for rated frequency f_r, the rotor current I_2 referred to stator is given as

$$I_2 = \frac{X_m I_1}{\sqrt{(\frac{R_2}{s_r})^2 + (X_2 + X_m)^2}} \qquad \ldots\ldots(6.22)$$

where s_r is the rated slip

$$S_r = \frac{\omega_{sr} - \omega_r}{\omega_{sr}}$$

where

ω_{sr} is the rated synchronous speed

ω_r is the rated speed of the motor

$$\text{Developed torque } T_d = \frac{3I_2^2 R_2}{S_r \omega_{sr}} = \frac{3}{S_r \omega_{sr}} \frac{I_1^2 X_m^2 R_2}{\left[(\frac{R_2}{S_r})^2 + (X_2 + X_m)^2 \right]} \qquad \text{.....(6.23)}$$

$$I_m = \left[\frac{(\frac{R_2}{S}) + X_2^2}{(\frac{R_2}{S}) + (X_2 + X_m)^2} \right]^{\frac{1}{2}} I_1$$

Equivalent circuit at operating frequency $f = K.f_r$ is shown in Fig. 6.15

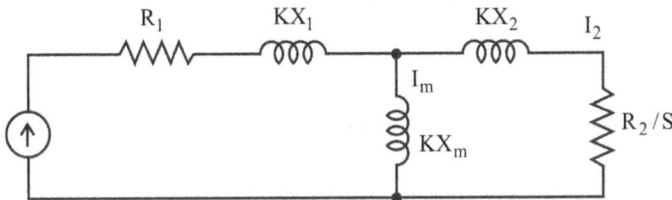

Fig. 6.15 Per phase equivalent circuit of induction motor fed from CSI at operating frequency $f = kf_r$

From the equivalent circuit

$$I_2 = \frac{I_1 KX_m}{\sqrt{(\frac{R_2}{S})^2 + K^2(X_m + X_2)^2}} \qquad \text{.....(6.24)}$$

where s is the slip at operating frequency and $s = \frac{\omega_s - \omega_m}{\omega_s}$

where ω_s is the synchronous speed at operating frequency f and ω_m is the actual speed of the motor.

Torque developed at slip s is given as

$$T_d = \frac{3I_2^2 R_2}{s\omega_s}$$

$$= \frac{3R_2}{s\,\omega_s} \cdot \frac{I_1^2 \, K^2 \, X_m^2}{\left(\frac{R_2}{s} \right)^2 + K^2(X_m + X_2)^2} \qquad \text{.....(6.25)}$$

$$= \frac{3\,R_2}{s\,\omega_s} \cdot \frac{I_1^2\,X_m^{\;2}}{\left(\dfrac{R_2}{s\,K}\right)^2 + \left(X_m + X_2\right)^2}$$

$$= \frac{3\,R_2}{\left(\omega_s - \omega_m\right)} \cdot \frac{I_1^2\,X_m^{\;2}}{\left(\dfrac{R_2\,\omega_{sr}}{\omega_s - \omega_m}\right)^2 + \left(X_m + X_2\right)^2}$$

$$= \frac{3\,R_2}{\omega_{sl}} \cdot \frac{I_1^2\,X_m^{\;2}}{\left(\dfrac{R_2\,\omega_{sr}}{\omega_{sl}}\right)^2 + \left(X_m + X_2\right)^2} \qquad .....(6.25)$$

$$I_m^{\;2} = \frac{\left(K^2 X_2^2 + \dfrac{R_2^2}{s^2}\right) I_1^2}{K^2\left(X_2 + X_m\right)^2 + \dfrac{R_2^2}{s^2}} = \frac{\dfrac{R_2^2}{s^2 K^2} + X_2^2}{\dfrac{R_2^2}{s^2 k^2} + \left(X_2 + X_m\right)^2} I_1^2$$

where $(\omega_s - \omega_m)$ is the slip speed denoted by ω_{sl}.

From equation 6.23 the torque developed at rated slip and rated speed is given as

Let the rated slip speed be equal to s_1.

$$T_d = \frac{3\,R_2}{s_r\,\omega_{sr}} \cdot \frac{I_1^2\,X_m^{\;2}}{\left(\dfrac{R_2}{\omega_{sr}}\right)^2 + \left(X_m + X_2\right)^2}$$

$$= \frac{3\,R_2}{\left(\omega_{sr} - \omega_{mr}\right)} \cdot \frac{I_1^2\,X_m^{\;2}}{\left(\dfrac{R_2\,\omega_{sr}}{\omega_{sr} - \omega_{mr}}\right)^2 + \left(X_m + X_2\right)^2}$$

$$= \frac{3\,R_2}{\omega_{slr}} \cdot \frac{I_1^2\,X_m^{\;2}}{\left(\dfrac{R_2\,\omega_{sr}}{\omega_{slr}}\right)^2 + \left(X_m + X_2\right)^2} \qquad .....(6.26)$$

where $(\omega_{sr} - \omega_{mr})$ is the slip speed at rated frequency and denoted by ω_{sr}.

$\omega_s - \omega_m$ = slip speed when operating at $f = \omega_{sl}$

$\omega_{sr} - \omega_{mr}$ = slip speed when operating at $f_r = \omega_{slr}$

Equations 6.25 and 6.26 show that the torque developed for a certain stator current will be same at all speeds, if the slip speed is maintained constant at rated slip speed ω_{slr}.

6.5.1 The General Features of an Induction Motor on a Current Source Inverter can be Summarized as follows

1. Load dependent commutation : As the load parameters form part of the commutation circuit, the inverter and motor must be matched.

2. The inverter has a simple configuration.

3. It can be used only for single motor operation.

4. The DC link contains only inductance. To maintain constant current this has to be very large. Two quadrant operation is straight forward.

5. Invariably, a phase controlled rectifier is required on the line side. The variable DC link voltage is converted to a constant current source by means of high link inductance.

6. The inverter is force commutated to give variable frequency currents to feed the motor.

7. The value of the capacitance is a compromise between the voltage spikes and highest operating frequency. Larger the capacitance smaller the voltage spikes. The highest operating frequency is also limited.

8. The leakage reactance of the motor influences the harmonic voltages. It is also responsible for spikes of voltages during commutation. The leakage reactance being a parameter of the commutation circuit, determines the time of commutation, and consequently the upper operational frequency is limited. A motor must have smaller leakage reactance to have reduced harmonic voltages and spikes of voltage and to increase the range of speed control. The spikes influence the rating of the thyristor and affect the insulation. The motor size becomes larger if leakage reactance is small.

9. Converter grade thyristors are sufficient. Thyristor utilization is good.

10. The inverter recovers from commutation failure. The link inductance causes a slow rise of fault current and by the time it reaches high value it can be suppressed.

11. There is a stability problem at light load. A minimum current should be there for commutation.

12. Open loop operation is not possible. Dynamic response is sluggish.

13. The line pf is poor due to phase control.

14. Finds application as medium to high power drive.

15. Torque pulsations cause speed oscillations at very low speeds. PWM strategies are being employed to eliminate the speed oscillations and make the running smooth.

16. Both constant torque and constant horse power operation are possible.

6.5.2 Closed Loop Speed Control of Induction Motor Driven by CSI

The block diagram of closed loop speed control scheme of 3 phase induction motor driven by current source converter is shown in Fig. 6.16. A large inductance is connected between controlled rectifier and current source inverter for keeping the input current at prescribed constant value. This current requirement depends on load torque. ω_r is the reference speed and ω_m is the actual speed of the motor. The signal corresponding to the difference between ω_r and ω_m is given as input to the speed controller. The output of the speed controller changes till the difference is zero. The output of speed controller sets a reference value for I_d and also for slip speed through slip regulator. The frequency of inverter will be such that the synchronous speed $\omega_s = \omega_{mf}^* \, \omega_{slip}$. This arrangement makes the slip speed constant so that the load of a particular torque can be driven at any speed without exceeding the input current I_d. There is an inner current loop which controls and keeps the value equal to the reference value of current by changing the firing angle of the controlled rectifier. Thus the induction motor speed can be controlled from low value to any value by changing the frequency of current source inverter. The value of input current I_d changes depending on the load torque requirement.

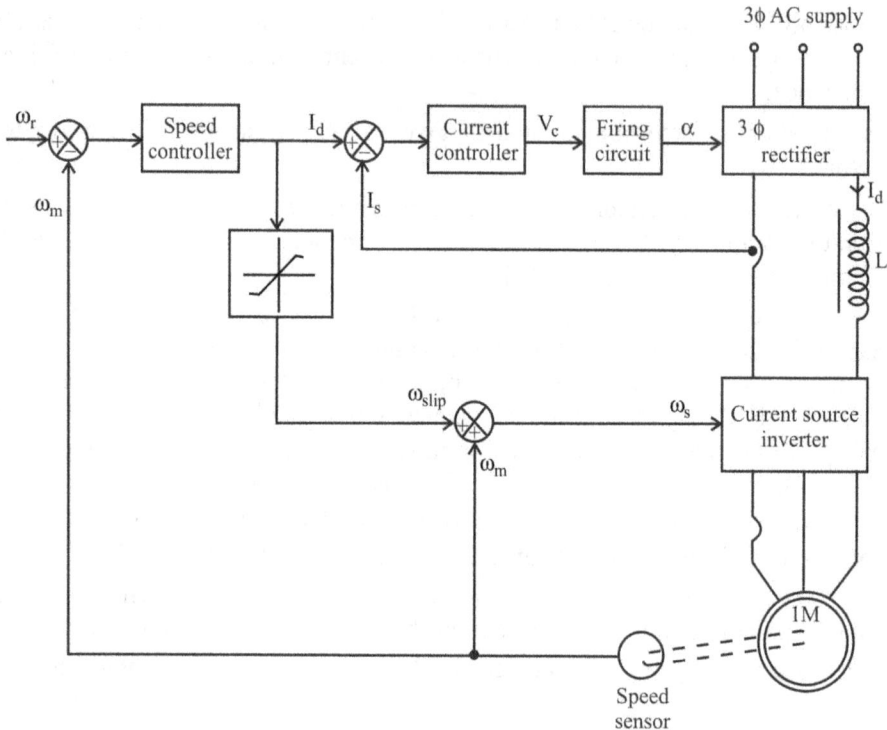

Fig. 6.16 Closed loop speed control of IM fed from current source inverter.

6.6   Speed Control of Three-phase Induction Motor with Cycloconverter

A Cycloconverter is a frequency changer that converts ac power at one input frequency to output power at a different frequency with a one-stage conversion process. Cycloconverters of high power capacity are very popular for driving induction motors and wound field synchronous motors. Applications of cycloconverters are in:

Cement and ball mill drives.

Rolling mill drives.

Slip-power recovery Scherbius drives.

Variable speed, constant frequency power generation for aircraft 400 Hz power supplies.

A half-wave cycloconverter is shown in the figure along with the nature of its output voltage wave form. Because of low harmonic content when operating at low frequencies, smooth motion is obtained at low speeds. Harmonic content increases with frequency, making it necessary to limit the maximum output frequency to 40% of the source frequency. The cycloconverter drive has regenerative braking capacity. Full four-quadrant operation is obtained by reversing the phase sequence of motor terminal voltage. Since the cycloconverter employs large number of thyristors, it becomes economically acceptable only in large power drives. The low speed operation is obtained by feeding a motor with large number of poles from a cycloconverter operating at low frequencies. These drives are called gearless drives because, unlike conventional drives, the low speed operation of load is obtained without reduction gear, thus eliminating the associated cost, space and maintenance.

Operation of the drive : A practical and commonly used cycloconverter uses the three-phase half wave configuration shown in the Fig. 6.17.

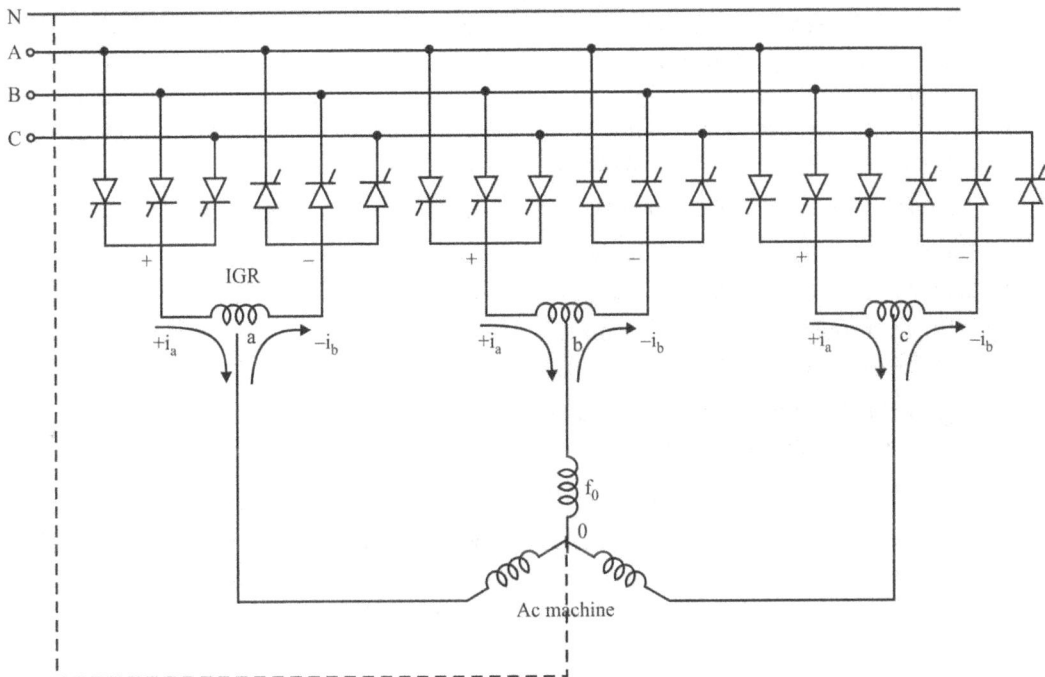

Fig. 6.17 Three-phase-to-phase cycloconverter with 18 thyristors.

It is known as 18 thyristor, three-pulse cycloconverter. The circuit consists of three identical half-wave, anti-parallel phase groups connected to a three-phase ac machine load. Each phase group functions as a dual converter, but the firing angle of each group is modulated sinusoidally with $2\pi/3$ phase angle shift so as to fabricate three phase balanced voltage at the machine terminals. An inter group reactor IGR is connected to each phase group to restrict circulating current.

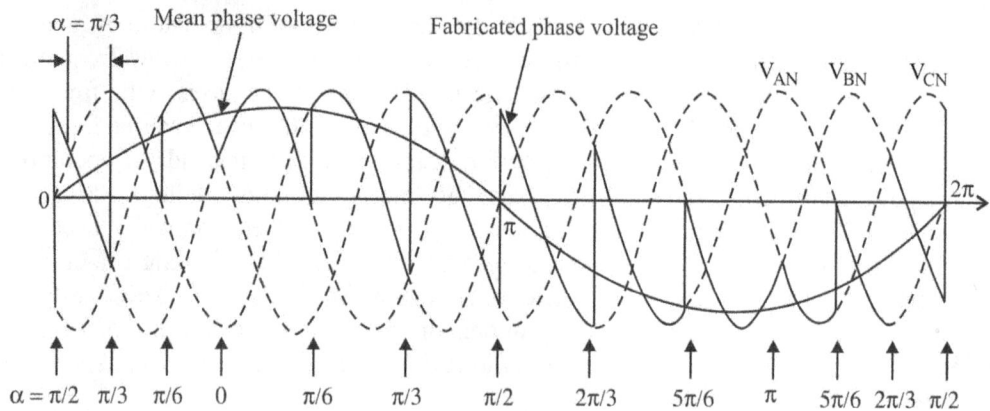

Fig. 6.18 Phase-controlled fabrication of output voltage wave in three-phase, half-wave cycloconverter

The figure explains the synthesis of output phase voltage wave by sinusoidal modulation of the firing angle α. The output frequency and depth of modulation can be varied to generate a variable frequency, variable voltage power supply to motor. The fabricated output voltage wave contains complex harmonics, which may be adequately filtered by the machine's effective leakage inductance.

Any arbitrary displacement power factor load can be supplied by a cycloconverter.

Fig.6.19(a) shows the phase voltage and current waves in active or motoring mode, when the current wave lags the voltage wave by DPF angle φ. The positive half-cycle of the current flows through the positive converter, where as the negative converter takes the negative half-cycle of current.

Fig. 6.19(b) shows the voltage and current waves in regenerative mode of the drive, where the polarity of the current wave is reversed.

6.6.1 Main Features of Cycloconverter Drive

1. The cycloconverter operates by means of line commutations. No forced commutation is required. Losses due to forced commutations are eliminated. The converter operates at lagging power factor. The line power factor is very poor at light loads.

2. The voltage control is possible in the converter itself, so that the machine operates at rated flux condition so that v/f can be held constant.

3. The firing angles of thyristors are modulated so that a mean sine wave voltage is synthesized from the input voltage wave form.

4. It is capable of power transfer in either direction between ac source and motor load. It can feed power to a load of any pf. Regeneration is possible in the complete speed range. A four-quadrant operation is simple and straightforward.

5. The output frequency of the cycloconverter is limited to 1/3 rd of input frequency.

6. A speed control range of 0 to $33\frac{1}{3}$% of base speed is possible.

7. It requires many thyristors.

8. The cost of cycloconverter is high because of large number of thyristors and the control is complex.

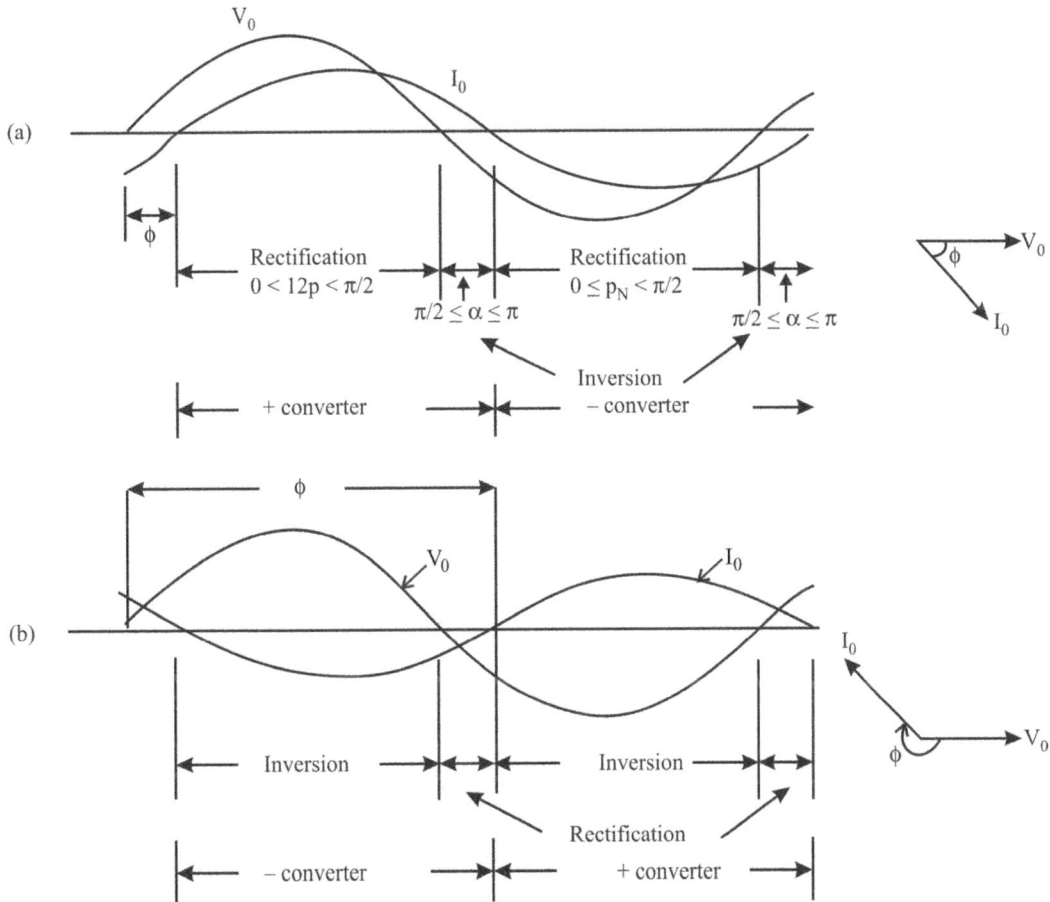

Fig. 6.19 Voltage and current waves for motoring and regenerative modes
(a) Motoring mode, (b) Regenerative mode.

Example 6.1

A star-connected squirrel cage induction motor has the following ratings and parameters referred to stator; 400 V, 50 Hz, 4 pole, 1370 rpm. $R_s = 2\,\Omega$, $R_r^{\,1} = 3\,\Omega$, $X_s = X_r^{\,1} = 3.5\ \Omega$; $X_m = 55\,\Omega$. It is controlled by VSI at constant flux.

Calculate; (i) Motor torque, speed and stator current when operating at 30 Hz and rated slip speed. (ii) Inverter frequency and stator current for rated torque and motor speed of 1200 rpm. Assuming motor speed-torque curves to be parallel straight lines in the region of interest, calculate motor speed when operating at (iii) 30 Hz and half the rated torque and (iv) 45 Hz and braking torque equal to rated torque.

Solution :

(i)   Synchronous speed $= \dfrac{120 \times 50}{4} = 1500$ rpm

Rated speed $= 1370$ rpm

Slip speed $= 1500 - 1370 = 130$ rpm at rated torque

$$\text{Slip} = \dfrac{130}{1500} = 0.0867$$

When operating at 30 Hz, Synchronous speed $= \dfrac{120 \times 30}{4} = 900$ rpm

Motor speed $= 900-130 = $ **770 rpm.**

Per phase voltage $= \dfrac{400}{\sqrt{3}} = 231\,V$

Equivalent circuit at 50 Hz & speed of 1370 rpm is shown below

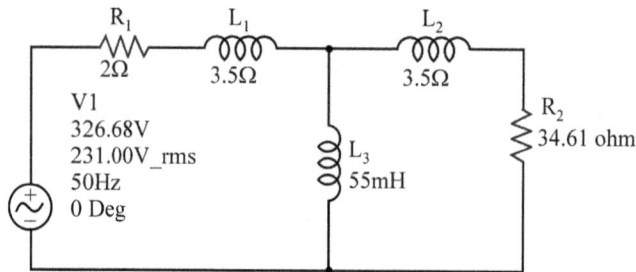

$R_2/s = 3/0.0867 = 34.61\ \Omega$

$$Z = 2 + j\,3.5 + \dfrac{j55(34.61 + j3.5)}{34.61 + j58.5} = 24.65 + j\,20.19$$

$$= 31.86\ \angle 39.3^{0}\,\Omega$$

$\therefore$ Full load stator current $= \dfrac{231}{31.86} = 7.2486\ A$

Full load rotor current $I_2 = 7.2486 \left[\dfrac{j55}{34.61 + j58.5} \right] = 5.865\ A$

$$T_d = \frac{3\, I_2^2\, R_2}{s\, \omega_s}$$

$\therefore$ Full load torque $= \dfrac{3}{50\pi}(5.865)^2 \times 34.61 = 22.73$ N-m

(i)   At rated slip speed, torque and I_S will have the same values as at 50 Hz operation. Thus at 30 Hz, Torque = 22.73 N-m and I_S = 7.2486 A

Synchronous speed $= \dfrac{30}{50} \times 1500 = 900$ rpm

Full load slip speed = 130 rpm

Motor speed = 900-130 = 770 rpm

(ii) At rated motor torque slip speed and I_s will be same as at 50Hz.

Hence I_s = 7.2486 A,   Slip speed = 130 rpm

Motor speed is given as 1200 rpm

$\therefore$ Synchronous speed = 1200 +130 =1330 rpm

Frequency $= \dfrac{1330}{1500} \times 50 = 44.33$ Hz.

(iii) Slip speed at full load Torque = 130 rpm

Slip speed at half the rated torque = 65 rpm

At 30 Hz the synchronous speed is 900 rpm

Motor speed = 900 – 65 = 835 rpm.

(iv) At rated braking torque slip speed = - 130 rpm

Synchronous speed $= \dfrac{45}{50} \times 1500 = 1350$ rpm

Motor speed = 1350 + 130 = 1480 rpm.

Examples 6.2(a)

A Y connected squirrel cage induction motor has the following ratings and parameters: 400 V, 50 Hz , 4 pole , 1370 rpm. R_S= 2 Ω; R_r^1 = 3 Ω; X_S= X_r^1 = 3.5 Ω. Motor is controlled by a voltage source inventor at constant V/f ratio. Calculate approximate values of the following:

(i)   Speed for a frequency of 30 Hz and 80% of full load torque.

(ii) Frequency for a speed of 1000 rpm and full load torque.

(iii) Torque for a frequency of 40 Hz and speed of 1100 rpm.

Solution :

(a)   Motor speed torque curves for various frequencies from full load motoring to full load braking can be assumed to be parallel straight lines each passing through corresponding synchronous speed without significant error.

(i)   At 50 Hz full load speed = 1370 rpm.

Drop in speed from no load to full load = 1500-1370 = 130 rpm

Drop in speed at 80% of full load = 0.8 ×130

$$= 104 \text{ rpm.}$$

Synchronous speed at 30 Hz = $\dfrac{120 \times 30}{4}$ = 900 rpm

∴ the speed of the motor = 900- 104 = **796 rpm.**

(ii)   At full load torque the drop in speed = 130 rpm.

Actual speed = 1000 rpm.

∴ Synchronous speed = 1000+130 = 1130 rpm.

$$f = \dfrac{4}{120} \times 1130 = \textbf{37.67 Hz.}$$

(iii)   Synchronous speed at 40 Hz = $\dfrac{120 \times 40}{4}$ = 1200 rpm

Drop in speed from no load to 1100 rpm = 1200 – 1100

$$= 100 \text{ rpm}$$

$$\text{Torque} = \dfrac{100}{130} \times T_{FL} = 0.769 \, T_F$$

$$\text{Full load torque} = \dfrac{3}{\omega_s} \cdot \dfrac{V^2 R_r^{\,1}}{s\left[(R_s + \dfrac{R_r^{\,1}}{s})^2 + (X_s + X_r^{\,1})^2 \right]}$$

$$V = \dfrac{400}{\sqrt{3}} = 230.94 \text{ V}$$

$$S = \dfrac{1500 - 1370}{1500} = 0.08667$$

$$\omega_s = \dfrac{1500}{60} \times 2\pi = 50 \, \pi \quad \text{rad/sec}$$

$$\therefore \text{ Full load torque } T_{FL} = \frac{3}{50\pi} \cdot \frac{230.94^2 \times 3}{0.08667\left[(2+\frac{3}{0.08667})^2 + (3.5+3.5)^2\right]}$$

$$= \frac{35258.9}{1389.58} = 25.37 \text{ N-m}$$

Hence torque $= 0.769 \times 25.37 = $ **19.51 N-m**

Example 6.2(b)

For the above drive calculate motor breakdown torque for a frequency of 60 Hz compared to that at 50 Hz: Above 50 Hz, V is held constant.

Solution:

At 60 Hz $K = 60/50 = 1.2$

$$T_{max} = \frac{3}{2K\omega_s} \cdot \frac{V^2}{R_s + \sqrt{R_s^2 + K^2(X_s + X_r^1)^2}}$$

$$= \frac{3}{2 \times 1.2 \times 50\pi} \cdot \frac{(\frac{400}{\sqrt{3}})^2}{2 + \sqrt{2^2 + 1.2^2(3.5+3.5)^2}} = \frac{424.41}{10.63}$$

$$= 39.90 \text{ N-m}.$$

At 50 Hz $T_m = \frac{3}{2 \times 50\pi} \cdot \frac{(\frac{400}{\sqrt{3}})^2}{2 + \sqrt{2^2 + 7^2}} = \frac{509.29}{9.28} = 54.88 \text{ N.m}$

$$\therefore \quad \frac{T_m \text{ at } 60Hz}{T_m \text{ at } 50Hz} = \frac{39.90}{54.88} = 0.727$$

Example 6.3

A 440 V, 50 Hz, 4 pole, 1420 rpm delta connected squirrel cage induction motor has the following parameters:

$R_s = 0.35\,\Omega$ $R_r^1 = 0.4\,\Omega$ $X_s = 0.7\,\Omega$ $X_r^1 = 0.8\,\Omega$

The motor is fed from a voltage source inventor. The drive is operated with a constant V/f control up to 50 Hz and at rated voltage above 50 Hz. Calculate the breakdown torques for a frequency of 75 Hz both for motoring and braking operations.

Solution :

The expression for breakdown torque is

$$T_{max} = \frac{3}{2\omega_s K} \frac{V^2}{R_s \pm \sqrt{R_s^2 + K^2(X_s + X_r^1)^2}}$$

$$V = 440; \quad K = \frac{75}{50} = 1.5 \quad \omega_s = \frac{120 \times 50 \times 2\pi}{4 \times 60} = 157.08 \text{ rad/sec}$$

$$R_s = 0.35 \quad X_s = 0.7, \quad X_r^1 = 0.8$$

$$\therefore \qquad T_{max} = \frac{3}{2 \times 157.08 \times 1.5} \cdot \frac{440^2}{0.35 \pm \sqrt{0.35^2 + 1.5^2(0.7 + 0.8)^2}}$$

$$= \frac{1232.5}{0.35 \pm 2.277}$$

$$= 469 \text{ N-m (or) } -639.6 \text{ N-m}$$

Example 6.4

A star connected squirrel cage induction motor has the following ratings and parameters:

400 V, 50 Hz, 4 pole, 1370 rpm, $R_s = 2\Omega$, $R_r^1 = 3\Omega$, $X_s = X_r^1 = 3.5\Omega$. The speed is controlled at constant V/f ratio from 10 Hz to 50 Hz. Plot the breakdown torque and frequency curve.

Solution :

Breakdown torque
$$T_m = \frac{3}{2\omega_s K} \frac{KV^2}{R_s \pm \sqrt{R_s^2 + K^2(X_s + X_r^1)^2}}$$

$$\omega_s = 50\pi$$

$$V = \frac{400}{\sqrt{3}}; \quad K = \frac{f}{50}$$

$$T_m = \frac{3}{2 \times 50\pi} \frac{f}{50} \times \frac{(\frac{400}{\sqrt{3}})^2}{2 \pm \sqrt{2^2 + (\frac{f}{50})^2(3.5 + 3.5)^2}}$$

$$= \frac{10.186f}{2 \pm \sqrt{4 + .0196f^2}}$$

f = 10	$T_m = 22.93$ N-m
f = 20	$T_m = 37.44$ N-m
f = 30	$T_m = 45.93$ N-m
f = 40	$T_m = 51.27$ N-m

f = 45 T_m = 53.23 N-m

f = 50 T_m = 54.88 N-m

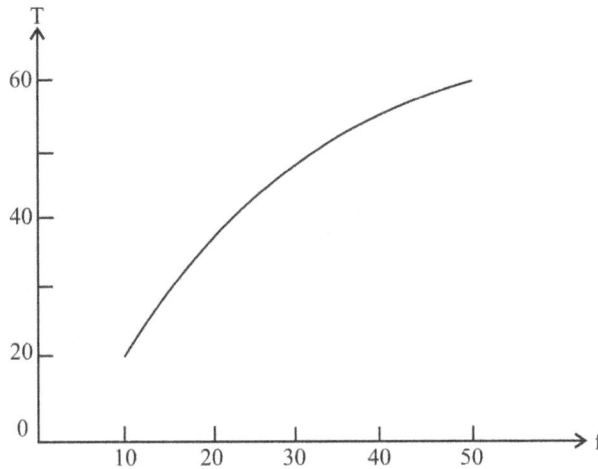

Example 6.5

A Three phase , star connected, 50 Hz, 4 pole induction motor has the following parameters in ohms per phase referred to the stator.

$R_1 = R_2 = 0.034$ and $X_1 = X_2 = 0.18$

The motor is controlled by the variable frequency control with a constant (V/f).

Determine the following for an operating frequency of 15 Hz.

(a)   The breakdown torque as a ratio of its value at the rated frequency for motoring and braking

(b)   The starting torque and rotor current in terms of their values at the rated frequency.

Solution :

Let $K = \dfrac{f}{f_r}$

If V/f is held constant,

$$T = \frac{3}{\omega_{sr}} \frac{V_r^2 \dfrac{R_2}{Ks}}{(\dfrac{R_1}{K} + \dfrac{R_2}{Ks})^2 + (X_1 + X_2)^2} \qquad \text{for K<1}$$

$$T_{max} = \frac{3}{2\omega_{sr}} \frac{V_r^2}{\frac{R_1}{K} \pm \sqrt{(\frac{R_1}{K})^2 + (X_1 + X_2)^2}}$$

$$K = \frac{15}{50} = 0.3$$

Breakdown torque for K = 0.3

$$T_{max} \ (K = 0.3) = \frac{3}{2\omega_{sr}} \cdot \frac{V_r^2}{\frac{0.034}{0.3} \pm \sqrt{(\frac{0.034}{0.3})^2 + 0.36^2}}$$

$$T_{max} \ (K=1) = \frac{3}{2\omega_{sr}} \cdot \frac{V_r^2}{0.034 \pm \sqrt{0.034^2 + 0.36^2}}$$

$$\therefore \qquad \frac{T_{max} \ (K = 0.3)}{T_{max} \ (K = 1)} = \frac{0.034 \pm \sqrt{0.034^2 + 0.36^2}}{\frac{0.034}{0.3} \pm \sqrt{(\frac{0.034}{0.3})^2 + (0.36)^2}}$$

For motoring we have to take +ve sign and for braking we have to take −ve sign.

$$\therefore \ \text{For motoring} \quad \frac{T_{max} \ (K = 0.3)}{T_{max} \ (K = 1)} = \frac{0.034 + \sqrt{0.034^2 + 0.36^2}}{\frac{0.034}{0.3} + \sqrt{(\frac{0.034}{0.3})^2 + 0.36^2}}$$

$$= \textbf{0.806.}$$

For braking $$\qquad \frac{T_{max} \ (K = 0.3)}{T_{max} \ (K = 1)} = \frac{0.034 - \sqrt{0.034^2 + 0.36^2}}{\frac{0.034}{0.3} - \sqrt{(\frac{0.034}{0.3})^2 + (0.36)^2}}$$

$$= \textbf{1.24}$$

Starting torque T_s

$$T_s = \frac{3}{\omega_s} \left[\frac{V_r^2(\frac{R_2}{K})}{(\frac{R_1 + R_2}{K})^2 + (X_1 + X_2)^2} \right]$$

For K = 0.3 $$\qquad T_s = \frac{3 V_r^2 \left[\frac{0.034}{0.3} \right]}{\omega_s \left[\left(\frac{0.068}{0.3} \right)^2 + (0.36)^2 \right]} = \frac{3V_r^2}{\omega_s} .0.626$$

For K = 1 $\qquad T_s = \dfrac{3\,V_r^2}{\omega_s} \cdot \dfrac{0.034}{(.068)^2 + (0.36)^2} = 0.253 \times \dfrac{3V_r^2}{\omega_s}$

$\therefore \qquad \dfrac{T_s(K=0.3)}{(K=1)} = \dfrac{0.626}{0.253} = 2.47$ $\qquad\qquad$ .....(i)

Starting rotor current I_{2s}

$$I_2 = \frac{V_r}{\sqrt{\left(\dfrac{R_1 + R_2}{K}\right)^2 + (X_1 + X_2)^2}}$$

$$\frac{I_2(K=0.3)}{I_2(K=1)} = \frac{\sqrt{0.068^2 + (0.36)^2}}{\sqrt{\left(\dfrac{0.068}{0.3}\right)^2 + (0.36)^2}} = 0.86 \qquad\qquad .....(ii)$$

From eqn. (i) & (ii) it can be seen that the starting torque is more with reduced starting current with constant V/f control.

Example 6.6

A Three phase, star connected, 50 Hz, 4 pole induction motor has the following parameters in ohms per phase referred to the stator.

$\qquad R_1 = R_2 = 0.034;\ X_1 = X_2 = 0.18$

It develops rated torque at 4% slip. Determine the motor speed at rated torque and

f = 25 Hz. The motor is controlled with a constant V/f ratio.

Solution :

Rated torque is developed at 4% slip & at 50 Hz.

$$T = \frac{3}{\omega_s} \cdot \frac{V_r^2 \cdot \dfrac{R_2}{s}}{\left(R_1 + \dfrac{R_2}{s}\right)^2 + (X_1 + X_2)^2}$$

$$= \frac{3}{\omega_s} \cdot \frac{V_r^2 \cdot \dfrac{0.034}{0.04}}{\left(0.034 + \dfrac{0.034}{0.04}\right)^2 + (0.36)^2} = \frac{3V_r^2}{\omega_s}(0.933) \qquad .....(1)$$

For 25 Hz $\qquad$ K = 0.5

$$T_d = \frac{3}{\omega_s} \frac{V_r^2 \dfrac{0.034}{0.5 \times s}}{(\dfrac{0.034}{0.5} + \dfrac{0.034}{0.5 \times s})^2 + 0.36^2} \qquad \ldots\ldots(2)$$

Equating (1) & (2)

$$0.933 = \frac{\dfrac{0.068}{s}}{(0.068 + \dfrac{0.068}{s})^2 + 0.36^2}$$

$$s^2 - 0.4742\, s + 0.0344 = 0$$

$$s = 0.384 \ (or) \ 0.089$$

The slip on the stable part will be 0.089

$$\frac{K\omega_s - \omega_r}{K\omega_s} = 0.089$$

$$\frac{KN_s - N_r}{KN_s} = 0.089$$

$$\frac{0.5 \times 1500 - N_r}{0.5 \times 1500} = 0.089$$

$$N_r = 750\ (1-0.089)$$

$$= 683.2 \ rpm$$

Problems

P.1 (a) An inverter supplies a 4 pole, 3 phase induction motor rated at 220 V, 50 Hz. Determine the approximate voltages required of the inverter for motor speeds of 300/400/800/1000/1200/1500/1800 rpm.

(b) Explain the principles of operation of VSI fed induction motor.

P.2 (a) Explain the general features of the induction motor on a current source inverter.

(b) Draw closed loop block schematic diagram of a slip controlled drive using CSI.

P.3 A three phase, 4 pole, 18 kW, 300 V, star-connected induction motor is driven at 50 Hz by a six step voltage source inverter supplied from a dc supply of 200 V. The motor equivalent circuit parameters for 50 Hz operation are $R_1 = 0.1\,\Omega$, $R_2 = 0.17\,\Omega$, $X_1 = 0.3\,\Omega$, $X_2 = 0.5\,\Omega$, X_m= large. Calculate the harmonic torques due to the 5th and 7th harmonic currents. Show that, for operation at 1450 rpm, 50 Hz, the harmonic torques are negligible.

P.4 (a) What is a voltage source inverter?

   (b) Explain with suitable block diagrams the various types of VSI controlled induction motor drives.

P.5 While explaining the principle of varying the speed of 3 phase induction motor by V/f method discuss it for the following two different modes.

   (i) Operation below rated frequency,

   (ii) Operation above rated frequency.

P.6 Discuss in detail the role of cyclo-converters for speed control of induction motor. Draw neat circuit diagram for speed control for 3 phase induction motor using cyclo-converters. Mention the merits and limitations of the above scheme.

P.7 For variable frequency control of induction motor explain the following points:

   (i) For speeds below base speed (V/f) ratio is maintained constant. Why?

   (ii) For speeds above base speed the terminal voltage is maintained constant. Why?

   (iii) Discuss in detail the merits, demerits of variable frequency control of induction motor.

P.8. (a) Draw and explain the speed torque curves with variable frequency control for two different modes.

   (i) Operation at constant flux.

   (ii) Operation at constant V/f ratio.

   (b) Explain the advantages of variable frequency drives.

P.9 (a) Compare CSI and VSI drives.

   (b) Show that a variable frequency induction motor drive develops at all frequencies the same torque for a given slip-speed when operating at constant flux.

P.10 With the help of circuit diagram and wave forms explain the induction motor with current source inverter. Draw the circuit diagram of the Auto-sequentially commutated inverter.

P.11 A three phase star connected 50 Hz, 4 pole induction motor has the following approximate per-phase equivalent circuit parameters referred to stator side.

$R_s = R_r' = 0.024\,\Omega$, $X_s = X_r' = 0.12\,\Omega$. The motor is controlled by variable frequency control with constant V/f ratio. For an operating frequency of 12 Hz calculate (i) the break down torque as a ratio of its value at the rated frequency for the motoring operation, (ii) The starting torque and rotor current in terms of their values at the rated frequency.

P.12 A three phase, 4 pole, 18 KW, 300 V, star-connected induction motor is driven at 50 Hz by a six step voltage source inverter supplied from a dc supply of 200 V. The motor equivalent circuit parameters for 50 Hz operation are $R_1 = 0.1\,\Omega$, $R_2 = 0.17\,\Omega$, $X_1 = 0.3\,\Omega$, $X_2 = 0.5\,\Omega$, X_m= large. Calculate the rms current and the harmonic copper losses when this operates at 1450 rpm, 50 Hz. Estimate the motor efficiency compared with sinusoidal operation.

P.13 With a block schematic diagram explain how the speed of induction motor can be controlled automatically (i.e., using closed loop scheme) with voltage source inverter. Mention the application of the above scheme

P.14 Explain the operation of a voltage source inverter (180-degree conduction mode), used for induction motor speed control. Draw neat wave forms of line voltages (V_{ab}, V_{bc}, V_{ca}) and hence show that the phase voltage, V_{ab} is six-step voltage wave form.

P.15 Explain the following for variable frequency control of induction motor.

    (a)   The motor has higher efficiency and better low speed performance when fed from a pulse-width modulated inverter instead of 6-step inverter

    (b)   The inverter has excellent low speed performance when fed from a Cyclo converter

    (c)   Cyclo-converter is suitable only for low speed drives.

P.16 A 460 V, 100 HP (74.6 KW), 1775 rpm, three phase, squirrel cage induction motor has the following equivalent circuit parameters.

R_s = 0.060 ohm, $R_r^{'}$ = 0.0302 ohm, L_{1s} = 0.638 mH, L_{ms} = 23.3 mH, $L_{1r}^{'}$ = 0.957 mH

The motor is to be driven from a current source inverter with the rotor frequency controlled at the rated value. Maximum output power is to be limited to 80% of the rated value. Motor friction, windage and core losses may be neglected. The load is to consist of a pump presenting a load characteristic described by the equation $T = \omega_m^2/110$ N-m. Determine the maximum values of motor speed, inverter frequency, rms motor line current, and fundamental line-to-line motor terminal pd at maximum power output.

P.17 For a 3 phase delta connected, 6 pole, 50 Hz, 400 V, 925 rpm, squirrel cage induction motor the parameters are:

$$R_s = 0.2 \text{ ohm}, R_r = 0.3 \text{ ohm}, X_s = 0.5 \text{ ohm}, X_r = 1 \text{ ohm}$$

The motor is fed from a voltage source inverter with a constant V/f ratio from 0 to 50 Hz and constant voltage of 400 V above 50 Hz frequency. Calculate

(a) Speed for a frequency of 35 Hz and half of full load torque,

(b) Torque for a frequency of 35 Hz and a speed of 650 rpm.

P.18 The equivalent circuit of a three phase induction motor having a delta connected stator has the following parameters at rated frequency of 50 Hz.

$$R_1 = 0.1\,\Omega, R_2 = 0.2\,\Omega, X_1 = 1.0\,\Omega, X_2 = 1.0\,\Omega, X_m = 20\,\Omega$$

The rated voltage of machine is 250 V. Determine the voltage to be applied as a function of stator frequency for rotor frequencies of 0, 2, 3 Hz. Consider only the fundamental.

Objective Type Questions

1. Speed of an induction motor can be varied by means of variable frequency supply from a static power converter. A simultaneous voltage variation is also effected in order to

   (a) avoid saturation and provide optimum torque capability

   (b) the torque pulsations decrease if supplied from variable supply

   (c) to limit the peak value of stator current

   (d) to minimize the additional losses   [   ]

2. A variable frequency variable voltage induction motor

   (e) can be accelerated at constant torque or constant current

   (f) suffers from poor starting characteristics as in the case of mains fed motor

   (g) has only stepped variation of speed

   (h) suffers from stability considerations   [   ]

3. If the supply voltage to the induction motor is held constant and the frequency is reduced to reduce the speed

   (a) The iron gets saturated and distorts the line current

   (b) The voltage is disturbed and increases core loss.

   (c) The copper loss increases & noise is produced.

   (d) All the above effects occur.   [   ]

4. For controlling the speed of induction motor below rated speed V /f should be held constant (True /False)   [   ]

5. When the induction motor is working in the regenerative mode, the flux and braking torque will have higher values at low frequencies compared to corresponding values of motor working in the motoring mode (True /False)   [   ]

6. For an induction motor if the supply voltage is held constant and frequency decreased the flux density will ....(Increase /Decrease)   [   ]

7. If the supply voltage to the induction motor is held constant and frequency increased beyond rated value the torque developed .... (Reduces / Increases)   [   ]

8. While controlling the speed of induction   motor at very low speeds E /f ratio is .... (Decreased / Increased) and thus the torque capability .... (Decreased /Increased/held constant)   [   ]

9. To make full use of motor torque capability at start and at low speeds the V/f ratio is increased. Why?

10. Name three solid state sources from which variable voltage variable frequency is obtained.

11. What are the two methods of controlling the speed of 3 ϕ squirrel cage induction motor ?

12. In a variable frequency method of speed control the v/f ratio should be held ....

13. Draw the torque speed characteristics of induction motor whose speed is controlled by variable frequency for the range of speed from zero to 1.5 times the synchronous speed.

14. Draw the schematic diagram, which can be used for getting variable frequency from 3-ϕ , 50 Hz AC supply for driving 3-ϕ induction motor. The voltage is controlled internal to the inverter by pulse width modulation.

15. What are the various methods of controlling the voltage internal to the inverter?

16. Draw the wave form of phase voltage applied to star connected induction motor, connected to 3-ϕ VSI operating with 120^0 conduction mode.

17. Draw the wave form of phase voltage applied to star connected induction motor connected to 3-ϕ VSI operating with 180^0 conduction mode.

18. In the output voltage of 3-ϕ VSI operating with 180^0 conduction mode, the triplen harmonics are .... (Present / Absent)

19. In a variable frequency variable voltage method of speed control the V/f ratio is normally to be held constant, but at low frequencies V/f ratio has to be slightly .... (Increased / Decreased)

20. A star connected 3-phase, 4 pole induction motor is driven by a three phase inverter operating in 180^0 conduction mode. The input voltage is 400 V DC. What is the rms value of the fundamental component of the voltage applied to a phase of the induction motor?

21. A star connected 3-phase, 4 pole induction motor is driven by a three phase inverter operating in 180^0 conduction mode. The input voltage is 400 V DC. What is the total rms value of the voltage applied to a phase of the induction motor?

22. A delta connected 3-phase, 4 pole induction motor is driven by a three phase inverter operating in 180^0 conduction mode. The input voltage is 400 V DC. What is the rms value of the fundamental component of the voltage applied to a phase of the induction motor?

23. A delta connected 3-phase, 4 pole induction motor is driven by a three phase inverter operating in 180^0 conduction mode. The input voltage is 400 V DC. What is the total rms value of the voltage applied to a phase of the induction motor?

24. A 400 V, 50 Hz, 3 phase, 4 pole star connected induction motor is connected to a 3-phase inverter for controlling the speed keeping V/f ratio as constant. The equivalent circuit parameters referred to stator are

$R_1 = 0.02\ \Omega$, $X_1 = 0.36\ \Omega$, $R_2 = 0.65\ \Omega$, $X_2 = 0.36\Omega$ & $X_m = 17.13\ \Omega$.

Draw equivalent circuit of the motor when the frequency is 20 Hz.

25. A 400 V, 50 Hz, 3 phase, 4 pole delta connected induction motor is connected to a 3-phase inverter for controlling the speed . The equivalent circuit parameters referred to stator are

   $R_1 = 0.02\ \Omega$, $X_1 = 0.36\ \Omega$, $R_2 = 0.65\ \Omega$, $X_2 = 0.36\Omega$ & $X_m = 17.13\ \Omega$.

   What is the speed of the motor running with a slip of 4% when the supply frequency is 40 Hz?

26. Draw the power circuit diagram of CSI driving 3 ϕ induction motor.

27. In three-phase current source inverter how many SCRs, diodes and capacitors are required?

28. In three-phase current source inverter how many SCRs will be conducting at any instant?

29. If I_d is the input current to 3 ϕ CSI driving 3 ϕ induction motor the rms value of the fundamental component of line current to induction motor is given by the equation…

30. A CSI is supplying power to delta connected induction motor. Draw the wave form of the phase current. The d. c current is I_d.

31. If I_d is the source current to CSI supplying power to delta connected induction motor the rms value of the fundamental component of phase current is given by equation……

32. Explain how regenerative braking is applied to CSI driven induction motor.

33. Draw the per phase equivalent circuit of induction motor fed from CSI.

34. If the air gap flux is to be held constant at all frequencies in a CSI driven induction motor what is to be done?

35. Give two merits of cyclo converter driving a 3ϕ IM.

36. What are the disadvantages of cyclo converter driving 3ϕ IM.

37. CSI can be used for controlling the speed of a number of 3ϕ induction motors connected in parallel (True/False).

38. In a CSI drive of a 3ϕ IM, ordinary diode rectifier can be used on line side (True/False).

39. Converter grade SCR s can be used in CSI (True/False).

40. Power factor on the line side of CSI drive is poor (True/False).

Speed Control of
Three-phase Induction Motors from Rotor Side

Control of induction motor from rotor side : State rotor resistance control - slip power recovery - static scherbius drive – static Kramer drive – their performance and speed torque characteristics – advantages – applications problems.

7.1   Rotor Resistance Control

One of the methods of controlling the speed of slip ring induction motor is by controlling the rotor resistance. A resistance is connected to the rotor windings through slip rings as shown in the Fig.7.1. By varying this resistance the effective resistance of the rotor is changed. The speed can be controlled by this method.

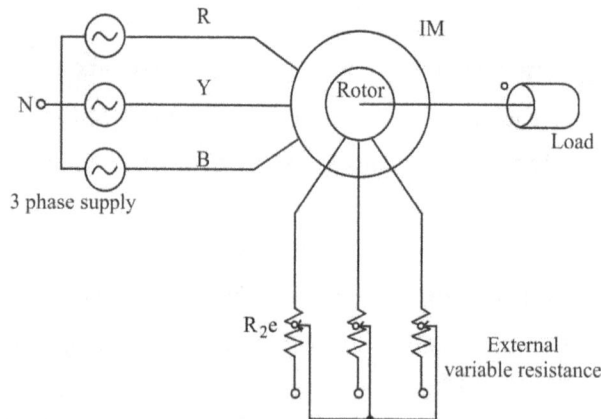

Fig. 7.1 Slip ring induction motor with variable rotor resistance.

Normally the rotor is designed with low resistance so that efficiency is high. But starting performance will be poor. The starting performance is improved by connecting external resistance in series with the rotor. The torque-speed characteristics for different values of rotor resistance is shown in the Fig. 7.2. The increase in rotor resistance does not affect the values of maximum torque but increases the slip at maximum torque. The external resistance can be chosen such that maximum torque is developed at starting. The resistance can be gradually cut off as the speed increases so that maximum torque is developed during acceleration.

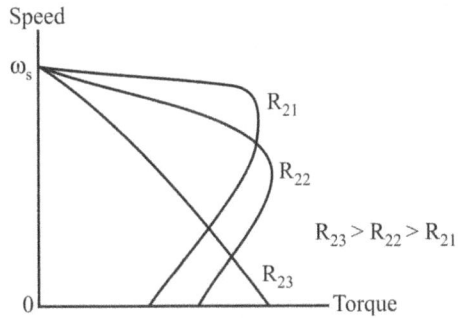

Fig. 7.2 Torque-speed characteristics.

The resistance can be varied mechanically by means of sliding contacts or by providing tapings and switches. By this method smooth variation of resistance is not obtained.

7.2 Static Rotor Resistance Control

Instead of mechanically varying the resistance, the rotor circuit resistance can be varied statically by using the principle of a chopper. This gives stable and smooth variation of resistance and consequently that of motor speed.

The rotor voltages at slip frequency are converted to DC by diode bridge rectifier. A GTO is connected across the external resistance R and is operated periodically with period T and it remains ON for an interval t_{ON} in each period. The effective value of R changes from 0 to R as t_{ON} changes from T to 0. The filter inductor L_d is provided to minimize the ripple in the current. The arrangement is shown in Fig.7.3.

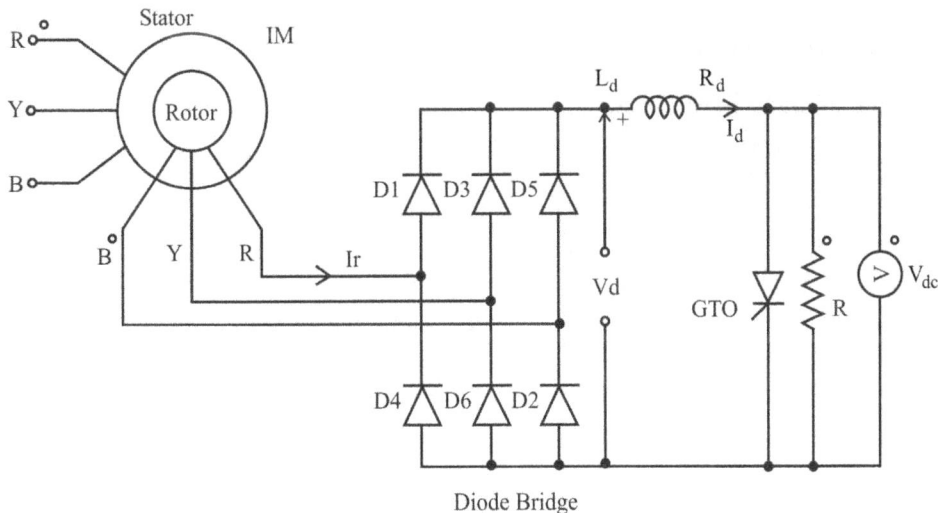

Fig. 7.3 Rotor resistance control with GTO.

The rotor phase current will be rectangular wave as shown in the Fig. 7.4. The fundamental component of phase current will be in phase with phase voltage.

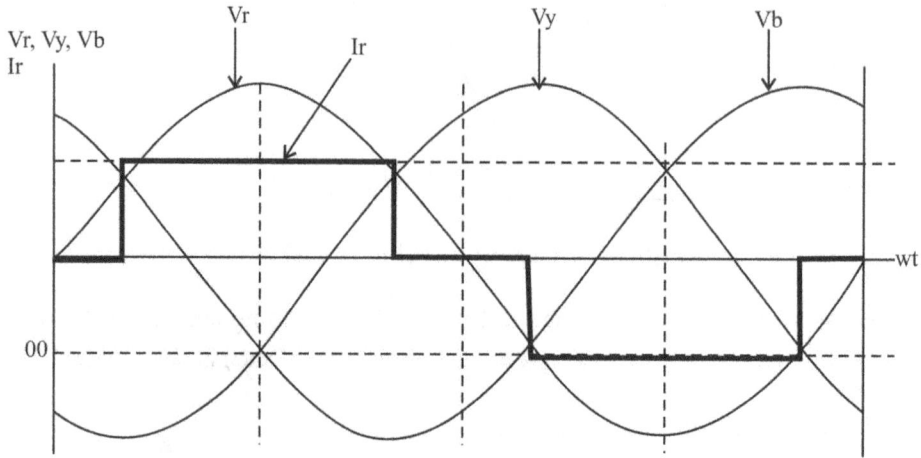

Fig. 7.4 Rotor phase voltage and current wave forms.

Let $\alpha = \dfrac{t_{ON}}{T}$ where t_{ON} is the period for which GTO switch is ON. During t_{ON} the resistance R is shorted. During T- t_{ON} the R is connected to the bridge. I_D passes through R during the period (T- t_{ON}). The energy absorbed by the resistance R during the period of operation of GTO is given by $E_R = I_d{}^2 R(T - t_{ON})$

Now the average power absorbed by R is given by $P_R = \dfrac{I_d^2 R(T - t_{ON})}{T} = I_d^2 R(1 - \alpha)$

Thus the effective value of resistance R is given by $R_e = R(1 - \alpha)$.

rms value of rotor phase current is given by

$$I_{rms} = \left[\sqrt{\frac{1}{\pi} \int_{\frac{\pi}{6}}^{\frac{5\pi}{6}} I_d^2 d(\omega t)} \right] = \sqrt{\frac{I_d^2}{\pi}\left(\frac{5\pi}{6} - \frac{\pi}{6}\right)} = I_d \sqrt{\frac{2}{3}} \quad \dots\dots(7.1)$$

The coefficient of fundamental component of Furrier series

$$a_1 = \frac{4}{\pi} \int_{\frac{\pi}{6}}^{\frac{\pi}{2}} I_d \sin \omega t\, d(\omega t) = \frac{2\sqrt{3}}{\pi} I_d$$

The fundamental component $i_1 = \dfrac{2\sqrt{3}}{\pi} I_d \sin \omega t$.

rms value of fundamental component of rotor current

$$I_{r1} = \frac{2\sqrt{3}\, I_d}{\pi} \frac{1}{\sqrt{2}} = \frac{\sqrt{6}}{\pi} I_d = \frac{\sqrt{6}}{\pi} \sqrt{\frac{3}{2}}\, I\text{ rms} = \frac{3}{\pi} I_{rms} \qquad \dots(7.2)$$

The equivalent circuit diagrams of schematic shown in Fig. 7.4 are shown in Fig. 7.5.

Now the total resistance across the diode bridge is $R_e = R_d + R\,(1-\alpha)$

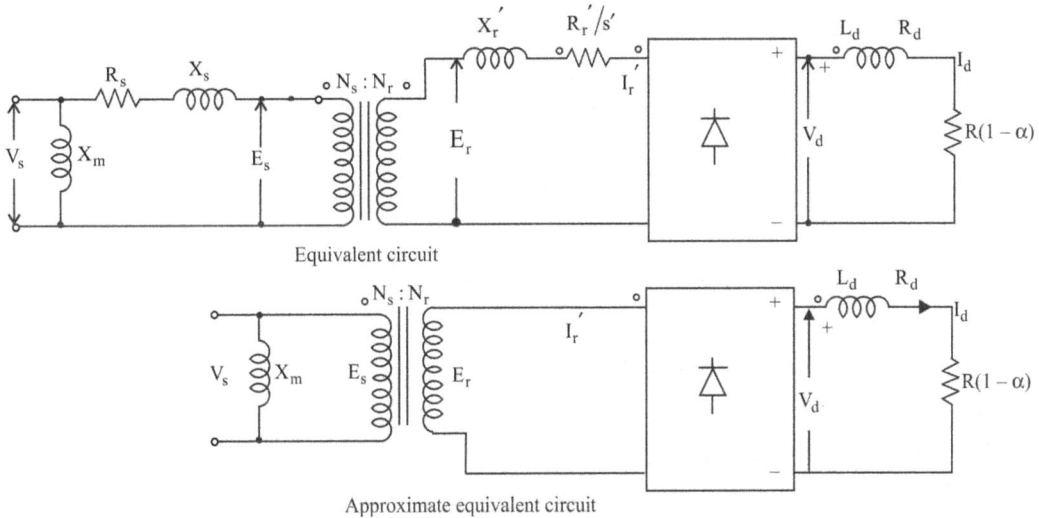

Equivalent circuit

Approximate equivalent circuit

Fig. 7.5 Equivalent circuit diagrams of Slip ring induction motor with rotor resistance control.

In the approximate equivalent circuit diagram R_s, X_s, R_r^1, X_r^1 are neglected compared to the reactance of L_d which is very large.

The dc voltage at the rectifier output

$$V_d = I_d R_e = I_d R\,(1-\alpha) \qquad \dots(7.3)$$

Per phase rotor voltage $E_r = s\, E_s \dfrac{N_r}{N_s} = sV_s\, n_m$, where n_m is the ratio of rotor turns to stator turns.

In a three phase diode rectifier the relationship between V_d and E_r is

$$V_d = 1.35 \times V_L = 1.35 \times \sqrt{3}\ E_r$$

$$V_d = 2.3394\ E_r = 2.3394 \times sV_s\, n_m \qquad \dots(7.4)$$

If P_r is the slip power i.e., power loss in the rotor circuit per phase

Total copper losses in the rotor = $3\ P_r$

Total air gap power developed $P_g = 3\ P_r\ /s$

Total mechanical power developed $P_d = P_g\ (1\text{-}s) = \dfrac{3P_r(1-s)}{s}$

Total slip power $= 3P_r = V_d\ I_d$

i.e., Electrical power lost in the rotor resistance.

$\therefore$ Mechanical power developed $= V_d\ I_d\ \dfrac{(1-s)}{s}$

If T_L is the load torque and ω_m is the motor speed in rad/sec

$$T_L\ \omega_m = V_d I_d\ \dfrac{(1-s)}{s}$$

i.e.,
$$T_L\ \omega_s\ (1\text{-}s) = V_d\ I_d\ \dfrac{(1-s)}{s}$$

$$T_L\ \omega_s = \dfrac{V_d\ I_d}{s}\ = 2.3394\ V_s n_m I_d$$

$\therefore$
$$I_d = \dfrac{T_L \omega_s}{2.3394 V_s n_m} \qquad \qquad \dots(7.5)$$

This shows that I_d is independent of speed.

$$V_d = 2.3394\ s V_s n_m$$

$$V_d = I_d\ R(1\text{-}\alpha) = \dfrac{T_L \omega_s R(1-\alpha)}{2.3394\ V_s n_m}$$

From eqn. 7.4   slip $s = V_d/2.3394\ V_s n_m$

Therefore slip
$$s = \dfrac{T_L \omega_s . R(1-\alpha)}{(2.3394\ V_s n_m)^2} \qquad \qquad \dots(7.6)$$

Motor speed
$$\omega_m = \omega_s\ (1\text{-}s) = \omega_s\left[1 - \dfrac{T_L \omega_s R(1-\alpha)}{(2.3394\ V_s n_m)^2}\right] \qquad \dots(7.7)$$

$$= \omega_s\left[1 - \dfrac{I_d R(1-\alpha)}{2.3394\ V_s n_m}\right] \qquad \qquad \dots(7.8)$$

Eqn. 7.7 shows that for a fixed duty cycle the speed decreases with increase of load torque.

The Torque-speed curves for different values of duty cycles α is given in Fig. 7.6.

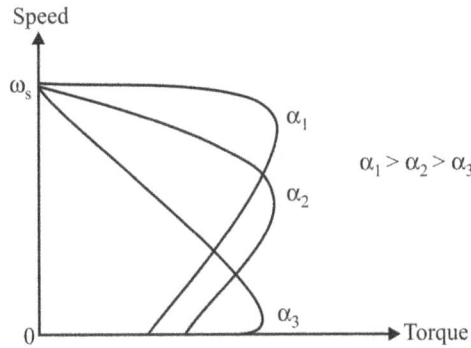

Fig. 7.6 Torque-speed characteristics.

7.3   Slip Power Recovery Scheme

If the speed of induction motor is controlled by varying the resistance of rotor circuit certain power is wasted in resistor connected external to the rotor. In other words slip power is wasted as heat loss. Instead of wasting slip power it can be pumped back into the mains supply. It is known that, if slip power is taken out from the rotor, the speed will decrease and if power is pumped into the rotor, the speed will increase. This can be done by injecting voltage into the rotor with appropriate polarity. If voltage is to be injected into the rotor, the frequency and magnitude should match with that of the rotor. The frequency and magnitude of rotor voltage continuously change with speed.

Equivalent circuit of induction motor with injected voltage in its rotor is shown in Fig. 7. 7. Vr is the voltage source connected to the rotor through slip rings. This source will either pump power into the rotor or take away power from the rotor.

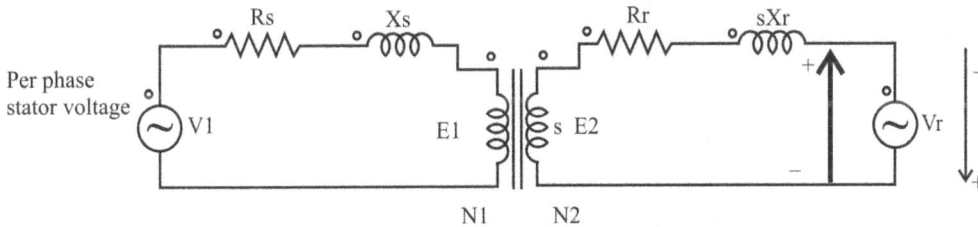

Fig. 7.7 Equivalent circuit of Slip ring induction motor with injection voltage.

Let

P_g be the power transferred from stator to rotor of induction motor through air gap,

P_m be the mechanical power developed,

P_r be the power absorbed by the source V_r,

Neglecting the Rotor copper losses, we have

$$P_m = P_g - P_r$$

The magnitude and sign of P_r can be controlled by controlling the magnitude and phase of V_r. If $P_r = 0$ the motor runs on its natural speed-torque curve.

If P_r is +ve, P_m will be reduced and the motor runs at lower speed. If P_r is –ve, the motor runs at higher speed. If $P_r = P_g$ the motor runs at zero speed. Polarity of V_r for speed less than synchronous speed is shown in thick line and for speed higher than synchronous speed is shown in thin line.

The injection of voltage in any polarity and frequency has become easy after the availability of SCRs. The voltage of variable frequency and magnitude at terminals of rotor is converted to DC and inverted to supply frequency and connected to source.

A schematic diagram by which this is achieved is shown in Fig. 7.8.

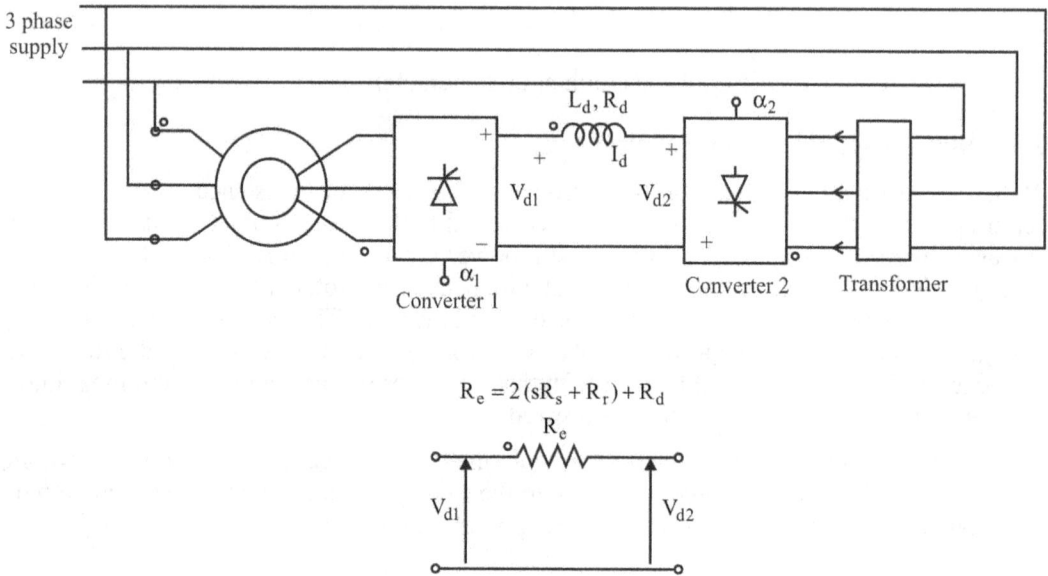

$$R_e = 2(sR_s + R_r) + R_d$$

Fig.7.8 Slip ring induction motor speed control schematic.

If P_r is + ve, speeds below synchronous speed are obtained. For this V_{d1} should be greater than V_{d2} Converter 1 work as Rectifier and $\alpha_1 < 90^0$

Converter 2 work as Inverter and $\alpha_2 > 90^0$

If P_r is – ve, speeds above synchronous speed are obtained. For this $V_{d1} < V_{d2}$

$V_{d1} < V_{d2}$ Converter 1 works as Inverter and $\alpha_1 > 90^0$

Converter 2 works as Rectifier and $\alpha_2 < 90^0$

Per phase equivalent circuit of induction motor is shown in Fig. 7.9.1

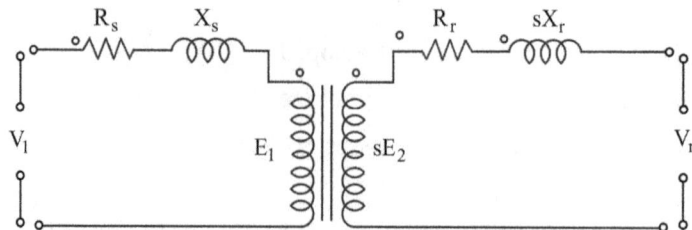

Fig. 7.9.1 Equivalent circuit.

V_1 – stator voltage per phase

E_2 - Rotor voltage per phase at standstill

$$\frac{E_2}{E_1} = a = \frac{N_2}{N_1} = \frac{\text{Rotor turns per phase}}{\text{Stator turns per phase}}$$

Equivalent circuit referred to secondary is shown in Fig. 7.9.2 in which

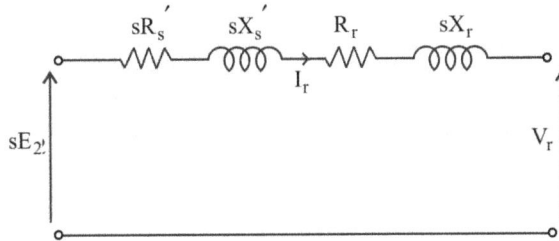

Fig.7.9.2 Equivalent circuit referred to rotor.

$$R'_s = a^2 R_s; \quad X_s^1 = a^2 X_s$$

The ac quantities I_r and V_r are converted to DC through converter 1 as I_d and V_{d1}

If I_d is the dc link current, rms value of rotor current $I_r = \sqrt{\frac{2}{3}} I_d$

Power loss in the equivalent rotor circuit $= 3 I_r^2 (sR_s + R_r) = 2(sR_s + R_r) I_d^2$ .....(7.9)

Now the equivalent circuit of the drive will be as shown in Fig. 7.9.3

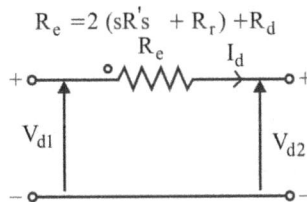

Fig.7.9.3 Equivalent circuit referred to dc side.

$$V_{d1} = 1.35 \times \sqrt{3} \times sE_2 \cos \alpha_1 = 2.339 \text{ s a } V_1 \cos \alpha_1 \quad .....(7.10)$$

$$V_{d2} = -2.339 \text{ } a_T V_1 \cos \alpha_2 \quad .....(7.11)$$

where α_T is ratio of transformer turns

$$I_d = (V_{d1} - V_{d2})/R_e \quad .....(7.12)$$

7.4   Slip Power Recovery Schemes :

There are two important schemes under slip power recovery schemes.

1. Static Scherbius drive.
2. Static Kramer drive.

7.4.1 Static Scherbius Drive

The schematic diagram is shown in Fig. 7.10(a)

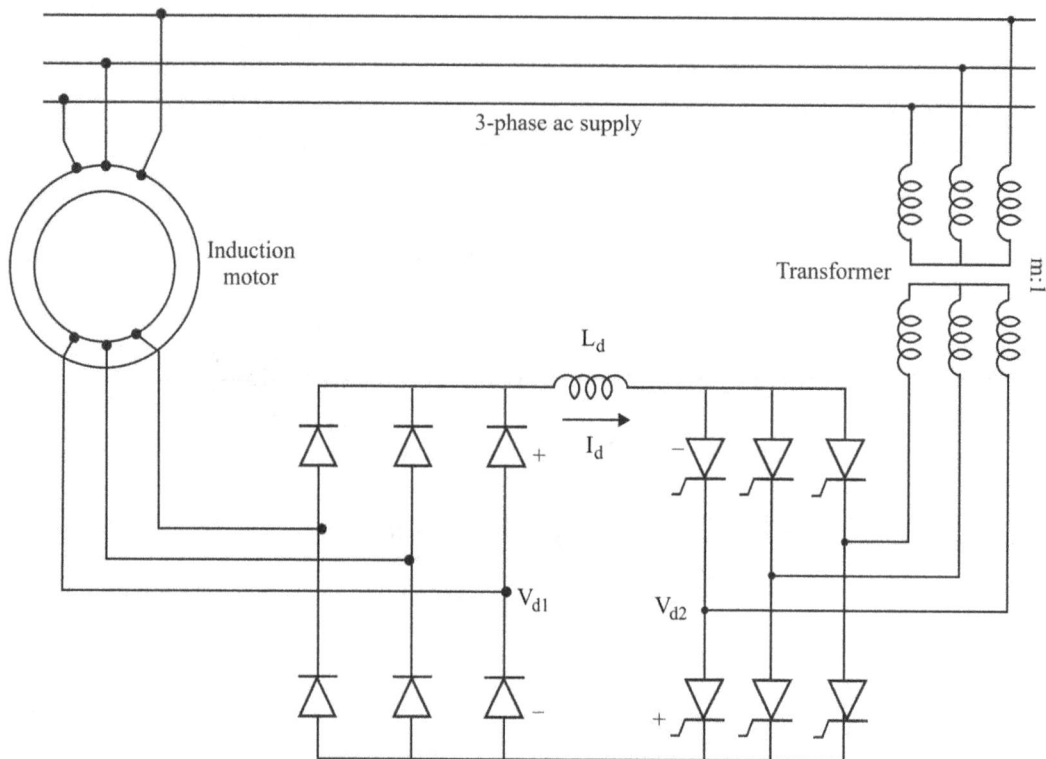

Fig. 7.10(a) Static Scherbius drive.

The slip frequency voltages of the rotor circuit are converted to dc voltage by diode rectifier. The dc voltage is converted to line frequency by line commuted inventor and the slip power is thus fed back into mains. In this scheme the slip power can flow only in one direction i.e., from rotor to mains, the speed control below synchronous speed only is possible.

Derivation of Speed Torque Expression

Let E_2 be the per phase rotor emf at stand still and s be the slip.

Then   Rotor voltage per phase = $s E_2$

V_{d1} is the diode bridge voltage.

$$V_{d1} = 1.35 \times \sqrt{3} \cdot sE_2 \ = 2.339 \ sE_2$$

If V_1 is the supply voltage per phase $E_2 = \dfrac{V_1 \times N_2^{1}}{N_1^{1}} = aV_1$

Where $\qquad\qquad a = \dfrac{\text{Effective rotor turns per phase} N_2^{1}}{\text{Effitive stator turns per phase} N_1^{1}}$

$\therefore \qquad\qquad V_{d1} = 2.339 \ s \ a \ V_1$

For a three phase line commutated inverter

$$V_{d2} = -1.35 \ V_L \cos\alpha$$

$$= -2.339 \times V_2 \cos\alpha$$

where V_2 is the per phase voltage of inverter

$$V_{d2} = -2.339 \ a_T \ V_1 \cos\alpha$$

where a_T is the turns ratio of transformer

The firing angle of inverter should be such that, $90^0 \le \alpha \le 180^0$

If the resistance of the link inductor is neglected $V_{d1} = V_{d2}$

$$2.339 \ a_s V_1 = -2.339 \ a_T V_1 \cos\alpha$$

$\therefore \qquad\qquad$ Slip $s = -\dfrac{a_T}{a} \cos\alpha \ ; \ 90 \le \alpha \le 180^0$

In order to develop motor torque a rotor current I_2 is required. V_d will force this current.

If the Resistance of the Rotor Circuits and Inductor L_d are Neglected

$$V_{d1} = V_{d2} = V_{dc}$$

Total slip power $= V_{dc} \cdot I_d$.

If T_d is the electro-magnetic torque developed and ω_s is the Synchronous speed in rad/sec.

$\therefore \qquad\qquad s\omega_s . T_d = $ slip power $= V_{dc} \ I_d$

$\therefore \qquad\qquad T_d = \dfrac{V_{dc} I_d}{s\omega_s} = \dfrac{2.339 \ s \ a \ V_1 \ I_d}{s \ \omega_s} = \dfrac{2.339 \ a \ V_1 I_d}{\omega_s}$ N-m

If a transformer is used $V_{d2} = V_{dc} = -2.339\, a_T V_1 \cos\alpha$; Since $s = -\dfrac{a_T}{a}\cos\alpha$

$\therefore$ $$T_d = \frac{2.339\, a_T\, V_1\, \cos\alpha.I_d}{\dfrac{a_T}{a}\cos\alpha\,\omega_s} = \frac{2.339\, a\, V_1 I_d}{\omega_s} \text{ N-m} \quad .....(7.13)$$

$\therefore$ whether a transformer is used or not expression for T_d is same and equal to
$$\frac{2.339\, a\, V_1 I_d}{\omega_s}.$$

From this it can be seen that torque developed is directly proportional to:

   1. dc link current I_d

   2. Stator line voltage V_1

   3. Turns ratio of induction motor and is inversely proportional to synchronous speed.

While deriving the above equations the resistance of inductance is neglected. Now if R_d is the resistance of the inductor

$$I_d = \frac{V_{d1} - V_{d2}}{R_d}$$

$$V_{d1} = 2.339\, s\, a\, V_1 = I_d\, R_d + V_{d2}$$

$\therefore$ $$s = \frac{I_d.R_d}{2.339\, aV_1} + \frac{V_{d2}}{2.339\, aV_1}$$

$$= \frac{I_d R_d}{2.339\, aV_1} + \frac{-2.339 a_T V_1 \cos\alpha}{2.339\, aV_1}$$

$$= \frac{I_d R_d}{2.339\, aV_1} - \frac{a_T}{a}\cos\alpha \qquad\qquad .....(7.14)$$

Motor speed $\omega_m = (1-s)\,\omega_s$

$$= \omega_s\left[1 + \frac{a_T}{a}\cos\alpha - \frac{I_d R_d}{2.339\, aV_1}\right] \qquad ......(7.15)$$

Under steady state condition $\quad T_d = \dfrac{V_{d1}.I_d}{s\omega_s}$;

$$I_d = \frac{sT_d\,\omega_s}{V_{d1}} = \frac{sT_d\,\omega_s}{s\,2.339\, a\, V_1} = \frac{T_d\omega_s}{2.339 aV_1} \qquad .....(7.16)$$

Substituting the value I_d in eqn. 7.15

$$\omega_m = \omega_s \left[1 + \frac{a_T}{a}\cos\alpha - \frac{T_d \omega_s R_d}{(2.339\ aV_1)^2} \right]$$

$$= \omega_s \left[1 + \frac{a_T}{a}\cos\alpha - K\ T_d \right] \qquad \qquad .....(7.17)$$

where $\qquad K = \dfrac{\omega_s R_d}{(2.339 a V_1)^2}$

The torque speed curves are shown in Fig. 7.10(b).

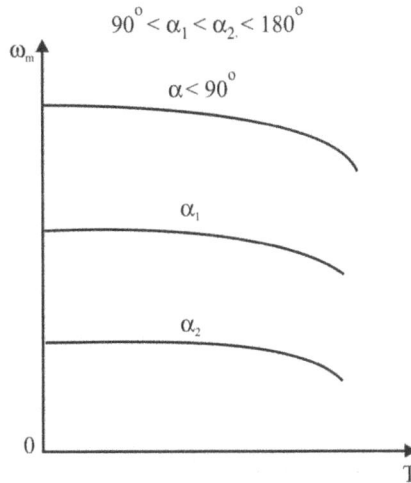

Fig. 7.10(b) Speed-torque curves.

With the static Scherbius drive shown in Fig.7.10, speed of slip ring induction motor can be controlled below synchronous speed only. For the speed control both below and above synchronous speed, static scherbius drive scheme is used with two possible configurations. These are:

1. dc Link static Scherbius drive, and

2. Cycloconverter static scherbius drive.

From the theory of induction motor it is known that the motor runs at speed less than synchronous speed if slip power is taken away from the rotor and it runs at speed above synchronous speed if power is pumped in to the rotor at slip frequency. The static schemes that do this job are described below.

7.4.3 DC Link Static Scherbius Drive Scheme

The power circuit diagram of this scheme is given in Fig. 7.11.

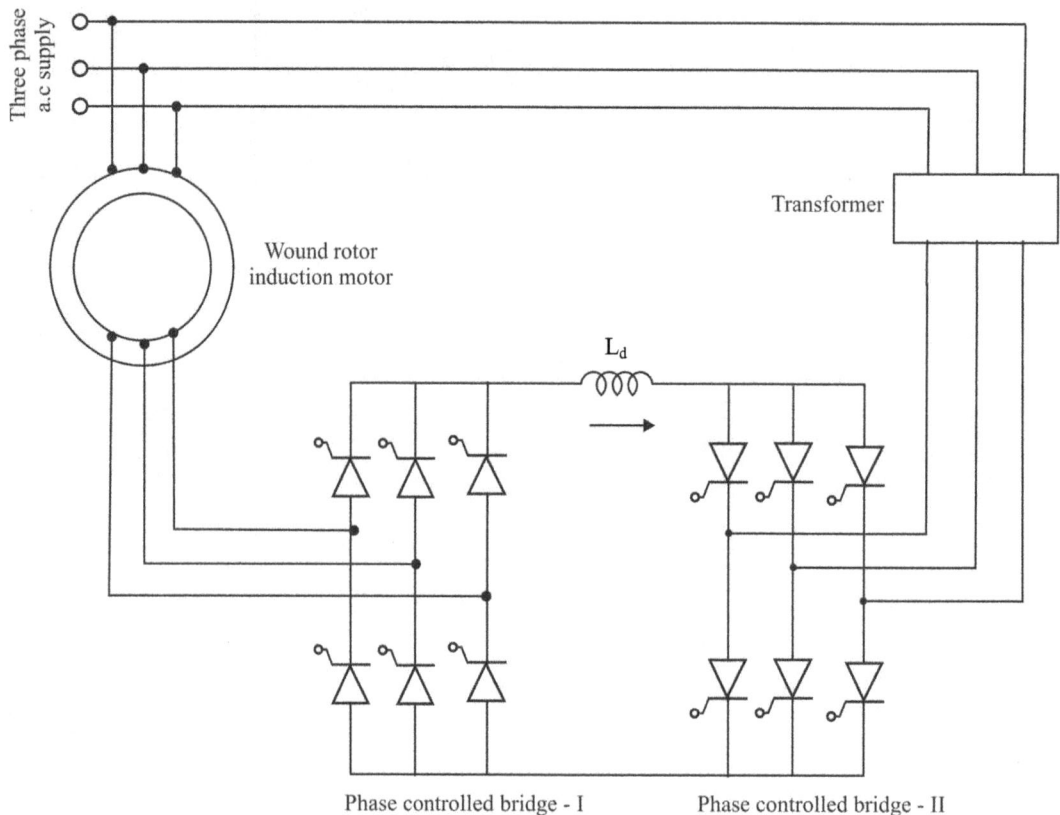

Fig. 7.11 DC scherbius drive.

The scheme has two phase controlled bridge converters, a smoothing inductor L_d and a 3 phase transformer coupling bridge II and ac supply mains. For sub synchronous speed control, bridge I has firing angle less than 90^0 and bridge II has firing angle greater than 90^0. Bridge 1 acts as rectifier converting ac slip power in to dc output thus drawing power from rotor. Bridge II converts this dc into ac and feeds ac power to ac mains. Thus the slip power flows from rotor to bridge 1, to bridge II, to the transformer and to the ac mains supply. For super synchronous speed bridge 1 has firing angle greater than 90^0 and bridge II has firing angle less than 90^0. Bridge II acts as rectifier converting ac mains power into dc. The bridge I acting as inverter converts the dc in to ac at slip frequency and feeds in to the rotor. Thus the slip power is taken from ac mains through transformer, bridge II, bridge I and to the rotor of slip ring induction motor. Thus additional power is pumped through rotor enabling the motor to run at speed higher than synchronous speed. The supply mains have to supply power to stator and to the rotor through transformer.

Near synchronous speed, the slip frequency is very small and the induced emfs in the rotor are very small. These emfs are insufficient for natural commutation of thyristors. This difficulty can, however, be overcome by using forced commutation. Thus the provision of both sub synchronous and super synchronous speed complicates system operation. The static converter of system nullifies the advantages of simplicity and economy which are inherent in a purely sub synchronous drive. In addition, Static Scherbius drive is costly compared to static Kramer drive.

7.4.3 Cycloconverter scherbius Drive

The power circuit diagram is given below. The two bridges in the previous circuit are replaced by a cyclo-converter.

An 18 SCR cycloconverter is shown in the Fig. 7.12, but other circuit configurations can be employed and a voltage-matching transformer may be introduced between the cycloconverter and the ac supply network. The cycloconverter must be controlled so that the frequency of the injected slip ring voltages tracks the rotor slip frequency. For sub synchronous motor speeds the slip frequency rotor emfs deliver slip power to the cycloconverter and this power is returned to the ac supply mains. At super synchronous speeds the cycloconverter delivers slip power to the rotor. Thus, for super synchronous motoring the ac mains must supply both the stator input power and the slip power.

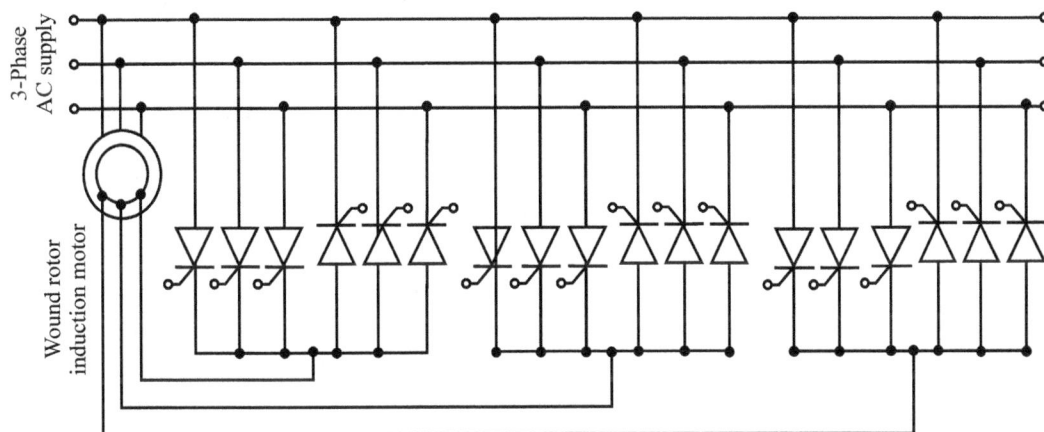

Fig.7.12 Slip ring Induction motor rotor connected to cycloconverter.

The line-commutated cycloconverter works with minimum harmonics if frequency ratio is less than one third. Thus for 50 Hz supply the maximum frequency of the voltages that can be injected in to the rotor circuit is $16 \dfrac{2}{3}$ Hz only. The speed range of the motor is from 67% to 133% of synchronous speed. At synchronous speed the rotor current should be dc and the cycloconverter has to supply voltages at zero frequency. That means it has to operate as rectifier.

The cycloconverter cascade drive is expensive and the control circuit is complex. But near sinusoidal rotor currents minimize harmonic heating effect and low frequency torque pulsations. Commutation problems near synchronous speed are eliminated. Since cycloconverter employs a large number of thyristors, the system is costly and the drive is suitable only for very large capacity drives.

7.5 Closed Loop Control of Static Sherbius Drive

The block diagram of closed loop control system for static Sherbius drive of induction motor is shown in Fig. 7.13. The torque current relation is linear. Hence a double loop control as shown in figure is suitable. Inner loop is current control and outer loop is for speed control. The actual speed is sensed by a sensor like taco-generator and compared with set speed. The error voltage is given to speed controller. The output of speed controller is fed to current limiter which sets the maximum allowable link current. The actual link current is sensed and compared with set current. The error is fed to current controller which in turn changes the firing angle of inverter. This will change link current which changes developed torque. Thus the speed error produces a motor torque which again reduces the error. The maximum current limit can be set to any desired value by setting the current limiter. The acceleration and deceleration is fairly smooth. The system is simple and stable. The speed accuracy of 0.1% from no load to full load can be achieved.

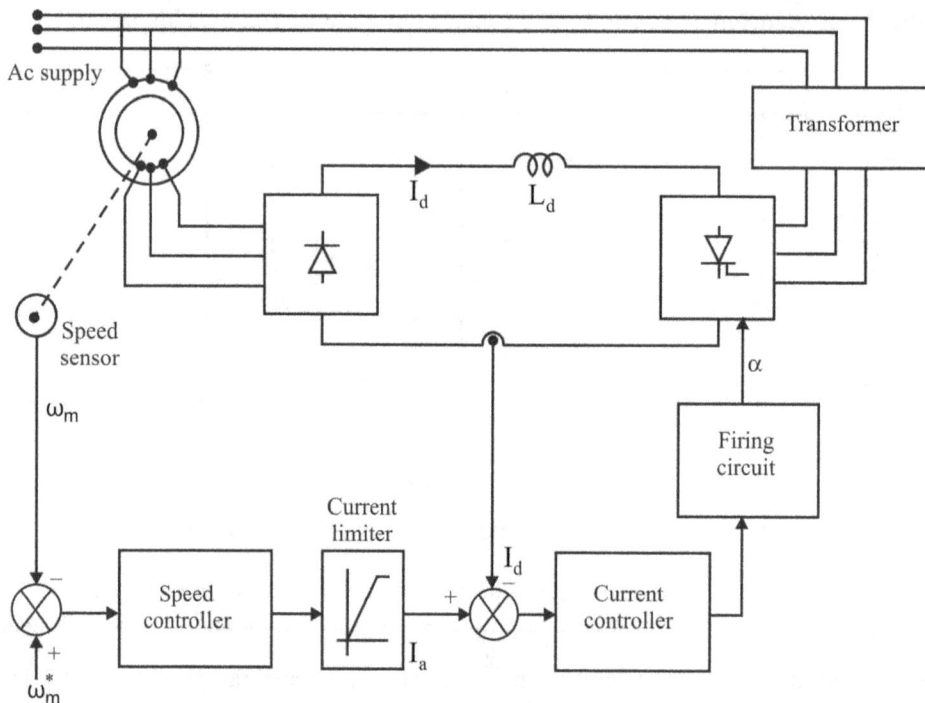

Fig. 7.13 Closed loop control of static Scherbius drive.

7.6   Static Kramer Drive of Slip Ring Induction Motor

This is also a slip recovery scheme. In this scheme the slip power is converted into mechanical power which is used to drive the rotor of induction motor in the same direction in which the rotor is already rotating. The schematic diagram is shown in Fig. 7.14(a). The diode bridge connected across the slip rings convert induced ac voltages of rotor into dc(V_{d1}). The output of rectifier is connected to the armature of a separately excited dc motor. The armature is mechanically coupled to the shaft of the induction motor rotor. An inductor is connected between rectifier and armature. Let the voltage at the terminals of armature be V_{d2}. Neglecting the armature resistance drop, V_{d2} is equal to induced emf in the rotor of dc machine. V_{d2} is directly proportional to speed of induction motor and field current. V_{d1} is proportional to slip. The variation of these two voltages is shown in Fig. 7.14(b). Whenever current flows through the armature of dc machine torque is produced which aids the torque produced by the induction motor. Under steady state condition V_{d1} is equal to V_{d2}. Therefore the speed will settle down at the point where the two characteristics of V_{d1} and V_{d2} intersect. This point is A for maximum field current. The speed is around 50% of synchronous speed. If the field current is decreased V_{d2} decreases shifting the operating point towards right. The speed thus increases. By this method the range of speed control is only from 50% to nearly 100% of synchronous speed.

If wider speed range of speed control is required the diode bridge is replaced by controlled rectifier. By controlling the firing angle the V_{d1} can be changed. Speed versus V_{d1} curves for two different values of firing angles are shown in Fig. 7.14(c). For $\alpha = 0$ and field current I_{f1} the speed corresponds to point A. For α_1 and I_{f1} the speed corresponds to point C. Thus by increasing the firing angle the speed can be reduced. Hence by controlling firing angle and field current the speed of induction motor can be varied over wide range.

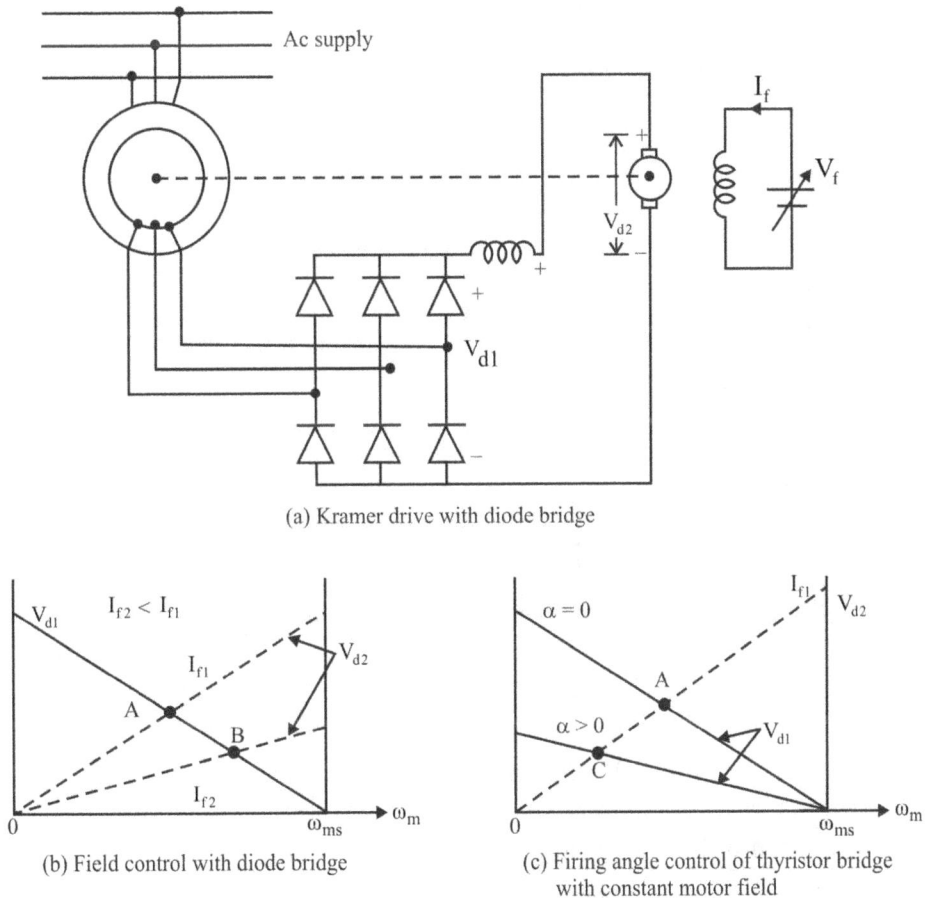

(a) Kramer drive with diode bridge

(b) Field control with diode bridge

(c) Firing angle control of thyristor bridge with constant motor field

Fig. 7.14 Static Kramer drive.

7.7   Modified Kramer Drive

In the Kramer drive described earlier a dc motor is employed to convert dc power to mechanical power. dc motor has inherent problems because of commutator, specially at higher power levels. This drive is slightly modified. The dc power is converted to ac power by having inverter as shown in Fig. 7.15. This ac power is used for driving a synchronous motor coupled mechanically to rotor of induction motor.

In this drive the dc motor is replaced by synchronous motor fed by load commutated inverter. The speed is controlled by varying the commutation lead angle. The speed can also be controlled by varying the field current. However it is not preferred because of the following problems:

1. To drive the system at synchronous speed the field current should be reduced to zero in order to reduce the inverter dc voltage to zero. Now the induced voltage will not be sufficient to obtain load commutation.

2. Slow response of field current.

3. Increased armature reaction at low field current.

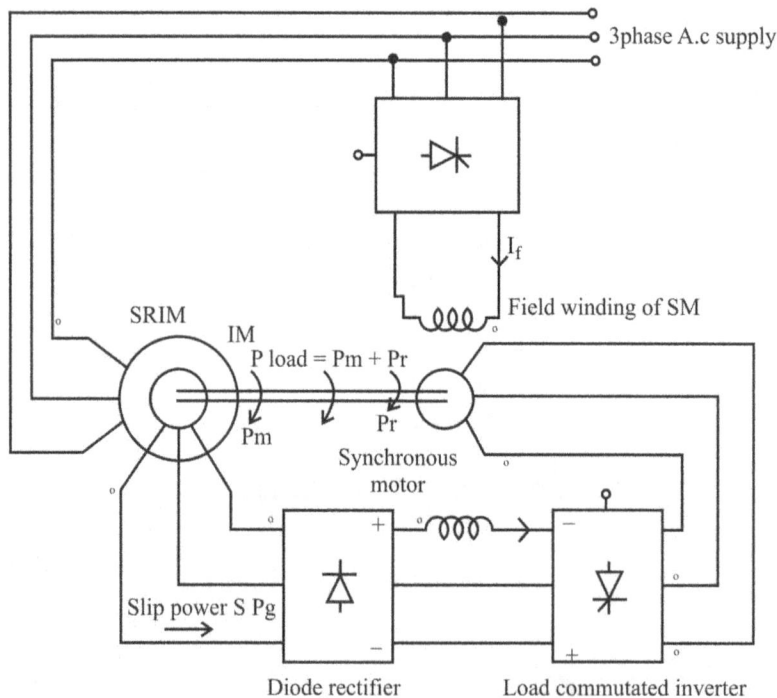

Fig. 7.15 Commutator less Kramer drive.

7.8   Comparision of Sherbius and Kramer Drives

Compared to Sherbius drive Kramer drive has better power factor and low harmonic content in line current. In Sherbius drive reactive power and harmonics are associated with the power fed back to the line. In Kramer drive since the power is not fed back to the lines, problems associated with the feedback power are also eliminated.

Worked Out Examples

Example 7.1

A Three phase, 420 V, 4 pole, 50 Hz star connected slip ring induction motor has its speed controlled by means of GTO chopper in its rotor circuit. The effective phase turns ratio from rotor to stator is 0.8. The filter inductor makes the inductor current ripple free. Losses in the rectifier, inductor, GTO chopper and no load losses of the motor are neglected. Load torque proportional to speed squared is 450 N-m at 1440 rpm.

(a) For a minimum motor speed of 1000 rpm, calculate the value of chopper resistance R. For the value of R obtained in part (a), if the speed is to be raised to 1320 rpm calculate (b) inductor current (c) duty cycle of the chopper (d) rectified output voltage (e) efficiency in case per phase, resistances for stator and rotor are 0.015 Ω and 0.02 Ω respectively.

Solution:

(a)   Per phase stator voltage $= \dfrac{420}{\sqrt{3}} = 242.5$ V

   Load torque T_L for 1000 rpm $= 450 \times (\dfrac{1000}{1440})^2 = 217.01$ N-m

   Synchronous speed $\omega_s = 2\pi \times \dfrac{50}{2} = 50\ \pi$ rad/sec

   Minimum motor speed $= 1000$ rpm $= \dfrac{1000 \times 2\pi}{60} = 104.72$ rad/sec

   Inductor current $I_d = \dfrac{T_L \omega_s}{2.3394 V_s n_m} = \dfrac{217.01 \times 50\pi}{2.3394 \times 242.5 \times 0.8} = 75.11$ A

   For minimum motor speed $\alpha = 0;\ \omega_m = \omega_s \left[1 - \dfrac{I_d . R(1-\alpha)}{2.3394 V_s n_m} \right]$

$$104.72 = 50\ \pi \left[1 - \dfrac{75.11 \times R}{2.3394 \times 242.5 \times 0.8} \right]$$

$$1 - 0.1655\ R = 0.666; \qquad R = \dfrac{1 - 0.666}{0.1655} = \mathbf{2.018\ \Omega}$$

(b)   The new speed $= 1320$ rpm $= \dfrac{1320 \times 2\pi}{60} = 138.23$ rad/sec

   load torque $T_L = 450 \times (\dfrac{1320}{1440})^2 = 378.125$ N-m

Inductor current $\quad I_d = \dfrac{T_L \omega_s}{2.3394 \times V_s \times n_m} = \dfrac{378.125 \times 50\pi}{2.3394 \times 242.5 \times 0.8} = \mathbf{130.87\ A}$

(c)
$$\omega_m = \omega_s \left[1 - \dfrac{I_d R(1-\alpha)}{2.3394 \times 242.5 \times 0.8} \right]$$

$$138.23 = 50\pi \left[1 - \dfrac{130.87 \times 2.018(1-\alpha)}{2.3394 \times 242.5 \times 0.8} \right]$$

$$0.88 = 1 - 0.5819(1-\alpha).$$

$$1-\alpha = \dfrac{1-0.88}{0.5819}$$

$$\alpha = 1 - \dfrac{1-0.88}{0.5819} = \mathbf{0.7938}$$

(d) Rectified output voltage

$$s = \dfrac{1500 - 1320}{1500} = 0.12$$

$$V_d = s \times 2.3394 \times V_s \times n_m$$

$$V_d = 0.12 \times 2.3394 \times 242.5 \times 0.8 = \mathbf{54.46\ V}$$

(e) Power loss in chopper resistance $= V_d I_d = 54.46 \times 130.87 = 7127.18$ W

Inductor current referred to rotor $I_r = \sqrt{\dfrac{2}{3}} I_d = \sqrt{\dfrac{2}{3}} \times 130.87 = 106.85$ A

Total rotor ohmic loss $= 3 I_r^2 R_r = 3 \times 106.85^2 \times 0.02 = 685.1$ W

Stator current $= 106.85 \times 0.8 = 85.48$ A

Stator ohmic loss $= 3 \times 85.48^2 \times 0.015 = 328.8$ W

Power output $= T_L \omega_m = 378.125 \times 138.23 = 52268.2$ W

Power input $= 52268.2 + 7127.18 + 685.1 + 328.8 = 60409$ W

Efficiency $= \dfrac{52268}{60409} \times 100 = \mathbf{86.52\%}$

Example 7.2

A three phase 460 V, 60 Hz, six pole star connected wound rotor induction motor whose speed is controlled by slip power as shown in the figure below. It has the following parameters.

$R_s = 0.041\ \Omega$, $R_r = 0.044\ \Omega$, $X_s = 0.29\ \Omega$, $X_r = 0.44\ \Omega$, and $X_m = 6.1\ \Omega$.

The turns ratio of the rotor to stator windings is $n_m = N_r/N_s = 0.9$. The inductance L_d is very large and its current I_d has negligible ripple. The values of R_s, R_r, X_s, X_r can be considered as negligible compared to effective impedance of L_d. The no load losses of the motor are negligible. The losses in the rectifier, inductor L_d and GTO chopper are also negligible. The speed torque, which is proportional to square of speed is 750 N-m at 1175 rpm.

(a) If the motor has to operate with a minimum speed of 800 rpm, determine the resistance R. With this value of R, if the desired speed is 1050 rpm, calculate

(b) the inductor current I_d

(c) the duty cycle of the chopper α

(d) the dc voltage V_d

(e) the efficiency

(f) the input power factor of the drive

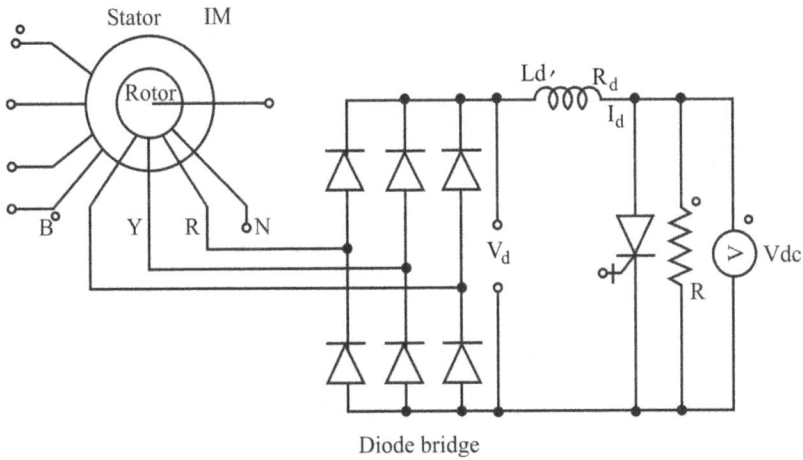

Diode bridge

Solution :

Per phase stator voltage $V_s = \dfrac{460}{\sqrt{3}} = 265.58$ V ; $\quad$ P = 6

Synchronous speed $\omega_s = 2\pi \dfrac{2f}{P} = \dfrac{4\pi \times 60}{6} = 125.66$ rad/sec

The equivalent circuit neglecting the motor parameter is shown below.

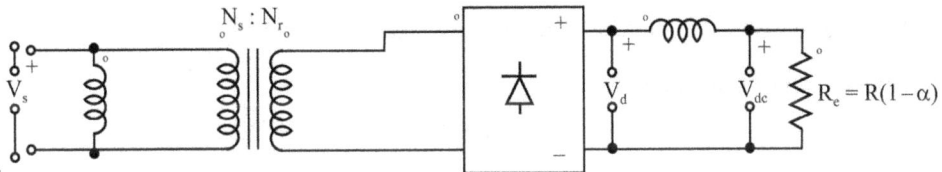

The dc voltage at the rectifier output $V_d = I_d R_e$

(a) Minimum speed = 800 rpm

$$\omega_m = \frac{800 \times 2\pi}{60} = 83.77 \text{ rad/sec}$$

Torque at 800 rpm $= 750 \times (\dfrac{800}{1175})^2 = 347.67$ N-m

The corresponding current $I_d = \dfrac{T_L \omega_s}{2.3394 V_s n_m} = \dfrac{347.67 \times 125.66}{2.3394 \times 265.58 \times 0.9} = 78.13$ A

The speed will be minimum when the duty cycle $\alpha = 0$

$$\omega_m = \omega_s \left[1 - \dfrac{T_L \omega_s R(1-0)}{(2.3394 \times 265.58 \times 0.9)^2} \right]$$

$$83.77 = 125.66 \left[1 - \dfrac{347.67 \times 125.66 \times R}{(2.3394 \times 265.58 \times 0.9)^2} \right]$$

$$R = 2.3856 \, \Omega$$

(b)   At 1050 rpm; $T_L = 750 \times (\dfrac{1050}{1175})^2 = 598.91$ N-m

$$I_d = \dfrac{598.91 \times 125.66}{2.3394 \times 265.58 \times 0.9} = \mathbf{134.\,6 \; A}$$

(c) $\omega_m = 1050 \times \dfrac{2\pi}{60} = 109.96$ rad/sec

$$\omega_m = \omega_s \left[1 - \dfrac{I_d R(1-\alpha)}{2.3394 V_s n_m} \right]$$

$$109.96 = 125.66 \left[1 - \dfrac{134.6 \times 2.3856(1-\alpha)}{2.3394 \times 265.58 \times 0.9} \right]$$

$$\boldsymbol{\alpha = 0.782}$$

(d)   Slip $s = \dfrac{125.66 - 109.96}{125.66} = 0.125$

$V_d = 2.3394 \, s V_s \, n_m = 2.3394 \times 0.125 \times 265.58 \times 0.9 = \mathbf{69.\,9 \; V}$

(e)   The power loss $P_r = V_d I_d = 69.9 \times 134.6 \quad = 9409$ W

Mechanical power output $P_m = T_L \, \omega_m$

$$= 598.91 \times 109.96 = 65856 \; W$$

The rms rotor current referred to stator $I_r^1 = \sqrt{\dfrac{2}{3}} \, I_d \, n_m \sqrt{\dfrac{2}{3}} \times 134.6 \times 0.9 = 98.9$ A

The rotor copper loss $= 3 \times 98.9^2 \times 0.044 = 1291$ W

The stator copper loss $= 3 \times 98.9^2 \times 0.041 = 1203$ W

The input power = 65856 + 9409 + 1291 + 1203 = 77759 W

$$\text{Efficiency} = \frac{65856}{77759} \times 100 = \textbf{85\%}$$

(f)   The rms value of fundamental component of rotor current referred to primary

$$I_{r1} = 0.7797 \, I_d \, \frac{N_r}{N_s} = 0.7797 \times 134.6 \times 0.9 = 94.45 \text{ A}$$

rms current through magnetizing branch is $I_m = \dfrac{V_s}{X_m} = \dfrac{265.58}{6.1} = 43.54 \text{ A}$

The r s fundamental component of input current $I_{i1} = \sqrt{94.45^2 + 43.54^2} = 104 \text{ A}$

Power factor angle $= - \tan^{-1} \dfrac{43.54}{94.45} = -24.74^0$

Input power factor = cos(-24.74) = **0.908 lagging**

If the resistances and inductances of stator and rotor are not neglected the input power factor will slightly change. The power factor is calculated as shown below.

Power consumed by chopper resistance $= I_d^2 \, R(1-\alpha)$

Power consumed by chopper per phase $= \dfrac{I_d^2}{3} R(1-\alpha) = \dfrac{1}{3}(\sqrt{\dfrac{3}{2}} \, I_r)^2 R(1-\alpha)$

$$= 0.5 \, I_r^2 R(1-\alpha)$$

If R_r is the rotor resistance per phase, its value is increased to $R_r + 0.5R(1-\alpha)$.

In this problem effective rotor resistance = 0.044 + 0.5 × 2.3856(1-0.782)

$$= 0.044 + 0.26 = 0.304 \, \Omega$$

Rotor resistance referred to stator $= \dfrac{0.304}{(0.9)^2} = 0.375 \, \Omega$

$$R_s = 0.041; \; \frac{R_r^1}{s} = \frac{0.375}{0.125} = 3.0; \; X_s = 0.29; \; X_r^1 = \frac{0.44}{0.81} = 0.54$$

$$Z = (0.04 + 3.0) + j(0.29 + 0.54) \, ; \; = 3.04 + j0.83 = 3.15 \angle 15.20^0$$

$$Z_m = j6.1 = 6.1 \angle 90$$

$$Z_{eq} = \frac{Z_1 Z_2}{Z_1 + Z_2} = \frac{3.15 \angle 15.27 \times 6.1 \angle 90}{3.04 + j(6.1 + 0.83)}$$

$$= \frac{19.215 \angle 105.27}{7.56 \angle 66.31} = 2.54 \angle 38.95$$

Therefore the input power factor = cos 38.95^0 = **0.7777**

Example 7.3

A 3-phase, 415 V, 50 Hz, 1470 rpm, star connected slip ring induction motor has the following per phase parameters referred to stator:

$$r_1 = 0.12\,\Omega\,,\ r_2 = 0.1\,\Omega\,,\ x_1 = x_2 = 0.4\,\Omega\,,\ X_m = 10\,\Omega$$

Per phase turn ratio from rotor to stator = 0.8

Speed of this motor is controlled by rotor ON-OFF control. For a speed of 1200 rpm, the inductor current is 100 A and chopper resistance is 1.8 Ω. Calculate

(i) The value of chopper duty cycle.

(ii) Efficiency for a power output of 25 kW and for negligible no-load losses.

(iii) The input power factor.

Solution :

(i)   Per phase stator voltage $= \dfrac{415}{\sqrt{3}} = 239.6$ V

Speed $N_m = 1200$ rpm

$$\omega_m = \frac{1200 \times 2\pi}{60} = 40\pi\ ;\quad \omega_s = 50\pi\ ;\ I_d = 100\ \text{A}\ ;\ R = 1.8\,\Omega$$

$$\omega_m = \omega_s \left[1 - \frac{I_d R(1-\alpha)}{2.339 V_s n_m} \right];\ 40\,\pi = 50\pi \left[1 - \frac{100 \times 1.8(1-\alpha)}{2.339 \times 239.6 \times 0.8} \right]$$

$$1\text{-}0.4014(1\text{-}\alpha) = 0.8;\ 1\text{-}\alpha = \frac{0.2}{0.4014} = 0.498;\ \alpha = 1\text{-}0.498 = \mathbf{0.501}$$

(ii)   Power output = 25 kW

Copper losses $= I_d^2\, R_e = 100^2 \times 1.8\,(1 - 0.501) = 8964$ W

RMS value of rotor current per phase $= \sqrt{\dfrac{2}{3}}\, I_d;\ \sqrt{\dfrac{2}{3}} \times 100 = 81.65$ A

Stator current $= \sqrt{\dfrac{2}{3}}\, I_d\, 0.8 = 81.65 \times 0.8 = 65.32$ A.

Stator copper loss + Rotor copper $= 3 \times 65.32^2 \times (0.12 + 0.1);\ = 2816$ W

Total power loss $= 8964 + 2816 = 11780$ W

Efficiency $= \dfrac{25000}{25000 + 11780} \times 100 = \mathbf{67.97\%}$

(iii) Fundamental component of the rotor current referred to stator $= \dfrac{\sqrt{6}}{\pi}\, a\, I_d$

$$= \frac{\sqrt{6}}{\pi} \times 0.8 \times 100 = 62.37\ \text{A}$$

Power input $= \sqrt{3} \, V_L \, I_L \cos\phi \, ; \, \sqrt{3} \times 415 \times 62.37 \times \cos\phi$

$\therefore \qquad \sqrt{3} \times 415 \times 62.37 \times \cos\phi = 36780$

$$\cos\phi = \frac{36780}{\sqrt{3} \times 415 \times 62.37} = \textbf{0.820 lag}$$

Example 7.4

A three phase, 400V, 50 Hz, 4 pole, 1400 rpm, star connected wound rotor induction motor has the following parameters referred to stator:

$R_1 = 2\,\Omega$, $R_2{}' = 3\,\Omega$, $X_1 = X_2{}' = 3.5\,\Omega$. The stator to rotor turns ratio is 2. The motor speed is controlled by static scherbius drive. The inverter is directly connected to the source. Determine,

(i)   The speed range of the drive when $\alpha_{max} = 165^0$

(ii)   The firing angle for 0.4 times the rated motor torque and speed of 1200 rpm.

(iii)   Torque for a speed of 1050 rpm and firing angle of 95^0.

Solution:

(i)
$$V_{d1} = 2.339 \, s \, a \, V_1 = 2.339 \times s \times 0.5 \times \frac{400}{\sqrt{3}} = 270 \times s$$

$$V_{d2} = -1.35 \times 400 \times \cos\alpha$$

$$V_{d1} = V_{d2}; \, \alpha = 165^0; \, V_{d2} = -1.35 \times 400 \times \cos165 = 521.6$$

$$270 \, s = 521.6; \, s = \frac{521.6}{270} = 1.93$$

In this drive the s cannot be greater than 1

If s $= 0$, Speed $= 1500$ rpm and $\alpha = 90^0$

If s $= 1$ i.e., zero speed

$$-1.35 \times 400 \times \cos\alpha = 270$$

$$\cos\alpha = -\frac{270}{1.35 \times 400}; \, \alpha = 120^0$$

$\therefore$ The speed range is from 0 to 1500 rpm for $90^0 < \alpha < 120^0$

(i)   Rated speed $= 1400$ rpm

$$\text{Rated slip} = \frac{1500 - 1400}{1500} = \frac{1}{15}$$

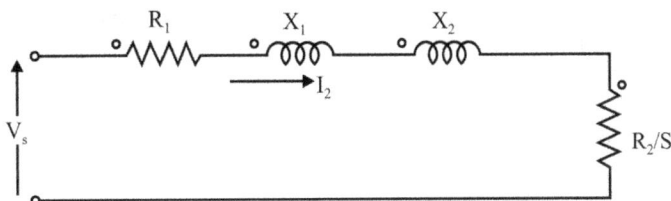

$$Z = R_1 + \frac{R_2}{s} + j(X_1 + X_2) = (2 + \frac{3}{1/15}) + j(3.5 + 3.5)$$

$$= (2 + 45) + j7 = 47 + j7 = 47.51\,\Omega$$

$$I_2 = \frac{400}{\sqrt{3} \times 47.51} = 4.86\ A$$

(Rotor copper losses $= 3I_2{}^2 R_2 = 3 \times 4.86^2 \times 3 = 212$ W

$$\text{Rated torque} = \frac{3I_2^2 R_2}{s\omega_s} = \frac{3}{s\omega_s} \frac{V^2 R_2}{(R_1 + \frac{R_2}{s})^2 + (X_1 + X_2)^2}$$

$$= \frac{3}{\frac{1}{15} \times 50\pi} \frac{400^2}{3} \frac{3}{(2 + \frac{3}{1/15})^2 + (3.5 + 3.5)^2}$$

$$= \frac{3}{2209 + 49} = \frac{3}{2258}$$

$$= \frac{45}{50\pi} \frac{400^2}{47^2 + 7^2} = 20.29\ \text{N-m}$$

0.4 times rated torque is $0.4 \times 20.29 = 8.12$ N-m; Speed is 1200 rpm

$$\text{Slip} = \frac{1500 - 1200}{1500} = \frac{300}{1500} = \frac{1}{5} = 0.2$$

$$V_{d1} = 2.339\ s\ a\ V_1 = 2.339 \times 0.2 \times 0.5 \times \frac{400}{\sqrt{3}} = 54.01\ V$$

$$V_{d2} = -1.35 \times 400 \cos\alpha = -540\cos\alpha$$

$$R = 2(sR_s{}^1 + R_r) + R_d; R_d = 0;$$

$$R_s{}^1 = 2 \times \frac{1}{2^2} = 2 \times 0.25; R_r = 3 \times \frac{1}{4}$$

$$R = 2(0.2 \times 2 \times 0.25 + 3 \times 0.25) + 0 = 2(0.1 + 0.75) = 1.7\,\Omega$$

$$I_d = \frac{V_{d1} - V_{d2}}{R} = \frac{54.01 + 540\cos\alpha}{1.7}$$

$$\text{Torque} = \frac{|V_{d2}|I_d}{s\omega_s} = \frac{540|\cos\alpha|(54.01 + 540\cos\alpha)}{0.2 \times 50\pi} = 8.12$$

$\therefore$

$$\left|\cos\alpha\left[54.01 + 540\cos\alpha\right]\right| = \frac{8.12 \times 0.2 \times 50\pi}{540} = 0.472$$

Let $\cos \alpha = -x$

$$X(54.01 - 540x) = 0.472$$

$$540x^2 - 54.01x + 0.472 = 0$$

$$X = \frac{54.01 \pm \sqrt{54.01^2 - 4 \times 0.472 \times 540}}{2 \times 540}$$

$$= \frac{54.01 \pm 43.56}{1080} = 0.090 \text{ or } 9.67 \times 10^{-3}$$

$$\cos \alpha = -0.090 \text{ or } -9.67 \times 10^{-3}$$

$$\alpha = 95.16^0 \text{ or } 90.55^0$$

(i)   Speed = 1050 rpm

$$S = \frac{1500 - 1050}{1500} = 0.3$$

$$V_{d1} = 1.35 \times 0.3 \times 0.5 \times 400 = 81 \text{ V}$$

$$V_{d2} = 1.35 \times 400 \cos 95^0 = -47.06$$

$$R = 2(sR_s{}^1 + R_r) + R_d$$

$$R = 2(0.3 \times 2 \times 0.25 + 3 \times 0.25) + 0 = 2(0.15 + 0.75) = 1.8\,\Omega$$

$$I_d = \frac{81 - 47.06}{1.8} = 18.85$$

$$T_d\, s\, \omega_s = V_{d2}\, I_d = 47.06 \times 18.85 = 887.08$$

$$T_d = \frac{887.08}{0.3 \times 50\pi} = 18.65 \text{ N-m}$$

Example 7.5

(a)   A 6 pole, 50 Hz slip ring induction motor is controlled by a static Scherbius drive. Determine the angle of firing advance in the inverter at (i) 600 rpm (ii) 800 rpm, if the open circuit standstill slip-ring voltage is 600 V, and the inverter is connected to a 415 V, 3 phase system. Neglect over lap and losses.

(b)   Recalculate the firing advance in part (a)(i)) if there is an overlap of 10^0 in the rectifier and 5^0 in the inverter. Allow for diode and thyristor voltage drops of 1.5 V and 0.7 V respectively.

(c)   Use the data of part(a), and taking the minimum required speed to be 600 rpm, estimate the voltage ratio of a transformer to be interposed between the inverter and the supply. Also, specify the power flow through the dc link as a ratio to the power input to the stator.

Solution :

(a) (i)   Speed = 600 rpm. ; Synchronous speed = $\dfrac{120 \times 50}{6}$ = 1000 rpm

∴   Slip s = $\dfrac{1000 - 600}{1000}$ = 0.4

Now when the speed is 600 rpm,

the rotor voltage across slip rings = 0.4 × 600 = 240 V

The d.c link voltage across rectifier V_d= 1.35 × V_L = 1.35 × 240 = 324 V

In the inverter side the relation between the dc voltage and ac voltage is given by

$$E_{dc} = 1.35 V_L \cos \alpha$$

$$-324 = 1.35 \times 415 \times \cos \alpha$$

$$\cos \alpha = -\dfrac{324}{1.35 \times 415} = -0.578; \alpha = 125.3^0$$

∴ Firing angle advance β = 180 - 125.3 = **54.7⁰**

(ii)   At 800 rpm   slip = $\dfrac{1000 - 800}{1000}$ = 0.2

Rotor voltage across slip rings = 600 × 0.2 = 120 V

$$V_{dc} = 1.35 \times 120 = 162$$

$$-162 = 1.35 \times 415 \times \cos \alpha$$

$$\alpha = \cos^{-1} \dfrac{-162}{1.35 \times 415} = 106.8^0$$

∴ Firing angle advance β = 180 - 106.8 = **73.2⁰**

(b)   For 3 phase fully controlled converter with firing angle α and overlap angle μ, the expression for average dc voltage is

$$V_{dc} = \dfrac{1.35 V_L}{2} (\cos \alpha + \cos(\alpha - \mu))$$

For uncontrolled rectifier $V_{dc} = \dfrac{1.35 V_L}{2} (1 + \cos \mu)$; as α = 0

For 600 rpm V_L= 240V; ∴ $V_{dc} = \dfrac{1.35 \times 240}{2} (1 + \cos 10^0)$; 321.5V

Taking drop 1.5 V in each diode dc voltage at the terminals of rectifier

$$= 321.5 - (2 \times 1.5) = 318.5 \text{ V}$$

Drop in inverter = 2 × 0.7 = 1.4 V

∴ The inverter dc voltage = 318.5 + 1.4 = 319.9 V

∴
$$319.9 = \frac{1.35 \times 415}{2}(\cos\beta + \cos(\beta - 5))$$

$$\cos\beta + \cos(\beta - 5) = \frac{319.9 \times 2}{1.35 \times 415} = 1.142$$

$$\cos\beta + \cos\beta\cos5 + \sin\beta\sin5 = \cos\beta + 0.996\cos\beta + 0.087\sin\beta = 1.142$$

$$1.996\cos\beta + 0.087\sin\beta = 1.142$$

dividing throughout with $\sqrt{1.996^2 + 0.087^2} = 1.998$

$$\frac{1.996}{1.998}\cos\beta + \frac{0.087}{1.998}\sin\beta = \mathbf{0.5716}$$

$$\cos a \cos\beta + \sin a \sin\beta = 0.5716 \; ; \; \cos(-a + \beta) = 0.5716$$

$$(-a + \beta) = \cos^{-1} 0.566 = 55.14^0$$

$$a = \cos^{-1}\frac{1.996}{1.998} = 2.56^0 \; ; \; \therefore\beta = 55.14 + 2.56 = \mathbf{57.7^0}$$

(c) As both rectifier and inverter are the same three phase configurations, then the same relationship exists between the ac side and dc side. At 600 rpm the rotor output voltage is 240 V across slip rings. This is the line voltage i.e., we must have 240 V line voltage as input to the inverter so that we get same dc voltage of 324 volts when $\alpha = 180^0$.

∴ the transformer ratio $V_2/V_1 = \frac{240}{415} = 0.578$

At 600 rpm the slip is 0.4

If the input power to the rotor is taken as 100, the rotor loss is 40. Balance 60 is converted to mechanical power neglecting the losses, the ratio of power flow through dc link to the power input to rotor is 40:100.

Example 7.6

A 3 phase, 420 V, 4 pole, 50 Hz star-connected slip ring induction motor has its speed controlled by means of static Scherbius drive. The effective phase turns ratio from rotor to stator is 0.8 and transformer has phase turns ratio from L.V to H.V as 0.4. The inductor current is ripple free. Losses in diode rectifier, inductor, inverter and transformer are neglected. The load torque is proportional to speed squared and its value at 1200 rpm is 450 N-m. For a motor operating speed of 1000 rpm, calculate (a) rotor rectified voltage (b) inductor current (c) delay angle of the inverter (d) efficiency in case the inductor resistance is 0.01 Ω and per phase resistances for stator and rotor are 0.015 Ω and 0.02 Ω respectively (e) for the firing angle obtained in part(c), the load torque is increased to 500 N-m, find the motor speed.

Solution:

(a)   Per phase stator voltage $V_1 = \dfrac{420}{\sqrt{3}} = 242.5$ V

slip $s = \dfrac{1500 - 1000}{1500} = \dfrac{1}{3}$

Turns ratio $a = 0.8$

Rotor rectified voltage $V_d = 2.339\, s\, a\, V_1$

$$= 2.339 \times \dfrac{1}{3} \times 0.8 \times 242.5 = 151.26 \text{ V}$$

(b)   Inductor current

Load torque $T_L = 450 \times (\dfrac{1000}{1200})^2 = 312.5$ N-m

Synchronous speed $= \dfrac{1500 \times 2\pi}{60} = 50\pi$ rad/sec

$$I_d = \dfrac{\omega_s T_L}{2.339 a V_1} = \dfrac{50\pi \times 312.5}{2.339 \times 0.8 \times 242.5} = 108.18 \text{ A}$$

(c) $s = \dfrac{-a_T}{a} \cos \alpha$

$$\dfrac{1}{3} = -\dfrac{0.4}{0.8} \cos \alpha$$

$$\cos \alpha = -\dfrac{2}{3} \qquad \alpha = \cos^{-1}(\dfrac{-2}{3}) = 131.81^0$$

(d)   Power output $= T_L \times \omega_m = 312.5 \times \dfrac{2\pi \times 1000}{60} = 32724.92$ W

Power loss in inductor $= I_d^2 . R_d = 108.18^2 \times 0.01 = 117.03$ W

rotor current $= \sqrt{\dfrac{2}{3}} I_d = \sqrt{\dfrac{2}{3}} \times 108.18 = 88.33$ A

rotor ohmic loss $= 3 \times 88.33^2 \times 0.02 = 468.13$ W

Stator current $= 0.8 \times 88.33 = 70.664$ A

Stator ohmic loss $= 3 \times 70.664^2 \times 0.015 = 224.7$ W

Power input $= 32724 + 117 + 468 + 224 = 33534$ W

$$\text{Efficiency} = \frac{32724}{33534} \times 100 = 97.58\%$$

(e)
$$\omega_m = \omega_s \left[1 + \frac{a_T}{a} \cos\alpha - \frac{\omega_s R_d T_L}{(2.339 a V_1)^2} \right]$$

$$= 50 \ \pi \left[1 + \frac{0.4}{0.8} \cos 131.81^0 - \frac{50\pi \times 0.01 \times 500}{(2.339 \times 0.8 \times 242.5)^2} \right]$$

$$= 104.121 \text{ rad/sec} = \textbf{994.3 rpm}$$

Example 7.7

A 440 V, 50 Hz, 970 rpm, 6 pole, Y connected, 3 phase wound rotor induction motor has following parameters referred to stator:

$R_s = 0.1\Omega$, $R_r^{1} = 0.08\ \Omega$; $X_s = 0.3\ \Omega$, $X_r^{1} = 0.4\ \Omega$. The stator to rotor turns ratio is 2.

Motor speed is controlled by static scherbius drive. Drive is designed for a speed range of 25% below the synchronous speed. Maximum value of firing angle is 165^0. dc link inductor resistance is 0.01Ω. Calculate,

   (a)   Transformer turns ratio

   (b)   Torque for a speed of 780 rpm at $\alpha = 140^0$

   (c)   Firing angle for half the rated motor torque and speed of 800 rpm.

Solution:

(a)   For a speed 25% below the synchronous speed maximum slip is $s_m = 0.25$

$$S_m = -\frac{a_T}{a} \cos\alpha_m = 0.25 = -\frac{a_T}{0.5} \cos 165$$

Transformer turns ratio is $a_T = \dfrac{N_2}{N_1}$

$\therefore$
$$a_T = \frac{0.25 \times 0.5}{-\cos 165^0} = 0.129$$

i.e., $\dfrac{N_2}{N_1} = 0.129$ or $\dfrac{N_1}{N_2} = \dfrac{1}{0.129} = 7.722$

(b)   For a speed of 780 rpm $\alpha = 140^0$

$$S = \frac{1000 - 780}{1000} = 0.22$$

Diode rectifier output $V_{d1} = 2.339s\ a\ V_1 = 2.339 \times 0.22 \times \dfrac{1}{2} \times \dfrac{440}{\sqrt{3}} = 65.36\ V$

Inverter output voltage $V_{d2} = -2.339a_T\,V_1\cos\alpha$

$$= -2.339 \times 0.129 \times \frac{440}{\sqrt{3}} \times \cos 140^0 = 58.94 \text{ V}$$

$$I_r = I_d\sqrt{\frac{2}{3}}$$

$\therefore$ The equivalent resistance referred to dc side is $2(R_r + s\,a^2 R_s)$

$$= 2(0.08 \times (0.5)^2 + (0.22)(0.5)^2 \times 0.1)$$

$$= 2\,(0.02 + 5.5 \times 10^{-3}) = 0.051$$

$$I_d = \frac{V_{d1} - V_{d2}}{2(sR_s^{\,1} + R_r) + R_d}$$

$$= \frac{65.36 - 58.94}{2\left[0.22 \times 0.1 \times (0.5)^2 + 0.08 \times (0.5)^2\right] + 0.01}$$

$$= \frac{6.42}{0.051 + 0.01} = 105.24 \text{ A}$$

$$\text{Torque} = \frac{V_{d2}I_d}{s\omega_s} = \frac{58.94 \times 105.24}{0.22 \times \dfrac{2\pi \times 1000}{60}} = 269.24 \text{ N-m}$$

(c)   Rated slip $= \dfrac{1000 - 970}{1000} = 0.03$

$$\omega_s = \frac{2\pi \times 1000}{60} = 104.72 \text{ rad/sec}$$

$$\text{Rated torque} = \frac{3V_1^{\,2}R_2}{\omega_s.s.[(R_1 + \dfrac{R_2}{s})^2 + (X_1 + X_2)^2]}$$

$$= \frac{3}{104.72} \times (\frac{440}{\sqrt{3}})^2\frac{0.08}{0.03}[\frac{1}{(0.1 + \dfrac{0.08}{0.03})^2 + (0.3 + 0.4)^2}]$$

$$= \frac{4930}{7.65 + 0.49} = 605.65 \text{ N-m}$$

Half the rated torque $= \dfrac{605.65}{2} = 302.82 \text{N} - \text{m}$

Speed $= 800$

Slip at this speed $s = \dfrac{1000 - 800}{1000} = 0.2$

$V_{d1} = 2.339 \, s \, a \, V_1$

$\qquad = 2.339 \times 0.2 \times 0.5 \times \dfrac{440}{\sqrt{3}} = 59.41V$

$V_{d2} = -2.339 \times 0.129 \times \dfrac{440}{\sqrt{3}} \cos \alpha$

$\qquad = -76.65 \cos \alpha$

$I_d = \dfrac{V_{d1} - V_{d2}}{2(sR_s^1 + R_r) + R}$

$\qquad = \dfrac{59.41 - (-76.65\cos\alpha)}{2(0.2 \times 0.025 + 0.02) + 0.1} = \dfrac{59.41 + 76.65\cos\alpha}{0.06}$

$\qquad = 990.2 + 1277.5 \cos \alpha$

$T = \dfrac{|V_{d2}|I_d}{s\omega_s} = \dfrac{+76.65|\cos\alpha|(990.2 + 1277.5\cos\alpha)}{0.2 \times 104.72}$

$\qquad = 3.66 \cos \alpha \, (990.2 + 1277.5 \cos \alpha)$

Let $\qquad \cos \alpha = -X$

$\qquad -3.66 \, X(990.2 + 1277.5X) = 302.82$

$1277.5X^2 + 990.2x = \dfrac{302.82}{-3.66} = -82.73$

$X = \dfrac{-990.2 \pm \sqrt{990.2^2 - 4 \times 82.73 \times 1277.5}}{2 \times 1277.5}$

$\qquad = \dfrac{-990.2 \pm 746.82}{2*1277.5} = \dfrac{-1998.12}{2555}$ or $\dfrac{7.72}{2555}$

$\qquad = -0.681$ or -0.097

$\qquad \alpha = 132.92$ or 95.56

$\alpha = 95.56$ is is unstable operation since $I_d = \dfrac{59.41 + 76.65\cos 95.56^0}{0.06} = 866 \, A$

if $\qquad \alpha = 132.92$

$I_d = \dfrac{59.41 - (-76.65 \times -0.681)}{0.06} = 120.2 \, A$

This is stable operation. Hence firing angle of converter should be 132.92^0

Problems

P.1   (a)   What are the effects of line side inductance in a slip energy recovery scheme?

(b)   Derive the relation of derating of an induction motor when it is having different harmonics under slip energy recovery scheme.

(c)   Why the rotor resistance of an induction motor operating under slip energy recovery scheme should have less rotor resistance.

P.2   What are the characters of a slip energy recovery scheme in (i) Speed range (ii) Rating of the drive (iii) Transformation ratio of the motor (iv) Line side power factor.

(a)   Explain the conventional Scherbius system with that of a solid state Scherbius drive.

(b)   Suggest certain modifications for improving the power factor of the SER scheme.

P.3   (a)   Why has the static Kramer drive cannot be used for high-speed ranges?

(b)   A 440 V, 50 Hz, 6 pole star connected, wound rotor induction motor has the following parameters referred to the stator:

$$R_1 = 0.08\,\Omega, R_2{}^1 =\text{-}\, 0.12\,\Omega, X_1 = 0.25\,\Omega, X_2{}^1 = 0.35\,\Omega, X_0 = 10\,\Omega.$$

An external resistance is inserted into the rotor circuit so that the T_{max} is produced at $S_m = 2.0$. The motor connections are now changed from motoring to single phase AC dynamic braking with three lead connection (one phase in series with other two phases in parallel). Calculate the braking current (line) and torque for a speed of 900 rpm.

P.4   A 440 V, 50 Hz, 50 kW, 3 phase induction motor is used as the drive motor in SER system. It is required to deliver constant (rated) motor torque over the full range from 100 rpm to the rated speed of 1420 rpm. The motor equivalent circuit parameters are:

$R_1 = 0.067\,\Omega, R_2{}^1 = 0.04\,\Omega, R_0 = 64.2\ \Omega, X_0 = 19.6\ \Omega, X_1 + X_2{}^1 = 0.177\,\Omega$. Calculate the rotor current, efficiency and power factor at 100 rpm.

P.5   A fan with a load characteristic $T_L = kN^2$ is to be driven by an SER drive incorporating a 440 V, 50 Hz, 100 kW induction motor. It is required to deliver rated speed of 1440 rpm and to provide smooth speed control down to 750 rpm.

The motor equivalent circuit parameters referred to primary turns are

$$R_1 = 0.052\,\Omega, R_2{}^1 = 0.06\,\Omega, X_m = 10\,\Omega, X_1 + X_2{}^1 = 0.29\,\Omega, R_0 = 100\,\Omega$$

Stator to rotor turns ratio: 1:2. Calculate the motor efficiency and power factor at 750 rpm. Friction and windage effects may be neglected. It is assumed that the motor is started from rest and up to 750 rpm by the secondary resistance method.

P.6   A 3 phase 400 V, 50 Hz, 6 pole 960 rpm delta connected wound rotor induction motor has the following constants referred to the stator:

(a) $R_1 = 0.3\ \Omega$, $R_2{}' = 0.5\ \Omega$ and $X_r{}' = 1.8\ \Omega$. The speed of the motor is reduced to 750 rpm at half full load torque by injecting a voltage in phase with the source voltage in to the rotor. Calculate the magnitude and the frequency of the injected voltage. Stator to rotor turns ratio is 2:2.

(b)   Why the static Kramer drive has a low range of speed control?

P.7   (a)   How do you explain the operation of an induction motor speed control using rotor resistance variation?

(b)   Explain the operation of an induction motor speed control using a chopper control.

P.8   (a)   Why the static Kramer drive has a low range of speed control?

(b)   A 3 phase 400 V, 50 Hz, 6 pole, 960 rpm delta connected wound rotor induction motor has the following constants referred to the stator:

$R_1 = 0.3\ \Omega$, $R_2{}' = 0.5\ \Omega$ and $X_r{}' = 1.8\ \Omega$. The speed of the motor is reduced to 750 rpm at half full load torque by injecting a voltage in phase with the source voltage into the rotor. Calculate the magnitude and the frequency of the injected voltage. Stator to rotor turns ratio is 2:2.

P.10   (a)   What are the assumptions made in the static resistance control of wound rotor induction motors.

(b)   Draw the speed-torque characteristics of a rotor resistance controlled induction motor and explain the effect of rotor resistance variation.

(c)   Why rotor resistance control is preferred in low power crane drives?

P.11   (a)   Compare the power flow diagram of a normal speed control method with that of slip energy recovery scheme SER for an induction motor.

(b)   In which way the SER can be made operatable for both super synchronous and sub-synchronous speed control.

P.12   A 3 phase, 50 Hz star connected, 970 rpm, 6 pole induction motor has the following parameters referred to the stator:

$R_1 = 0.2\ \Omega$, $R_2{}' = 0.15\ \Omega$, $X_1 = X_2{}' = 0.4\ \Omega$. Stator to rotor turns ratio $= 3.5$. The motor is controlled by the static Kramar drive. The drive is designed for a speed range of 30% below the synchronous speed. The maximum value of firing angle is 170^0. Calculate,

(a)   Turns ratio of the transformer

(b)   Torque for a speed of 750 rpm and $\alpha = 140^0$

(c)   Firing angle for half the rated motor torque and a speed of 850 rpm.

P.13   A 3 phase 400 V, 4 pole 50 Hz, star connected induction motor has the following parameters referred to the stator: $R_2{}' = 0.2\ \Omega$, $X_2{}' = 0.35\ \Omega$. Stator impedance and the magnetizing branch can be ignored. When driving a load with its torque proportional to speed, the motor runs at 1450 rpm. Calculate the magnitude and phase of the voltage (referred to the stator) to be impressed on the slip rings in order that the motor may operate at 1200 rpm and unity power factor.

P.14 The speed of a 3 phase slip ring induction motor is controlled by variation of rotor resistance. The full load torque of the motor is 50 N-m at a slip of 0.3. The motor drives load having a characteristics $T \alpha \, N^2$. The motor has 4 poles and operates on 50 Hz, 400 V supply. Determine the speed of the motor for 0.8 times the rated torque. The operating condition is obtained with additional resistance in the circuit. The resistance is controlled by chopper in the rotor circuit. Determine the average torque developed for a time ratio of 0.4.

P.15 A 3 phase, 420 V, 50 Hz, star connected induction motor has the following parameters:

$R_1 = 2.95\,\Omega$, $R_2' = 2.08\,\Omega$, $X_1 = 6.82\,\Omega$, $X_2' = 4.11\,\Omega$ per phase. Neglect core loss.

The motor draws a current 6.7 A at no load and controlled by rotor resistance controller. A resistance R_e has been controlled by chopper. Determine the value of R_e to get a speed range of 1500 to 500 rpm, assuming a turns ratio of two between stator and rotor. The torque and speed of the load are related by $T \, \alpha \, N$. Determine the characteristics giving the speed versus time ratio of the chopper.

P.16 An 8 pole, 50 Hz, 380 V, star connected induction motor has a star connected slip ring rotor. The stator/ rotor turns ratio is 1.25. The speed of the motor is controlled by a converter cascade in the rotor circuit. Determine the firing angles of the inverter to get 600 rpm and 400 rpm at no load. The inverter is connected to a 380 V, 3 phase system. Assume no over lap in the rectifier as well as in the inverter. What is the minimum possible speed.

P.17 A 3 phase, 420 V, 50 Hz, star connected induction motor has the following parameters:

$R_1 = 2.95\,\Omega$, $R_2' = 2.08\,\Omega$, $X_1 = 6.82\,\Omega$, $X_2' = 4.11\,\Omega$ per phase. Neglect core loss.

The motor draws a current of 6.7 A at no load and controlled by rotor resistance controller. A resistance R_e has been controlled by chopper. Determine the value of R_e to get a speed range of 1500 to 500 rpm assuming a turns ratio of two between stator and rotor. The torque and speed of the load are related by $T \alpha \, N$. Determine the characteristics giving the speed versus time ratio of chopper.

P.18 A 600 V, 50 Hz, 30 kW, 3 phase induction motor is used as the drive motor in an SER system. It is required to deliver constant (rated) motor torque over the full range from 100 rpm to the rated speed of 1000 rpm. The motor equivalent circuit parameters are:

$R_1 = 0.05\,\Omega$, $R_2' = 0.07\,\Omega$, $R_0 = 53\,\Omega$, $X_0 = 23\,\Omega$, $X_1 + X_2' = 0.153\,\Omega$.

Stator to rotor turns ratio is 1.3. Calculate the motor currents, efficiency and power factor at 300 rpm

P.19 A SER controlled slip ring motor drives a fan load having a characteristic $T_L = kN^2$. The motor is rated at 440 V, 50 Hz, 100 kW. The speed is to be controlled from a rated value of 1420 to 750 rpm. The equivalent circuit has the following parameters: $R_1 = 0.052\,\Omega$, $R_2' = 0.06\,\Omega$, $X_1 + X_2' = 0.29\,\Omega$, $X_0 = 10\,\Omega$. Stator /rotor turns ratio = 1.2. Determine the firing angle and draw the speed-torque characteristic.

Objective Type Questions

1. State the disadvantage of speed control of induction motor by rotor resistance control.

2. What is the advantage of rotor resistance control?

3. What are the applications of rotor resistance control?

4. What are the advantages of static rotor resistance control compared to conventional rotor resistance control?

5. Draw the wave forms of the rotor phase voltage and current of slip ring induction motor operating with static rotor resistance control.

6. If I_d is the current through the inductor in the rotor chopper circuit the total rms value of the rotor phase current is ………………..

7. In static rotor resistance control if I_d is the current through inductor what is the expression for fundamental rotor phase current?

8. What is the value of rms fundamental component of rotor current in terms of total rms current I_{rms}

9. If α is the duty cycle of the chopper and R is the actual resistance across chopper, what is the effective resistance R_e?

10. In static rotor resistance control the d.c voltage V_d at the output of rectifier is given as………………………

11. In static rotor resistance control the inductor current I_d is independent of speed (True/False)

12. In static rotor resistance control the inductor current I_d in terms load torque T_L is given by expression ……………..

13. Neglecting copper losses in the stator and rotor write the equation for speed of slip ring induction motor driving a load T_L and operating with duty cycle α.

14. Draw the circuit diagram of 3 ϕ, slip ring induction motor with static rotor resistance control.

15. Draw torque speed curves of slip ring induction motor with static rotor resistance control for different values of duty cycle.

16. Total rotor copper loss in 3 ϕ slip ring induction motor with static rotor resistance control is 600 Wand rotor resistance per phase $R_r = 0.03\,\Omega$. What is the value of current I_d through smoothing inductor?

17. A 3 ϕ induction motor having speed control using chopper-controlled resistance is characterized by

    (a)   Poor power factor and good efficiency

    (b)   Poor efficiency and poor power factor

    (c)   Good power factor and good efficiency

    (d)   Good power factor and poor efficiency   [   ]

18. A Three-phase induction motor with chopper-controlled resistance has its torque proportional to

    (a)   Rotor current   (b)   Square of rotor current

    (c)   Stator resistance   (d)   Square root of rotor current   [   ]

19. In static Scherbius drive of 3 phase slip ring induction motor, converter1 is a diode rectifier connected to the rotor and converter 2 is a thyristor converter connected to 3 phase mains. This drive can give

    (a)   Speeds above Synchronous speeds only

    (b)   Speeds below Synchronous speeds only

    (c)   Speeds both above and below Synchronous speeds only

    (d)   Speeds varying from 0 to 50% of Synchronous speeds only   [   ]

20. In dc link Scherbius drive of 3 phase slip ring induction motor the converter 1 is connected to rotor and converter 2 is connected to ac mains. This drive can give speeds both above and below synchronous speeds. [True/ False]

21. In dc link scherbius drive, converter 1 is connected rotor and converter 2 is connected to ac mains. For getting speeds below synchronous speed the firing angle of converter 2 should lie between ......(a).........degrees and that of converter 2 should lie between. ....(b)......... Degrees.

22. In dc link static Scherbius system converter 1 is connected to the rotor and converter 2 is connected to the mains supply. State the range of firing angles of each converter when the induction motor is running above synchronous speed.

23. In static Kramer drive the slip energy is converted to mechanical energy through dc motor and used to drive the main motor shaft in the same direction [True/False]

24. In static Kramer drive the speed of slip ring induction motor can be below synchronous speed only [True/False]

25. In a static scherbius drive system cycloconverter is used for driving 4 pole, 50 Hz slip ring induction motor. What is the range of speed over which effective control is achieved?

26. A 3 phase slip ring induction motor has rotor to stator turns ratio of 0.8. The transformer turns ratio secondary to primary is 0.5. The inverter is operating at firing angle of 150^0. What is the value of slip?

27. A 3 phase slip ring induction motor has rotor resistance of $0.02 \,\Omega$ and total rotor copper losses of 300 Watts. What is the value of dc link current in the scherbius drive.

28. The torque-speed curve of slip ring induction motor with Scherbius drive is given below for firing angle of 90^0. Draw the curve for firing angle of 120^0.

29. Write equation relating the speed and torque of 3 phase slip ring induction motor using Scherbius drive.

30. In basic static Kramer drive the slip power is fed to the armature of a separately excited dc motor which is mechanically coupled to induction motor [True/False]

31. In the static Kramer drive the speed changes with field current. If the field current is increased what will happen to the speed of slip ring induction motor?

32. In the static Kramer drive a diode bridge and dc motor are used. What is the speed range possible?

33. In the static Kramer drive if the diode bridge is replaced by Thyristor bridge what will happen to the speed range?

34. What is the problem with Kramer drive at high power level and how they are eliminated in modified Kramer drive?

35. What is the advantages of Kramer drive compared to Scherbius drive?

Control of Synchronous Motors

Control of Synchronous Motors : Separate control & self control of synchronous motors – Operation of separately controlled synchronous motors by VSI, CSI and Cyclo converters – Operation of self controlled synchronous motors by VSI, CSI and Cyclo converters – Load commutated CSI fed Synchronous motor- Operation– Wave forms–speed torque characteristics – Applications - Advantages and Numerical Problems – Closed loop control operation of synchronous motor drives (Block Diagram Only) variable frequency control, Cyclo converter, PWM, VSI, CSI.

8.1   Synchronous Motor Drive

A synchronous motor is a constant speed motor. It runs at synchronous speed given by N_s=120 f/P rpm. For a given synchronous motor the speed can only be changed by changing the supply frequency. Variable frequency supply can be obtained by using either an inverter or cyclo-converter.

A synchronous motor has a stator having three phase winding similar to induction motor, wound for certain number of poles. It has a rotor excited by dc which produces flux which depends on the excitation current. The synchronous machine can operate either as motor or generator. The power factor can be controlled by varying the field current. Over excitation results in leading power factor and under excitation draws lagging pf currents. There are different types of synchronous motors. They are:

1. Round or cylindrical rotor motors
2. Salient pole rotor motors
3. Reluctance motor
4. Permanent magnet motors
5. Switched reluctance motors
6. Brush less dc or ac motors

First five types are classified as separately controlled synchronous motors where as sixth type is classified as self controlled motor. First two types will have field windings on the rotor which are excited by direct current. By varying the magnitude of field current, the flux produced by the rotor can be varied. These two types of motors will have two slip rings on the rotor and the excitation current is sent through two brushes placed on slip rings.

Types 3 to 5 i.e., Reluctance motors, Permanent magnet motors, and Switched reluctance motors do not have any windings on the rotor and hence no slip-rings and no brushes.

A permanent magnet synchronous motor has its rotor of different construction. Permanent magnets are mounted on the rotor. It is a brush less motor.

The synchronous reluctance motor is a brush less motor that uses an unexcited rotor having salient or projecting poles. It always operates at lagging power factor.

The permanent magnet and reluctance machines are restricted to low power ratings but the wound field motors can be of very high power rating.

8.2 Round (cylindrical) Rotor Motor

This machine has stator made of iron laminations with slots milled parallel to the axis and has 3 phase winding similar to that of a 3 phase induction motor. It has a rotor in cylindrical form with slots milled on its surface. A single field winding is placed in the rotor slots. For synchronous motor operation the three phase armature winding is supplied with balanced three phase currents, which sets up rotating magnetic field in the air gap. The air gap in this machine is uniform. A direct current is sent through the winding on the rotor, which set up magnetic field of constant magnitude. If the rotor rotates at synchronous speed, the magnetic fields of stator and rotor are stationary with respect to each other and develop steady electro magnetic torque. The two magnetic fields tend to align themselves. The synchronous motor has no starting torque. It must be brought to synchronous speed by using any of the available starting methods.

8.2.1 Equivalent Circuit and Vector Diagram

The per phase equivalent circuit of synchronous motor is shown in Fig. 8.1.

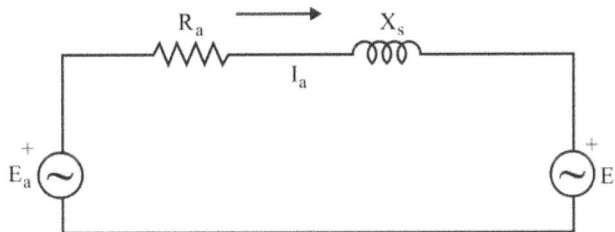

Fig. 8.1 Per phase Equivalent circuit of Synchronous motor.

E_a – Armature applied voltage per phase in Volts

E_f – Induced emf due to excitation flux in Volts

X_s – Synchronous reactance per phase in Ohms

R_a – Armature resistance per phase in Ohms

I_a – Armature current per phase in Amps

E_f depends on field current and number of turns on the stator per phase and speed of rotor.

The vector diagram is shown in Fig. 8.2.

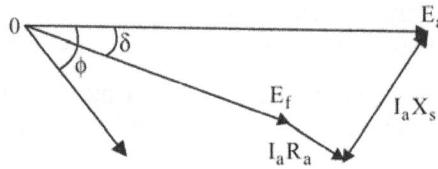

Fig. 8.2 Phasor diagram.

$$\overline{I_a} = \frac{\overline{E_a} - \overline{E_f}}{(R_a + jX_s)} = \frac{\overline{E_a}}{Z_s} - \frac{\overline{E_f}}{Z_s}$$

Taking E_a as the reference vector

$$\overline{E_a} = E_a\angle 0 \qquad\qquad \overline{E_f} = E_f\angle -\delta$$

$$\overline{Z_s} = Z_s\angle 0 \qquad \overline{I_a} = I_a\angle -\phi$$

$$\overline{E_f} = E_a\angle 0 - \overline{I_a}(R_a + jX_s)$$

$$= E_a - (I_a\cos\phi + jI_a\sin\phi)(R_a + jX_s)$$

$$= E_a - I_a R_a \cos\phi + I_a X_s \sin\phi + j(-I_a R_a\sin\phi - I_a X_s \cos\phi) \qquad .....(8.1)$$

L.H.S $= E_f\angle\delta$

where

$$\delta = \tan^{-1}\frac{(-I_a R_a\sin\phi - I_a X_s\cos\phi}{E_a - I_a R_a\cos\phi + I_a X_s\sin\phi}$$

$$E_f = \left[(E_a - I_a R_a \cos\phi + I_a X_s \sin\phi)^2 + (I_a R_a \sin\phi + I_a X_s \cos\phi)^2\right]^{\frac{1}{2}}$$

$$\overline{I_a} = \frac{\overline{E_a} - \overline{E_f}}{R_a + jX_s} = \frac{E_a - E_f(\cos\delta + j\sin\delta)}{R_a + jX_s}$$

$$= \frac{\left[E_a - E_f\cos\delta - jE_f\sin\delta\right]\left[R_a - jX_s\right]}{R_a^2 + X_s^2}$$

$$= \frac{(E_a - E_f\cos\delta)R_a - E_f X_s\sin\delta - j\left[E_f R_a\sin\delta + X_s\left[E_a - E_f\cos\delta\right]\right]}{R_a^2 + X_s^2}$$

$$.....(8.2)$$

$$I_a\cos\varphi = \frac{(E_a - E_f\cos\delta)R_a - E_f X_s\sin\delta}{R_a^2 + X_s^2} - \qquad .....(8.3)$$

8.2.2 Torque Versus Load–angle Characteristic

Power input to the motor $P_i = 3E_a I_a \cos\phi$

$$= \frac{3\left[E_a^2 - E_a E_f\cos\delta\right]R_a - E_a E_f X_s\sin\delta}{R_a^2 + X_s^2}$$

Stator copper loss $P_c = 3I_a^2 R_a$

Mechanical power developed $P_d = \rho_i - \rho_c$

If ω_s is the synchronous speed in rad/sec

Torque developed $T_d = \dfrac{P_d}{\omega_s}$ N-m

If R_a is neglected $\quad T_d = \dfrac{-3E_a E_f X_s \sin \delta}{\omega_s X_s^2} = \dfrac{-3E_a E_f \sin \delta}{X_s \omega_s}$ $\qquad$ .....(8.4)

For motoring action E_f will be lagging E_a as shown in the figure and δ will be $-$ve.

$\therefore$ $\quad$ Torque developed will be $+$ve.

Maximum torque developed $= \dfrac{3E_a E_f}{X_s \omega_s}$ N-m $\qquad$ .....(8.5)

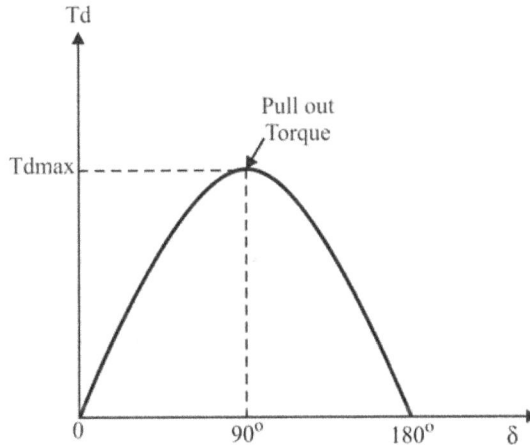

Fig. 8.3 Load angle versus torque.

The torque developed by the motor increases as δ increases from zero to 90^0 and then decreases as δ increases further. For $\delta = 90^0$ torque developed is maximum and this torque is called pullout torque.

8.3   Salient Pole Synchronous Motor

The armature winding on the stator of a salient pole motor is similar to that of a cylindrical rotor motor. But field winding on the rotor is a concentrated winding on the salient poles. In this type of motor the air gap is not uniform.

For analysis purposes its armature current I_a is resolved into two components called direct axis current I_d and quadrature axis current I_q. Like wise there are two reactances

X_d =direct axis synchronous reactance and X_q = quadrature axis synchronous reactance. The quadrature axis reactance is less than the direct axis reactance as the air gap along quadrature axis is larger than that along direct axis. $X_q < X_d$.

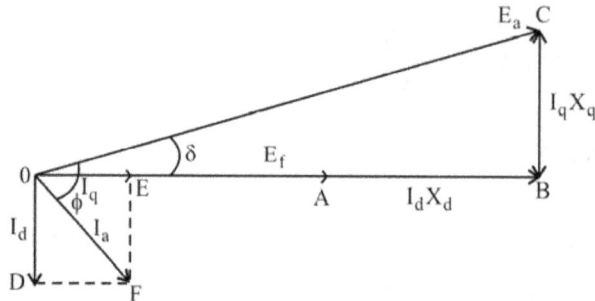

Fig. 8.4 Phasor diagram of salient pole synchronous motor.

For negligible resistance the phasor diagram is as shown above in Fig. 8.4.

$$\overline{E_f} = \overline{E_a} - j\overline{I_q}X_q - jX_d\overline{I_d} \qquad \qquad .....(8.6)$$

$$I_d = I_a \sin(\phi - \delta) \qquad \qquad .....(8.7)$$

$$I_q = I_a \cos(\phi - \delta) \qquad \qquad .....(8.8)$$

$$I_dX_d = E_a \cos \delta - E_f \qquad \qquad .....(8.9)$$

$$I_qX_q = E_a \sin \delta \qquad \qquad .....(8.10)$$

Substituting eqn. 8.8 in eqn. 8.10, for value of I_q.

$$E_a \sin \delta = I_aX_q \cos(\phi - \delta) = I_aX_q \cos \phi \cos \delta + I_aX_q \sin \phi \sin \delta$$

$$E_a = I_aX_q \cos \phi \cot \delta + I_aX_q \sin \phi \qquad \qquad .....(8.11)$$

$$\cot \delta = \frac{E_a - I_aX_q\sin\phi}{I_aX_q\cos\phi}$$

$$\delta = \tan^{-1}\frac{I_aX_q\cos\phi}{E_a - I_aX_a\sin\phi} \qquad \qquad .....(8.12)$$

Now resolving E_a along I_d axis & I_q axis

$$E_{ad} = -E_a \sin \delta$$

$$E_{aq} = E_a \cos \delta$$

The input power becomes

$$P_i = 3(E_{aq} I_q + E_{ad} I_d)$$

$$= 3\left[E_aI_q\cos\delta - E_aI_d\sin\delta\right]$$

$$= 3\left[E_a \frac{E_a \sin\delta}{X_q} \cos\delta - E_a \frac{(E_a \cos\delta - E_f)}{X_d} \sin\delta \right]$$

$$= 3\left[\frac{E_a^2 \sin 2\delta}{2X_q} - \frac{E_a^2 \sin 2\delta}{2X_d} + \frac{E_a E_f \sin\delta}{X_d} \right]$$

$$= 3\left[\frac{E_a E_f \sin\delta}{X_d} + \frac{E_a^2 \sin 2\delta}{2}\left[\frac{1}{X_q} - \frac{1}{X_d} \right] \right] \qquad .....(8.13)$$

Torque developed $$T_d = \frac{3}{\omega_s}\left[\frac{E_a E_f \sin\delta}{X_d} + \frac{E_a^2 \sin 2\delta}{2}\left[\frac{1}{X_q} - \frac{1}{X_d} \right] \right] \qquad .....(8.14)$$

The power in eqn. 8.13 has two components. The first component $\frac{3E_a E_f \sin\delta}{X_d}$ is similar to that of cylindrical rotor synchronous motor and is called electromagnetic power. The other component $\frac{E_a^2}{2}\left(\frac{1}{X_q} - \frac{1}{X_d}\right)\sin 2\delta$ is called reactance power, as it is present due to different reluctances along direct axis and quadratic axis. Similarly torque also has two components called electromagnetic torque and reluctance torque. From graph between developed torque and load angle δ, the maximum torque occurs at load angle slightly less than 90^0.

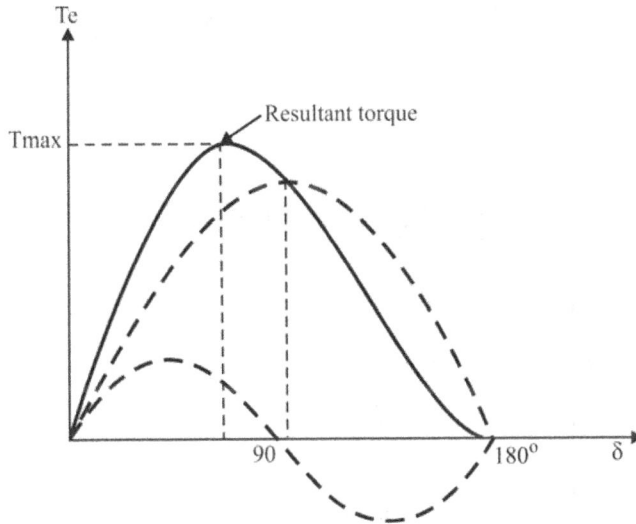

Fig. 8.5 Torque versus load angle characteristic of salient pole synchronous motor.

8.4   Reluctance Motor

Its construction is similar to salient pole synchronous motor but there is no field winding i.e., excitation current is zero. Hence $E_f = 0$. Because of presence of salient poles, having low reluctance path, they will align with rotating poles and follow the rotating magnetic field. The phasor diagram is shown in Fig 8.6.

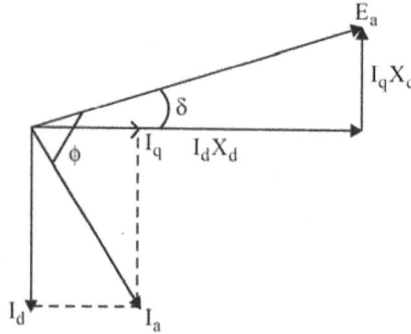

Fig. 8.6 Phasor diagram.

In the absence of field current $E_f = 0$

Power input

$$P = 3\left[E_a \cos\delta\, I_q - E_a \sin\delta\, I_d \right]$$

and

$$I_d = \frac{E_a \cos\delta}{X_d} \ ; \ I_q = \frac{E_a \sin\delta}{X_q}$$

$\therefore$

$$p = 3\left[E_a \cos\delta \frac{E_a \sin\delta}{X_q} - \frac{E_a \sin\delta E_a \cos\delta}{X_d} \right]$$

$$= \frac{3}{2} E_a^2 \sin2\delta \left[\frac{1}{X_q} - \frac{1}{X_d} \right] \qquad\qquad .....(8.15)$$

$$T_d = \frac{3}{2\omega_s} E_a^2 \left[\frac{1}{X_q} - \frac{1}{X_d} \right] \sin2\delta \qquad\qquad .....(8.16)$$

$$E_a \sin\delta = I_q X_q = X_q I_a \cos(\phi - \delta)$$

$$= I_a X_q \cos\phi \cos\delta + I_a X_q \sin\phi \sin\delta$$

$$(E_a - I_a X_q \sin\phi) \sin\delta = I_a X_q \cos\phi \cos\delta$$

$\therefore$

$$\tan\delta = \frac{J_a X_q \cos\phi}{E_a - I_a X_q \sin\phi} ;$$

$$\delta = \tan^{-1} \frac{I_a X_q \cos\phi}{E_a - I_a X_q \sin\phi} \qquad\qquad .....(8.17)$$

Torque developed is maximum when $\delta = 45°$

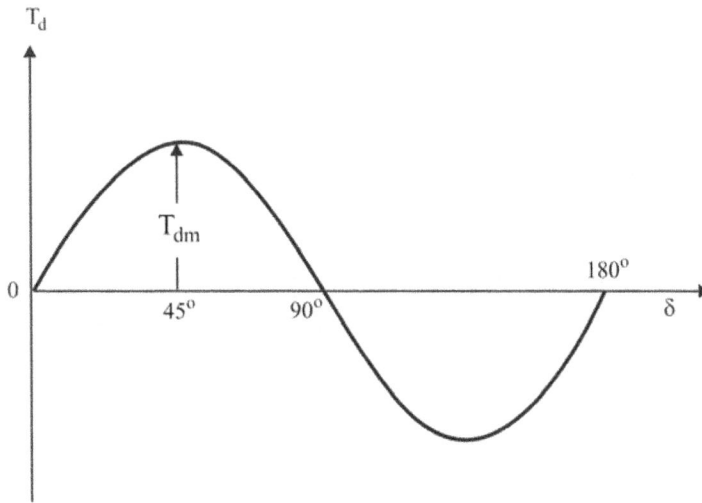

Fig. 8.7 Load angle versus torque characteristic of reluctance motor.

8.5   Permanent Magnet Motors

The permanent magnet motors are similar to salient pole motors, except that there is no field winding on the rotor and the field is provided by permanent magnets mounted on the rotor. Different types of mounting are shown in Fig. 8.8.

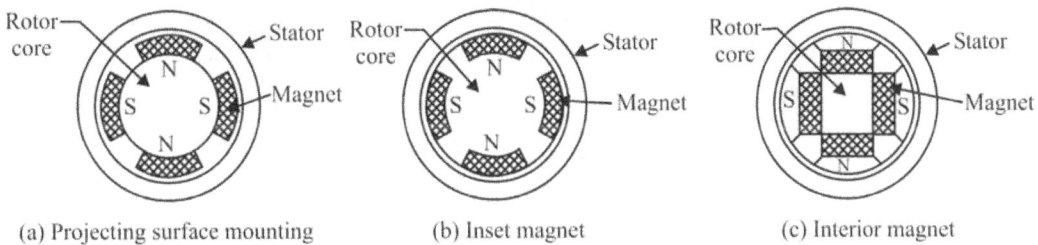

(a) Projecting surface mounting   (b) Inset magnet   (c) Interior magnet

Fig. 8.8 Permanent magnet synchronous motors.

The excitation voltage cannot be varied. For the same frame size, permanent magnet motors have higher pullout torque. The equations for the salient pole motor are applicable for permanent magnet motors. But V_f is to be taken as constant. The elimination of dc supply, field coil, brushes and slip rings reduce the motor loss. These motors are also known as brush less motors and finding increased applications in robots and machine tools. A permanent magnet motor can be fed either by rectangular current or sinusoidal current. The rectangular current fed motors have concentrated winding on the stator and are normally used for low power drives. The sinusoidal current fed motors have distributed winding on the stator, provide smoother torque and are normally used in high power drives.

8.6   Control of Synchronous Motor

There are two types.

1. Separate control

2. Self control

Separate control : In this type of control, the motor is fed from a variable frequency supply, the frequency being controlled externally from a crystal oscillator. The motor has the normal synchronous motor operation with all its stability and hunting problems.

Self control : In this method the frequency of the input voltage is decided by the rotor speed or stator voltages. A rotor position sensor is used to sense the position of the rotor and this sensor controls the inverter frequency (firing pulses). By the time the rotor moves two pole pitches, all the thyristors in the inverter are fired once in the sequence. Thus the input frequency and the rotor speed are related. The firing pulses may be derived by sensing the stator voltages also. The motor with self-control has good stability as well as good dynamic response. The inverter with rotor position sensing or induced voltage sensing is equivalent to a six-segment commutator. The synchronous motor with self control behaves like a dc motor.

8.6.1 Separately Controlled Synchronous Motor

Speed of the motor is determined from an external frequency obtained from a crystal oscillator. Variable voltage variable frequency supply can be obtained from constant voltage constant frequency ac supply by three methods already explained.

1. Controlled rectifier – DC link – 3 phase VSI either 120^0 or 180^0 conduction

2. Diode rectifier – DC link – PWM VSI inverter.

3. Diode rectifier – DC chopper – 3 phase VSI (120^0 or 180^0 conduction)

The performance of the motor is similar to a conventional synchronous motor with problems of hunting for sudden changes of load. Multi motor operation is possible.

A voltage source square wave inverter (VSI) feeding a synchronous motor is shown in Fig. 8.9.

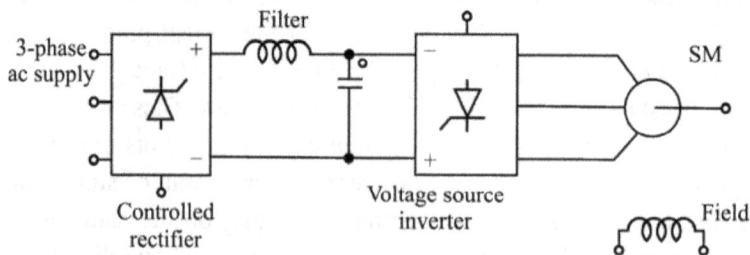

Fig. 8.9 Separate control of synchronous motor fed from square wave inverter.

A PWM voltage source inverter (VSI) feeding a synchronous motor is shown in Fig. 8.10

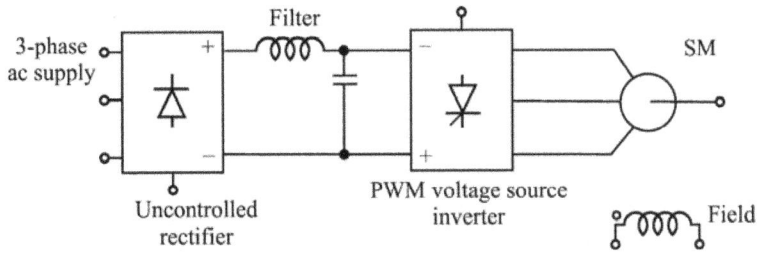

Fig. 8.10 Separate control of synchronous motor fed from a PWM inverter.

In the above two schemes open loop control is possible. Motor has instability problems such as hunting. The output voltage of the inverter is non sinusoidal. The behavior of the motor is different from the behavior of the motor supplied with sinusoidal voltages. Additional losses and pulsating torques occur due to harmonics present in the stator current. When PWM inverter is used the effect due to harmonics is minimized.

In the case of square wave inverter as we use controlled rectifier on the line side, at low speeds the firing angle will be near 90^0 and the pf is poor. In the case of PWM inverter the pf is unity.

The synchronous motor is operated at unity pf by controlling the field current. This will minimize the size of the inverter and losses. VSI drive has reasonably good efficiency. This finds applications as a general purpose industrial drive for low and medium powers.

The speed is changed by changing the frequency. For speeds below rated speed V/F is held constant. For speeds above rated speed V is held constant at its rated value and frequency is increased. Frequency from its initial value is changed very slowly to its desired value, so that the rotor speed tracks the speed of stator field and follows it. This mode of control can be used for starting the motor also.

A block diagram of drive operating in this mode is shown in Fig. 8.11. Frequency required for the desired speed is calculated and this frequency f^* is given as command signal to a delay circuit. Delay circuit changes the frequency f slowly and gives to inverter and to flux control block. The flux control block generates a signal corresponding to stator voltage V^* so that V/f is held constant up to rated speed and V is held constant above rated speed. This reference voltage V^* generates a signal which changes the firing angle α of controlled rectifier so that required DC voltage is given as input to inverter.

This type of control is used in spinning mills for driving spindles because all spindles are to be rotated at same speed simultaneously.

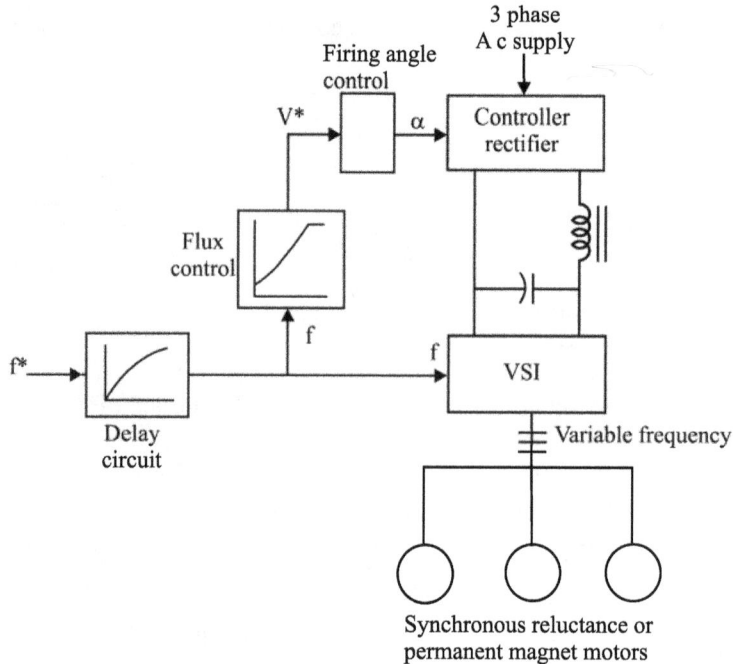

Fig. 8.11 Block diagram of separate control of synchronous motor.

8.6.2 VSI Fed Synchronous Motor Self Controlled Drives

A machine is said to be self-controlled if it gets its variable frequency from an inverter whose thyristors are fired in a sequence, using the information of rotor position or stator voltages.

Rotor position with respect to stator is sensed by a sensor. The sensor sends pulses to the thyristors. Thus the frequency of the inverter output is decided by the speed of the rotor. This kind of control makes it possible to control the angle between the stator and rotor mmf or load angle δ. The machine behavior is decided by the torque angle, and voltage/current. Such a machine can be looked upon as a dc motor having its commutator replaced by a converter connected to stator. The rotor of the synchronous motor may be conventional type having dc power through slip rings or may be of special construction to avoid slip ring and brushes. The schematic diagram of self controlled synchronous motor driven by VSI is shown in Fig. 8.12.

If the synchronous motor has its rotor fed from dc source and operating in self-control, it is called commutator less motor (CLM). If the synchronous motor has its rotor with permanent magnets or variable reluctance it is called Brush less and commutator less motor(BLCLM). The output voltage of the inverter is non sinusoidal (square wave). It contains lot of harmonics and introduces additional losses and heating. They also produce pulsating torques, which are objectionable at low speeds.

If PWM inverter is used harmonics are minimum and their effects are less.

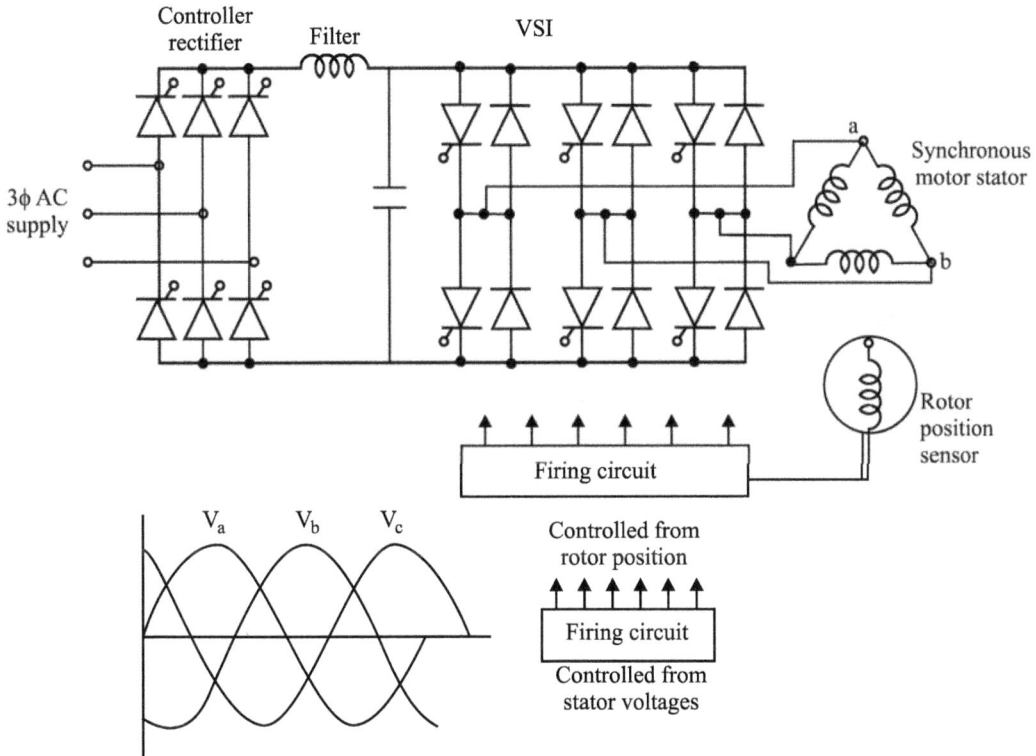

(a) Inverter output is square wave

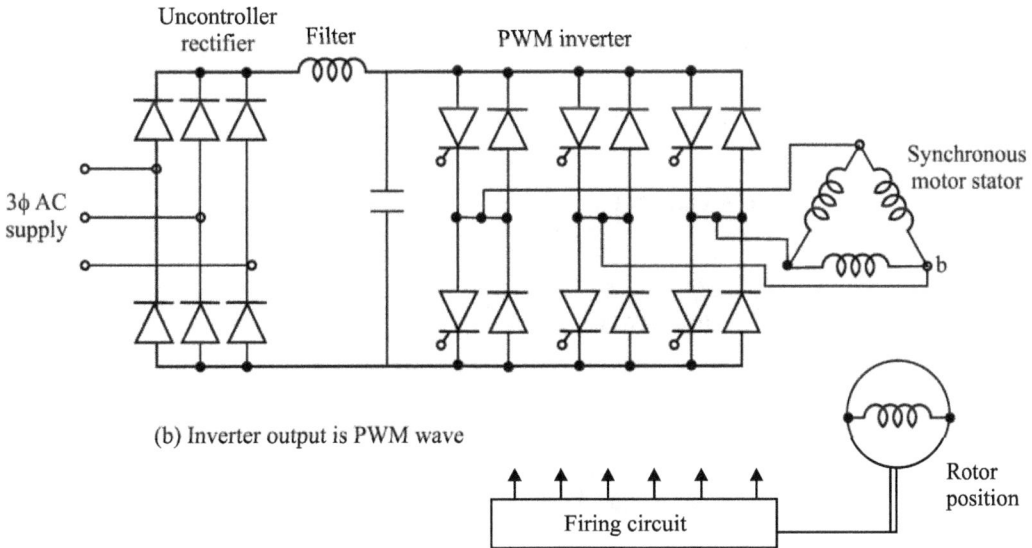

(b) Inverter output is PWM wave

Fig. 8.12 Self controlled synchronous motor with VSI.

When fed from square wave inverter the drives have speed range from medium to high speeds. This is because of low voltages at small frequencies, which are insufficient for commutation. When fed from PWM inverter a wide range of speeds down to zero speeds is possible because inverter input voltage is constant at all frequencies.

8.7   Braking of Synchronous Motor with VSI

8.7.1 Dynamic Braking

Dynamic braking can be employed by disconnecting the inverter supply and connecting external resistance to the stator. The per phase equivalent circuit during dynamic braking is shown in Fig. 8.13. By controlling the field current the braking torque can be controlled as induced emf depends on field current also apart from speed.

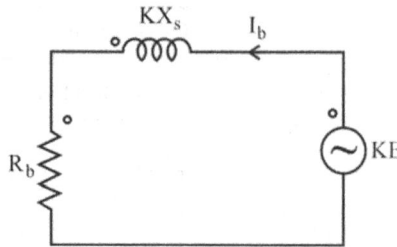

Fig. 8.13 Per phase equivalent circuit during dynamic braking.

Let   E = Per phase induced emf in the stator at rated synchronous speed ω_s and

$$K = \frac{\omega_m}{\omega_s} \quad \text{where } \omega_m \text{ is the actual speed of motor in rad/sec.}$$

E at speed ω_m is equal to KE.

$$\text{The stator current } I_{sb} = \frac{KE}{\left[R_b^2 + (K^2 X_s^2)\right]^{\frac{1}{2}}}$$

$$\text{Braking power } P_b = 3 I_{sb}^2 R_b$$

$$= 3 \frac{K^2 E^2 R_b}{R_b^2 + (KX_s)^2}$$

$$\text{Braking torque } T_b = \frac{P_b}{K\omega_s} = \frac{3KE^2 R_b}{\omega_s (R_b^2 + K^2 X_s^2)}$$

Usually R_b is very small compared to X_s, hence the braking torque will remain almost constant over wide speed range. The synchronous motor is brought to zero speed by dynamic braking.

8.7.2 Regenerative Braking

Regenerative braking can be applied but another phase controlled rectifier is required on the line side, apart from VSI connected to motor. Regenerative braking can be applied only at synchronous speed as shown in Fig. 8.14. The machine side converter is VSI. During regenerative braking, its firing angle is made less than 90^0. It works as rectifier and converts ac voltages available at terminals of synchronous motor into DC voltage with polarity shown in Fig. 8.14.

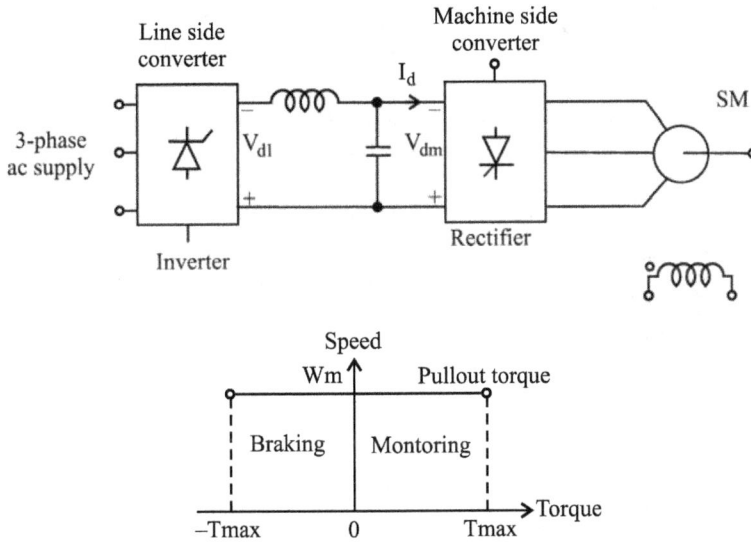

Fig. 8.14 Speed torque characteristic – regenerative braking.

The firing angle of line side converter is made greater than 90^0. It works as inverter, converting dc voltage V_{dl} into ac and feeds power to ac mains supply. Thus the kinetic energy of motor is converted to electrical energy and fed back to ac source. The difference between V_{dl} and V_{dm} is voltage drop in the resistance R_d of inductor. Regenerative braking is usually applied to limit the speed of moving object.

8.8   Current Source Inverters for Feeding 3 phase Synchronous Motors

In CSIs, the controlled quantity is the current in the dc link. The current from dc source remains constant at the controlled value, irrespective of the load and events in the inverter. The link current is maintained constant by means of a large inductance. The control of the link current is achieved by means of phase controlled rectifier on the line side. As the current is the controlled quantity feedback diodes are not required. Two types of CSIs are shown in Fig. 8.15(a) and 8.15(b). Both are force commutated. For changing the speed the frequency of the inverter has to be changed.

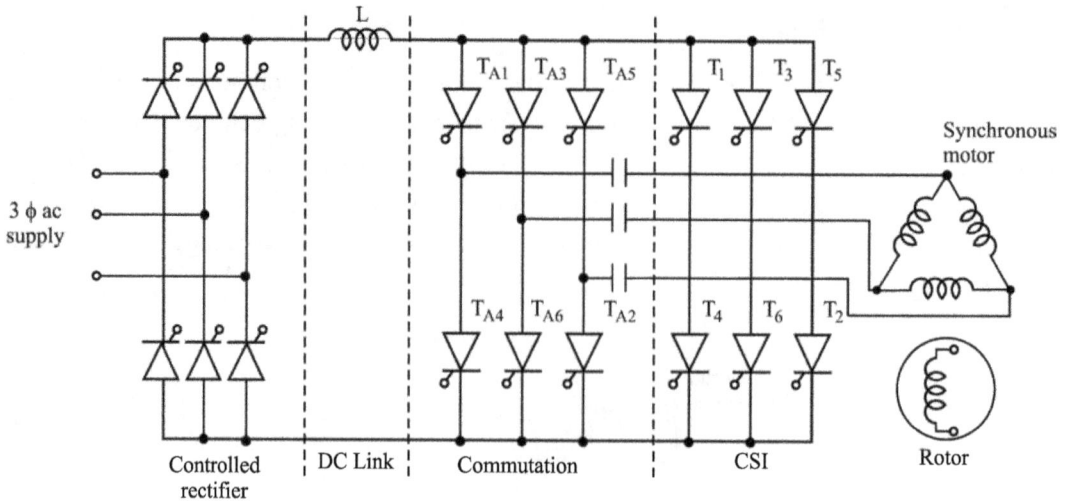

Fig 8.15(a) CSI with individual commutation.

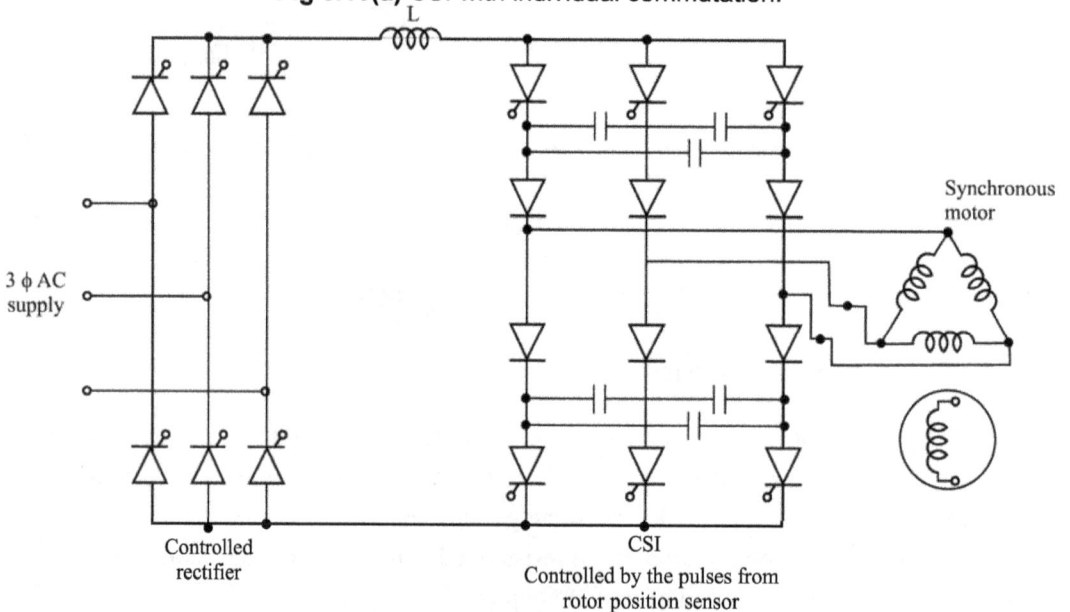

Fig. 8.15(b) CSI auto sequancial commutation.

CSI with independent commutation. Auxiliary Thyristors T_{A1} to T_{A6} are required for commutation.

Auto sequential commutation. Diodes D1 to D6 are used for retaining the charges on the capacitor.

8.9   Special Features of CSI

(i) They have load dependent commutation.

(ii) As the load circuit elements form a part of commutation circuit, the inverter and the load (motor) must be matched with each other.

(iii) These are suitable for single motor operation only.

(iv) Two and four quadrant operation is straightforward, and no additional converter on the line side is required.

(v) Inverters employ forced commutation to give variable frequency currents.

8.10 Control of Synchronous Motor Using Current Source Inverter (CSI)

CSI supplies constant current to synchronous motor. The amplitude of current can be controlled by line side converter. A Synchronous motor can have either separate control or self-control. Due to stable operation self-control is normally employed, by using either rotor position sensing or induced voltage sensing. The CSI operating on machine commutation (self control) is shown in Fig. 8.16.

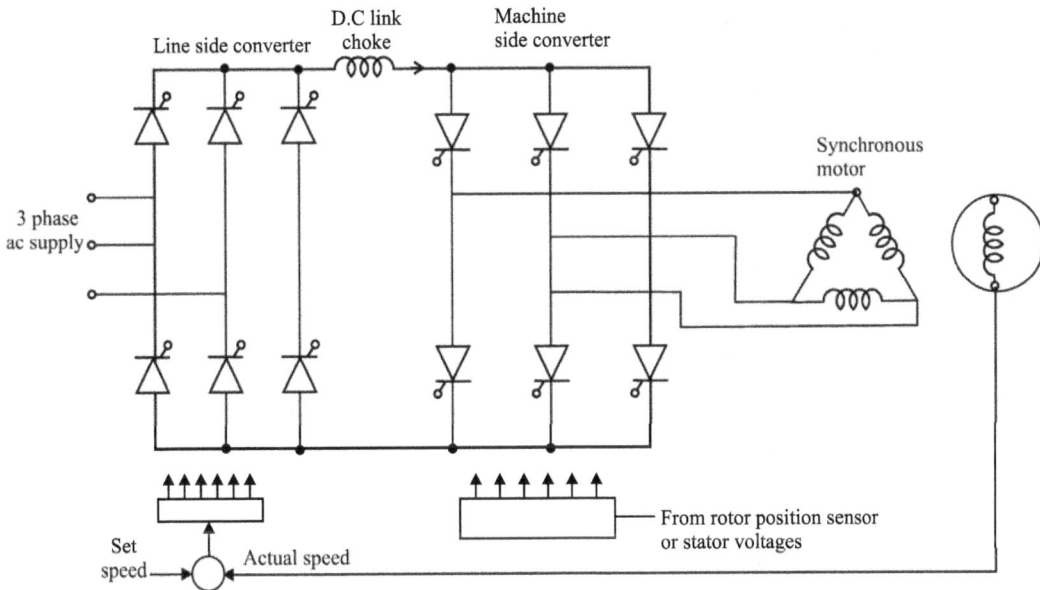

Fig. 8.16 CSI with machine commutation.

The Synchronous motor operates at leading power factor when over excited. The armature voltages can be used to commutate the thyristors of inverter on machine side. This is possible over a reasonably wide range of speeds. However at very low speeds up to 10% of the base speed the machine commutation is in-effective and therefore forced commutation is required. The inverter (ordinary) may be equipped with forced commutation circuit, which operates up to 10% of base speed, and above this speed the machine commutation takes over. Refer Fig. 8.15(a).

To accelerate the rotor to the speed where machine commutation can take over, forced commutation of the inverter thyristors is necessary. This can be achieved by an auxiliary circuit in the inverter with the facility of being cut off when natural commutation takes over. Refer Fig 8.15(a). The firing pulses to six auxiliary SCRs are to be stopped whenever the inverter has to work on machine commutation.

8.10.1   Closed Loop Speed Control with CSI

A block diagram showing speed control of synchronous motor with current source inverter using self control is shown in Fig. 8.17. The motor in the diagram is either normal synchronous motor with separate field winding or permanent magnet synchronous motor. The speed of rotor is sensed by the voltages at terminals of stator and firing angles to the inverter are generated accordingly. Up to 10% of rated speed forced commutation is used. Above this speed natural commutation with voltages available at stator terminals is used. ω_m^* is the reference speed .The actual speed ω_m signal is derived from terminal voltage sensor and compared with reference speed. The error signal is used for generating necessary phase delay and firing pulses to load side converter and for generating dc link current required for driving load. The actual link current is sensed and compared with set value. The difference between these two currents is used for controlling firing angle of source side converter till the difference becomes zero. As usual the inner loop is for current control and outer loop is for speed control.

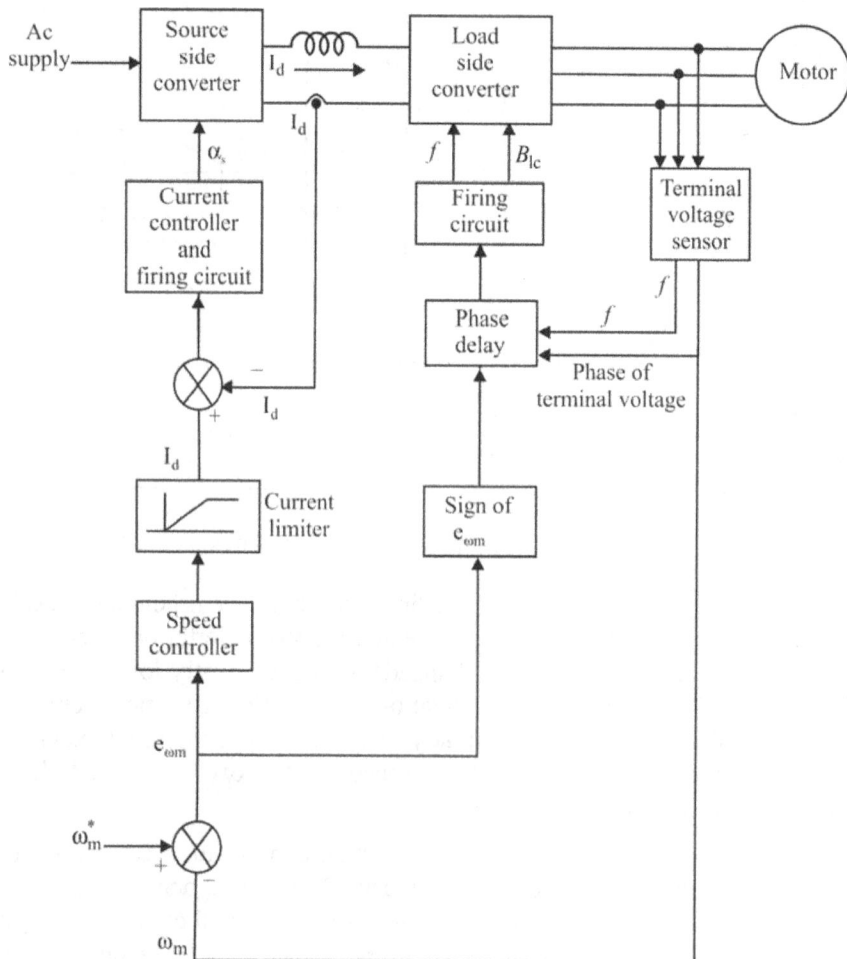

Fig. 8.17 Block diagram of self controlled synchronous motor with CSI.

8.10.2 Closed Loop Speed Control of Motor by DC Link Current Interruption

Block diagram for the control of CSI fed synchronous motor is shown in the Fig. 8.18. In this scheme a different method is used for commutating the thyristors. At very low speeds the induced voltages in armature windings are very small and they cannot be sufficient to commutate the thyristors. The assistance in commutation is provided by means of interrupting the dc link current. A thyristor TH across the dc link is provided as shown in Fig. 8.18.

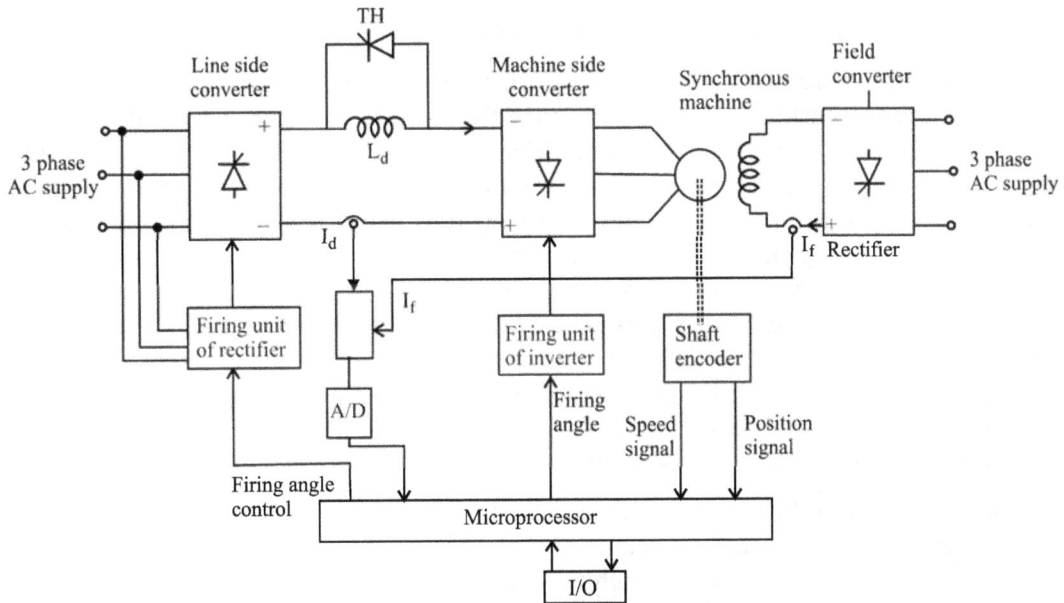

Fig. 8.18 Block diagram of load commutated CSI fed synchronous motor control.

The system consists of dc link converter which is made up of two six pulse bridge converters inter connected by a high smoothing inductance. The field of synchronous motor is fed from rectifier. In the normal operation, the line side converter operates as a rectifier and the machine side converter operates as line commutated inverter. The synchronous motor is fitted with shaft encoder. This provides information of the shaft position and speed.

The information regarding the field current, dc link current, shaft position, shaft speed are fed to the micro processor in digital form. Necessary software is developed for:

(i)   controlling the dc link current by the line side converter,

(ii)   triggering the suitable thyristors in the machine side converter operating as inverter,

(iii)   for using forced commutation of inverter up to 10% of base speed,

(iv)   for changing the operation from forced commutation to natural commutation using the voltages available at the terminals of synchronous motor,

(v)   to control the speed of synchronous motor,

(vi)   to apply braking,

(vii)   for protection, diagnosis and display.

Starting of Synchronous Motor with the Above Scheme

From zero to 10% of base speed natural commutation using machine voltage is not possible. To accelerate the rotor to the speed where machine commutation can take over, forced commutation of the inverter thyristors is necessary. It is done by interrupting dc link current with a Thyristor TH connected across the link inductance as shown.

The interruption of dc link current is explained now : The dc link current is interrupted at the instant of commutation and at the same time the line side converter is controlled so that it goes to inversion from rectification (i.e., firing angle goes to 90 to 180^0 range from 0 to 90^0 range. The rotor position sensor sends information to the control unit of the machine side converter to block firing pulses to the **out going thyristor** and provide them to the **incoming one**. Though firing pulse to outgoing SCR is stopped it will not go to OFF state because of DC link current. The firing angle of line side converter is made greater than 90^0 so that dc link voltage has changed its polarity due to transition. Consequently the dc link current decays to zero and is maintained at zero value for a time greater than turn off time of the thyristor making it go to OFF state. After this dead zone the line side converter is again made a rectifier. The dc link current builds up and flows through the machine phases via new thyristor. A similar sequence of operations take place in the other commutations.

The interruption of link current to zero at the instant of commutation and its build up to the reference value after commutation are delayed by the link inductance. When the inductance is large and frequency is high the current may not even reach its reference value by the time the next commutation starts. To make current variation faster, the thyristor TH, connected across the link inductor is fired just at the instant when the zero current should exist. At the end of commutation when the link voltage changes its sign, the thyristor automatically ceases to conduct. The link current flows through the inductor and builds up again and flows through the new phases via the thyristors.

The commutation process is explained in detail now. Please refer to Fig. 8.18(a). Let us suppose that at t = 0, thyristers T1 and T2 are conducting and current is I_d. Now T1 is to be turned OFF T3 is to be turned ON. Firing pulses to T1 are stopped and firing pulses to T3 are given. By doing this T1 will not go to OFF state and hence T3 will not come to ON state. We have to make the current through T1 zero. The firing angle of line side converter is made greater than 90^0 simultaneously and TH across link inductor is triggered with gate pulse. This negative voltage will reduce the current through T1 & T2. As TH is forward biased it comes to ON state, shorting the link inductor and reduces current faster. At t = t_1 the current has become zero. After a short time equal to turn off time of thyrister the firing angle of line side converter is made less than 90^0 making it operate as rectifier. Since firing pulses to T2 & T3 are already there, the thyristers come to ON state and current starts increasing from zero at t = t_1. As TH is now reverse biased it goes to OFF state automatically. The current through T3 & T2 increases to I_d at $t = t_2$. Thus the current I_d is transferred from T1 to T3 in time t_2 completing commutation process.

The process of starting the synchronous motor will be explained. Refer Fig 8.18(b). A rotor position sensor is essential. Let us assume that at time $t = 0$ the rotor is at position shown in (I) of the Fig. 8.18(b). Thyristors T1 & T2 are switched ON. Current flows through phases A & C and produces mmfs fA & fC. The resultant mmf vector is 'f$_{ar}$'. This mmf axis is 60^0 ahead of mmf axis of rotor. Therefore the rotor will move in anti-clock wise direction and try to align itself with axis of flux produced by armature. The rotor rotates by 60^0 in one sixth of periodic time T. Now the position of rotor is as shown in (II). The sensor senses this position and makes T2 & T3 come to ON state. Now the current flows through B & C phase windings, producing mmfs fB & fC as shown in (II). The resultant flux vector 'f$_{ar}$' is now 60^0 ahead of rotor position. The rotor again moves by 60^0 to be in line with stator flux. This happens in T/6 seconds. Sensor senses this new position and triggers T3 & T4. Consequently stator flux moves by another 60^0 making the rotor to follow. This process repeats and the rotor moves continuously. The sequence of operations and positions of rotor for one period are shown. In one period rotor makes one revolution. By reducing the periodic time slowly the speed is increased. The wave forms of currents in each phase winding and thyristors conducting in each one-sixth of a periodic time are shown in Fig. 8.18(c).

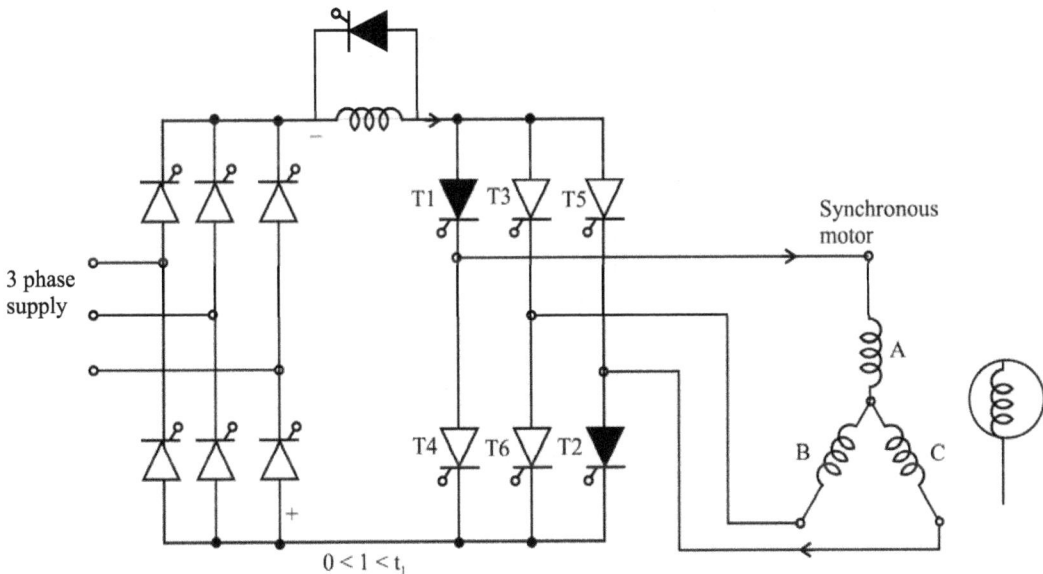

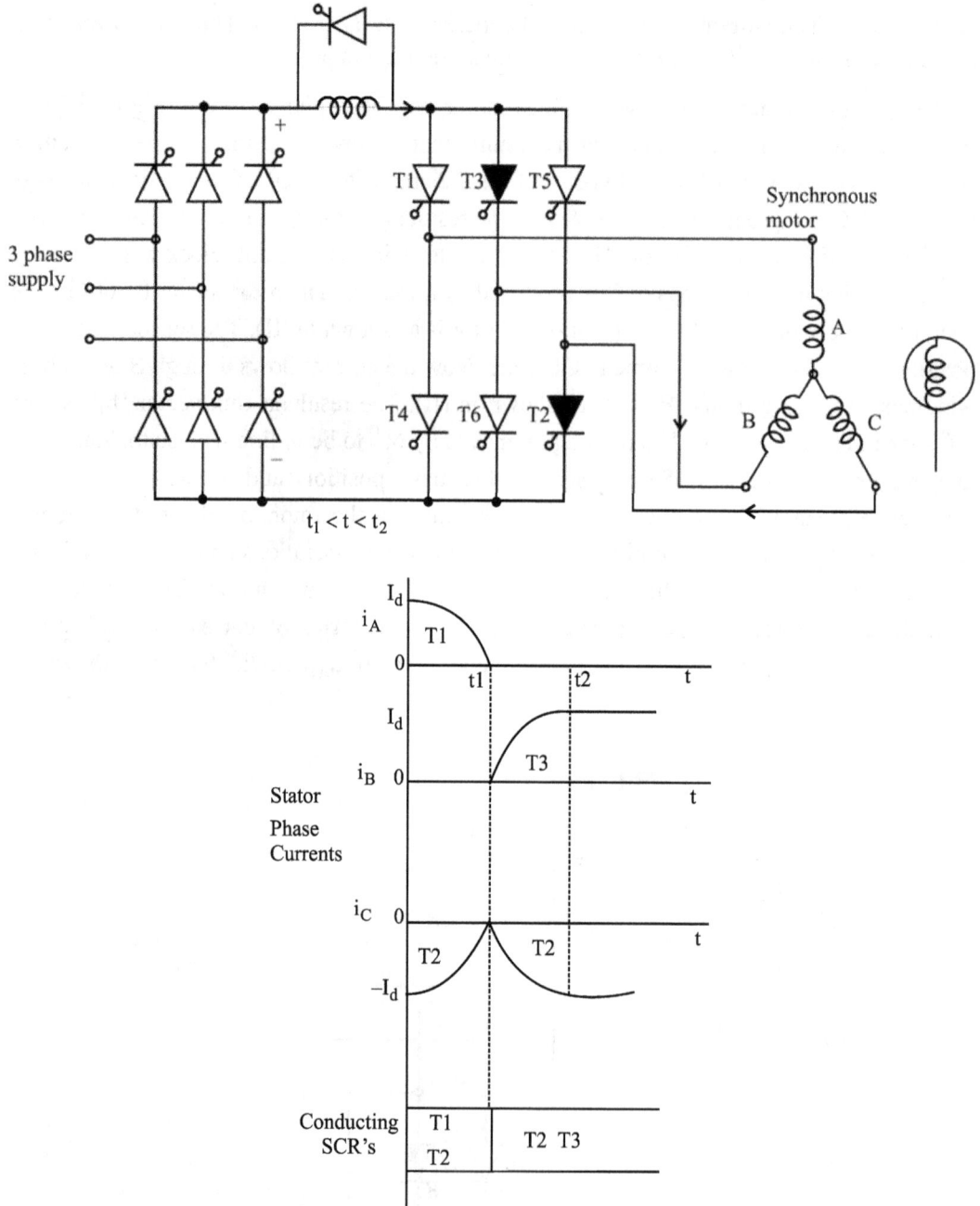

$t_1 < t < t_2$

Fig. 8.18(a) Wave forms of currents during commutation.

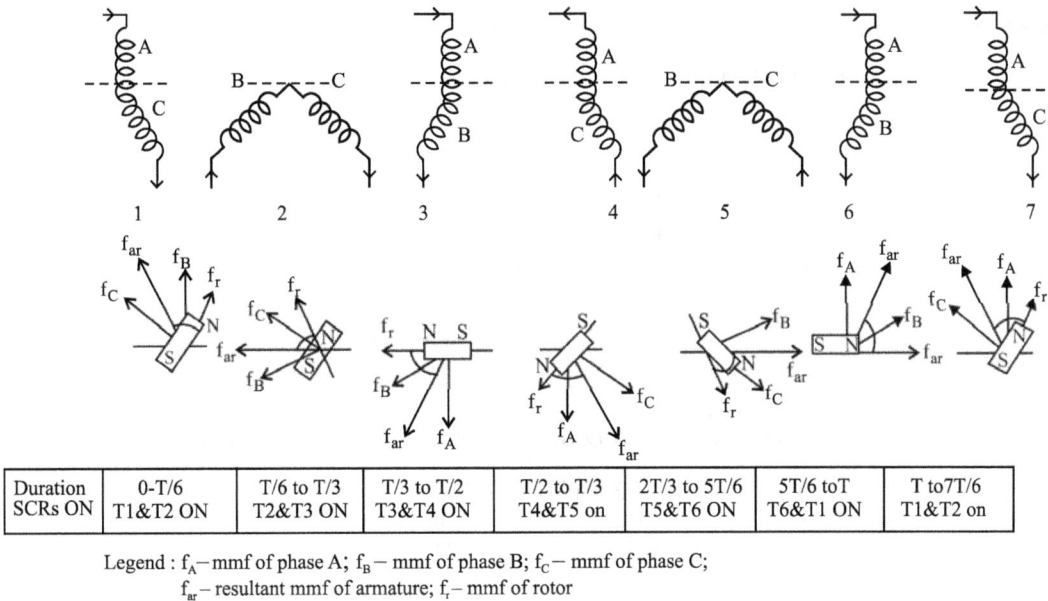

Duration SCRs ON	0-T/6 T1&T2 ON	T/6 to T/3 T2&T3 ON	T/3 to T/2 T3&T4 ON	T/2 to T/3 T4&T5 on	2T/3 to 5T/6 T5&T6 ON	5T/6 toT T6&T1 ON	T to7T/6 T1&T2 on

Legend : f_A– mmf of phase A; f_B – mmf of phase B; f_C – mmf of phase C;
f_{ar} – resultant mmf of armature; f_r – mmf of rotor

Fig. 8.18(b) Flux vectors and rotor positions of synchronous motor.

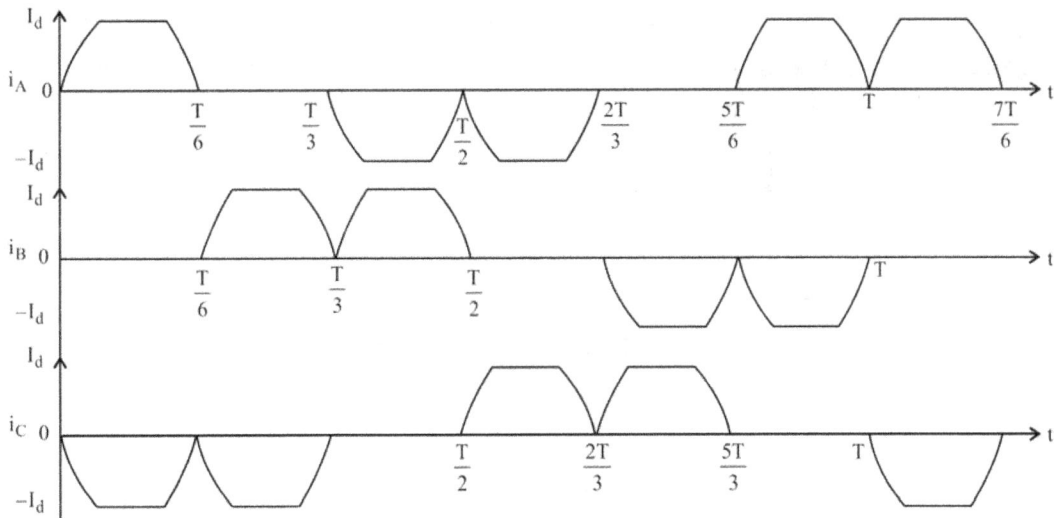

Fig. 8.18(c) Wave forms of currents in phase windings.

8.10.3   Applications of this Drive

Compressors, blowers, conveyors, steel rolling mills, main line traction, and air craft test facilities.

8.11 Synchronous Motor Fed from a Cycloconverter

A cycloconverter fed synchronous motor is shown in the Fig. 8.19.

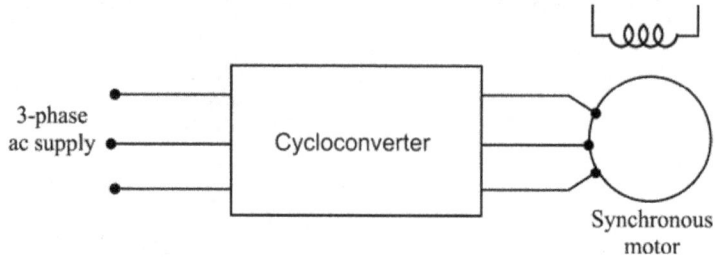

Fig, 8.19 Block diagram of cycloconverter connected to a synchronous motor.

The cycloconverter may be 18 thyristor converter or bridge type converter having 36 SCRs. In the block diagram of Fig. 8.20 cycloconverter with 18 thyristors is shown. Cycloconverter with 36 thyristors is shown in Fig. 8.21.

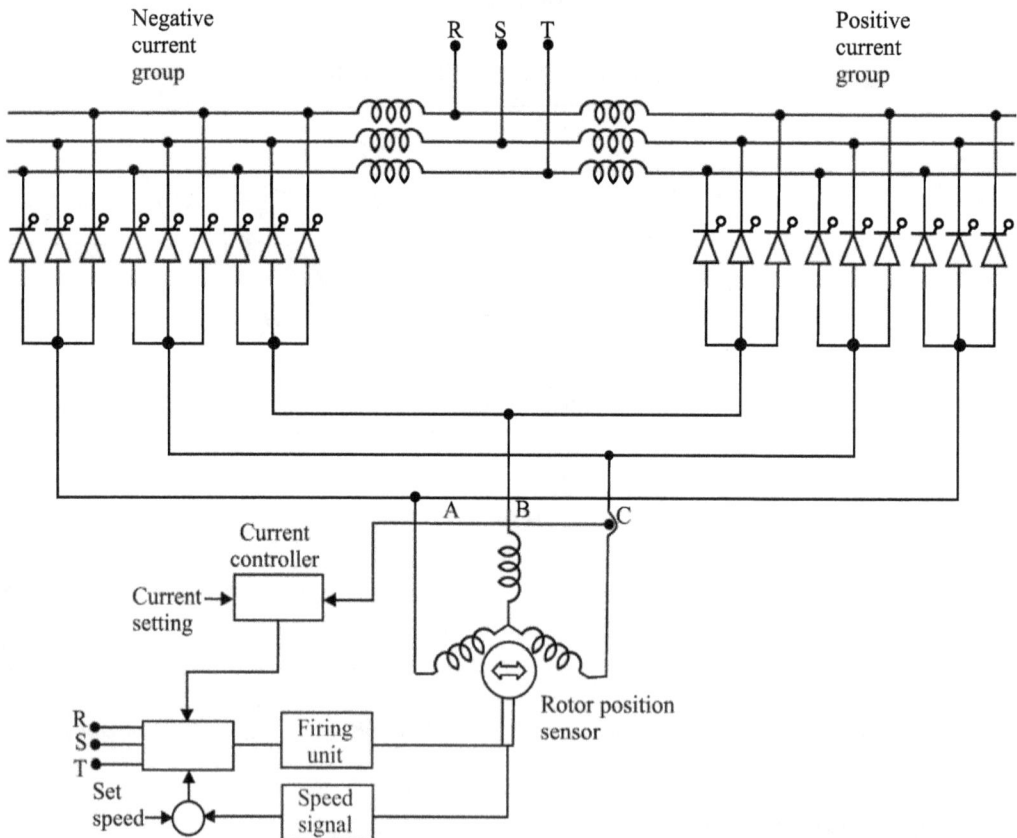

Fig. 8.20 Cycloconverter drive of synchronous motor operating in self control mode.

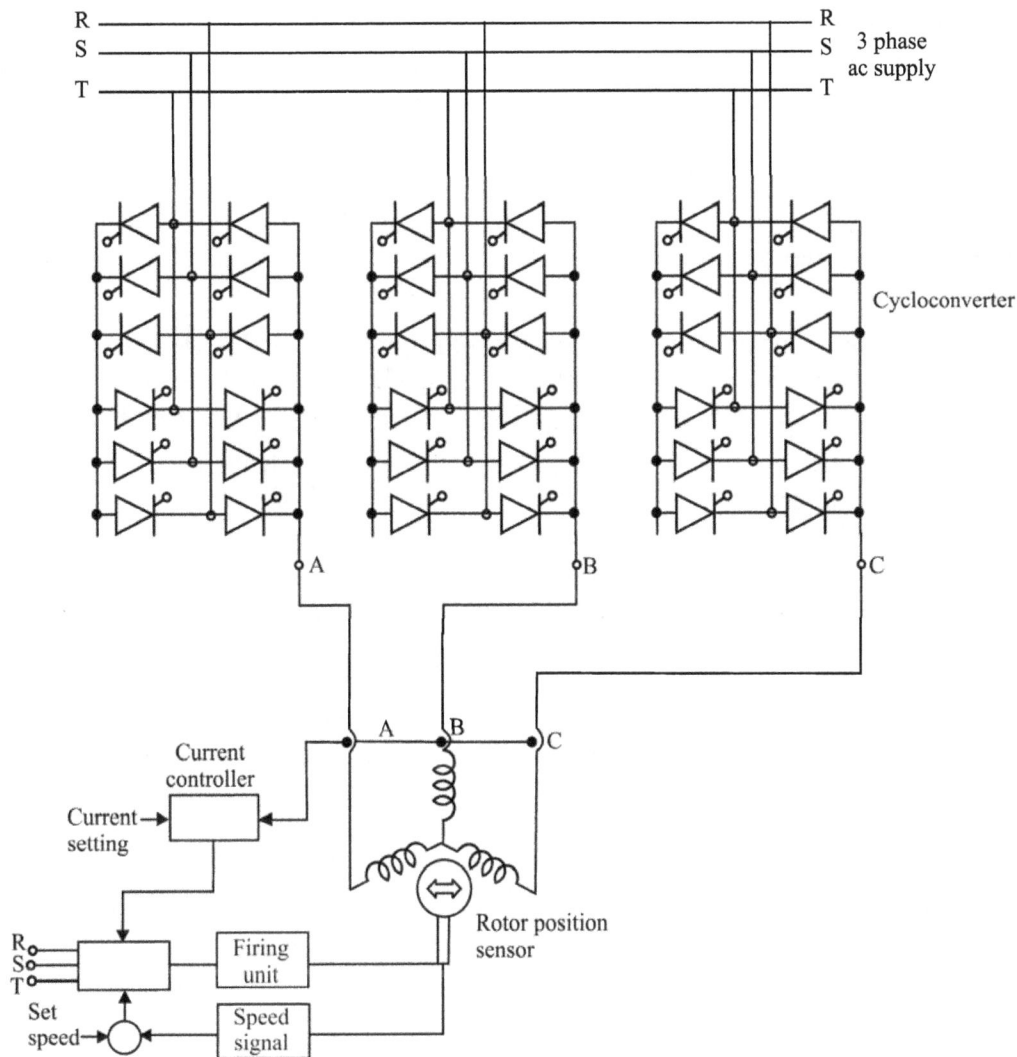

Fig. 8.21 Cyclo converter with 36 SCRs drive of synchronous motor with self control.

The motor operates in self-control mode. RST is 3-phase ac supply. A B C are the terminals of the synchronous motor. The position and speed of the rotor are to be sensed and fed to a microprocessor, which sends corresponding signals for controlling the frequency and voltage.

Variable voltage variable frequency supply is required for driving synchronous motor at different speeds. dc link converter is a two-stage conversion which provides variable voltage variable frequency supply. But this supply can be obtained from a cycloconverter, which is a single stage converter. Supply line voltages are made use of to commutate the thyristors of a cycloconverter. The output frequency can be varied from 0 to 1/3 of the input frequency. The range of speed is therefore limited, extending from 0 to 1/3 base speed.

Cycloconverters are capable of power transfer in both directions. Four-quadrant operation is simple. For the speeds below 1/3 base speed output voltages are nearly sinusoidal. The resulting currents are also nearly sinusoidal thus the harmonics are minimal. Losses, heating and torque pulsations are minimal. The line power factor is better because the machine power factor can be made unity.

Converter grade thyristors are sufficient but the cost of the converter is high because large number of thyristors are required. The efficiency is good and the drive has a good dynamic behaviour. The operation in CLM (Commutator less motor) mode is popular.

A cycloconverter drive is attractive for low speed operation and is frequently used in large, low speed reversing mills requiring rapid acceleration and deceleration. Typical applications are large gear less drives, e.g., drives for reversing mills, mine hoists, etc.

A cycloconverter can also be commutated using the voltage if the load is capable of providing the necessary reactive power for the inverter. An over excited synchronous motor can provide the necessary reactive power. Hence a cycloconverter feeding such a motor can be load commutated. With load commutation the range of speed control is from medium to base speed. At very low speeds load commutation is not possible. The speed range can be extended to zero if line commutation is used at low speeds. Four-quadrant operation is simple. The problem associated with harmonics is minimal due to high quality of the output. The line power factor depends on the angle of firing and is poor at low output voltages. The cost of the converter is high with complex control. Its efficiency is good and the drive has a fast response. It finds application in high power pump and blower type drives.

8.12 Braking of Synchronous Motors

Methods are:

1. Regenerative braking while operating on available frequency supply
2. Rheostatic braking
3. Plugging.

Regenerative braking : This is possible if the synchronous speed is less than the rotor speed. The input frequency is gradually reduced to achieve this at every instant. The kinetic energy of the rotating parts is returned to the mains. The braking takes place at constant torque with a current source inverter and cycloconverter. Regeneration is simple and straightforward. With voltage source inverter, an additional converter is required on the line side.

Rheostatic or Dynamic braking : The synchronous motor is disconnected from the mains and connected to a three phase resistive load, keeping the excitation constant. The machine now acts like a generator and its speed decreases. As the speed decreases the induced emf decreases and hence the current and hence the braking torque also decreases.

Plugging : This type of braking has serious disadvantages. Very heavy currents flow during plugging. Plugging is not used in ordinary synchronous motors.

8.13 Applications of Synchronous Motors

(a) Converter fed synchronous motors:
   (i) As start up converter for large synchronous machines
   (ii) Boiler feed pumps
   (iii) Turbo compressors
   (iv) Continuous rolling mills

(b) Cycloconverter fed synchronous motors:-
   Gear less drive for large tube & bowl mills
   Mine winders
   Reversing rolling mills
   Ship propeller
   Large reciprocating compressors

Worked Out Examples

Example **8.1**

A 500 KW, 3 phase, 3.3 KV, 50 Hz, 0.8(lagging) pf, 4 pole, star connected synchronous motor has following parameters:

$X_s = 15\Omega$, $R_s = 0$. Rated field current is 10 A. Calculate,

(i) Armature current and power factor at half the rated torque and rated field current

(ii) Field current to get unity power factor at the rated torque.

(iii) Torque for unity power factor operation at field current of 12.5 A

Solution:

Since $R_s = 0$, stator copper losses = 0 , power input = power output

At rated torque power input = 500 KW

At half rated torque power input = 250 KW

$$E_a = \frac{3300}{\sqrt{3}} = 1905.2 \text{ V}$$

E_f at rated power = ?

$$3 E_a I_a \cos \phi = P_m$$

$\therefore$
$$I_a = \frac{500 \times 10^3}{3 \times 1905.2 \times 0.8} = 109.3 \text{ A}, \phi = \cos^{-1} 0.8$$

$$\overline{E_f} = E_a - j\overline{I_a} \times X_s$$

$$= 1905.2 - j15(109.3 \times 0.8 - j109.3 \times 0.6) = 1905.2 - 983.7 - j1311.6$$

$$= 921.5 - j1311.6 = 1603 \angle -54.9^0$$

For field current of 10 A, induced emf $E_f = 1603$ V

At half the rated torque and rated field current

$$P_m = 250 \times 10^3 \text{ W}, E_f = 1603 \text{ V}$$

$$\frac{3E_a E_f \sin\delta}{X_s} = p_m \; ; \; \sin\delta = \frac{250 \times 10^3 \times 15}{3 \times 1905.2 \times 1603} = 0.409 \; ; \; \delta = 24.16^0$$

$$\overline{I_a} = \frac{\overline{E_a} - \overline{E_f}}{jX_s} = \frac{1905.2\angle 0 - 1603\angle -24.16}{15\angle 90}$$

$$= 127 \angle -90 - 106.86 \angle -114.16$$

$$= -j127 - 43.73 + j97.47$$

$$= 43.73 - j29.5$$

$$= 52.75 \angle -34^0$$

$$= 52.75 \text{ A}$$

Power factor $= \cos(-34^0) = \textbf{0.829 lagging}$.

(ii)   At unity power factor and rated torque $3E_a I_a \cos\phi = P_m$

$$I_a = \frac{500 \times 10^3}{3 \times 1905.2 \times 1} = 87.48 \text{ A}$$

$$\overline{E_f} = E_a - \overline{I_a} jX_s$$

$$= 1905.2 \angle 0 - 87.48 \angle 0 \times 15 \angle 90^0 = 1905.2 - j\,1312.2$$

$\therefore$ $E_f = 2313.4$ V

Field current $= \dfrac{2313.4}{1603} \times 10 = \textbf{14.43 A}$

(iii) $I_f = 12.5$ A, $\cos\phi = 1$

$$E_f = \frac{12.5}{10} \times 1603 = 2003.75 \text{ V}$$

$$E_f^2 = E_a^2 + (I_a X_s)^2$$

$$(I_a X_s)^2 = E_f^2 - E_a^2 = 2003.75^2 - 1905.2^2 = 385227$$

$$I_a X_s = 620.66$$

$$I_a = \frac{620.66}{15} = 41.37 \text{ A}$$

$$P_m = 3E_a I_a \cos\phi = 3 \times 1905.2 \times 41.37 \times 1$$

Torque $= \dfrac{P_m}{\omega_s} = \dfrac{3 \times 1905.2 \times 41.37}{50\pi} = 1505.6$ N-m

Example 8.2

A 6 MW, 3 phase, 11KV, star connected 6 pole, 50 Hz synchronous motor operating at 0.9 leading pf has $X_s = 9\ \Omega$ and $R_s = 0$. Rated field current is 50 A. The motor is controlled by variable frequency drive at constant V/f ratio up to the base speed and at constant voltage above the base speed. Determine,

(a) armature current and pf for regenerative braking power output of 4.2 MW at 750 rpm and rated field current

(b) Torque and field current for regenerative braking operation at rated armature current, 1500 rpm and unity power factor

Solution :

(a) At 750 rpm.

$$E_a = \frac{750}{1000} \times 6350.8 = 4763.1\ V$$

$$X_s = \frac{9 \times 750}{1000} = 6.75\ \Omega$$

Power output = 4.2 MW

Rated field current = 50 A

E_f at rated field current & 750 rpm

$$= \frac{8227 \times 750}{1000} = 6170.25\ V$$

Power output P_m at 750 rpm $= \dfrac{3 E_a E_f \mathrm{Sin}\delta}{X_s}$

$$= \frac{3 \times 4763.1 \times 6170.25 \times \mathrm{Sin}\delta}{6.75} = -4.2 \times 10^6$$

$\mathrm{Sin}\ \delta = -0.32, \qquad \delta = 18.757$

$$\overline{I_s} = \frac{\overline{E_f} - \overline{E_a}}{\overline{X_s}}$$

$$= \frac{6170\angle 18.25 - 4763\angle 0}{6.75\angle 90^0} = 293.98 - j159 = 334.66\ \angle -28.55$$

Thus $I_s = 334.66\ A$

pf $= \cos(-28.55^0) = $ **0.878 Lagging**.

(b) At 1500 rpm

$$X_s = 13.5\ \Omega, \qquad E_a = 6350.8\ V$$

Rated armature $I_a = 349.9$ A,   pf $= 1$

$$\overline{E_f} = \overline{E_a} + \overline{I_a X_s} = 6350.85\angle 0 + 349.9\angle 0 \text{X} 13.5\angle 90$$

$$= 6350.85 + j4723 = 7915 \angle 36.64$$

Field current $= \dfrac{7915}{12340.8} \times 50 = 32.07$ A

$$P_m = \dfrac{3E_a E_f \sin\delta}{X_s} = \dfrac{3 \times 6350.8 \times 7915 \sin 36.64}{13.5} = 6666300$$

$$\omega_s = \dfrac{1500 \times 2\pi}{60} = 50\pi \text{ rad/sec}$$

$$\text{Torque} = \dfrac{6666300}{50\pi} = 42438.9 \text{ N-m}$$

Example 8.3

A 6 MW, 3 phase, 11KV, star connected 6 pole, 50 Hz synchronous motor operating at 0.9 leading pf has $X_s = 9\ \Omega$ and $R_s = 0$. Rated field current is 50 A. The motor is controlled by variable frequency drive at constant V/f ratio up to the base speed and at constant voltage above base speed. Determine (i) the torque and field current for the rated armature current, 750 rpm and 0.8 leading pf. (ii) Armature current and pf for half the rated motor torque, 1500 rpm and rated field current.

Solution :

(i)   Rated armature current $= I_a$

Rated power input$= 6 \times 10^6$ W

$$\sqrt{3} \times V_L I_L \cos\phi = p$$

$$I_L = I_a = \dfrac{6 \times 10^6}{\sqrt{3} \times 11 \times 10^3 \times 0.9} = 349.9 \text{ A}$$

$$\overline{E_f} = \overline{E_a} - \overline{I_a X_s}$$

$$= \dfrac{11000}{\sqrt{3}} - 349.9\angle \cos^{-1} 0.9 \times 9 \angle 90^0 = 6350.8 - 3149.1\ \angle 115.84$$

$$= 6350.8 - (1372.66 + j2834.19) = 7723.46 - j2834.19 = 8227$$

Induced emf at 50 Hz and rated current $= 8227$ V,

corresponding field current $= 50$ A

Now $I_a = 349.9$A, speed is 750 rpm, and pf $= 0.8$ leading

For operation at 750 rpm

$$E_a = \frac{6350.8 \times 750}{1000} = 4763.1$$

$$X_s = \frac{9 \times 750}{1000} = 6.75$$

$\therefore$

$$E_f = 4763 - 349.9 \angle \cos^{-1} 0.8 \times 6.75 \angle 90^0$$

$$= 4763 - (-1417 + j1889)$$

$$= 6180 - j1889.2 = 6462.3 \angle -17^0$$

At rated field current of 50 A and at speed of 750 rpm

The induced emf $= \dfrac{8227 \times 750}{1000} = 6170.25$ V

$\therefore$ At 750 rpm, the field current for generating 6463.3 volts

$$= \frac{6463.5}{6170.25} \times 50 = \mathbf{52.37\ A}$$

Power input $= 3\ E_a\ I_a \cos \phi$

$$= 3 \times 4763 \times 349.9 \times 0.8 = 3999776\ W$$

Motor speed $\omega_m = \dfrac{2\pi \times 750}{60} = 78.53$ rad/sec

Torque $= \dfrac{P_m}{\omega_m} = \dfrac{3999776}{78.53} = \mathbf{50926.7\ N\text{-}m}$

(ii)   Speed = 1500 rpm,   rated field current = 50A

armature current I_a = ?   pf   = ?

Rated power input $= 6 \times 10^6$ W

Rated speed $\omega_m = \dfrac{2\pi 1000}{600}$

$\therefore$ Rated torque $T_R = \dfrac{P_m}{\omega_m} = \dfrac{6 \times 10^6 \times 60}{2\pi \times 1000} = 57.295 \times 10^3$

Half the rated torque $= \dfrac{57.295 \times 10^3}{2} = 28.647 \times 10^3$ N-m

At 1500 rpm $E_a = \dfrac{11000}{\sqrt{3}} = 6350.8$ V

At 1500 rpm frequency $= \dfrac{1500}{1000} \times 50 = 75$ Hz

$$E_f \text{ at 1500 rpm \& field current of 50 A} = \frac{8227 \times 1500}{1000} = 12340.5 \text{ V}$$

$$X_s \text{ at 1500 rpm} = \frac{9 \times 75}{50} = 13.5 \ \Omega$$

Mechanical power developed at half the rated torque & 1500 rpm

$$= 28.647 \times 10^3 \times \frac{2\pi \times 1500}{60} = 4499 \times 10^3 \text{ W}$$

$$P_m = \frac{3 E_a E_f \sin \delta}{X_s}$$

$$\therefore \sin \delta = \frac{4499 \times 10^3 \times 13.5}{3 \times 6350.8 \times 12340.5} = 0.2583$$

$$\therefore \delta = 14.97^0$$

$$\overline{I_s} = \frac{\overline{E_a} - \overline{E_f}}{jX_s} = \frac{6350.8 - 12340.8 \angle - 14.97}{13.5 \angle 90^0} = \textbf{475.5} \ \angle 60.2$$

$$\text{pf} = \cos 60.2^0 = \textbf{0.5 leading}$$

Example 8.4

A 415 V, 50 Hz, 4 pole, star connected synchronous motor has X_s = 1.5 Ω. Load torque, proportional to speed is 300 N-m at synchronous speed. The speed of the motor is lowered by keeping V/f constant and maintaining 0.8 pf leading by field control. For the motor operation at 840 rpm, calculate

(i) supply voltage (ii) the armature current (iii) the excitation voltage (iv) load angle (v) the pull out torque.

Neglect rotational losses.

Solution :

Synchronous speed at 50 Hz = 1500 rpm,   operating speed = 840 rpm

(i)   The supply voltage for this speed $= \dfrac{415 \times 840}{1500} = \textbf{232.4 V}$

(ii)   Power input at 840 rpm is

$$\sqrt{3} V_L I_L \cos\phi = T_L \omega_m$$

$$= \left(\frac{300 \times 840}{1500} \right) \times \frac{2\pi \times 840}{60} = 14778 \text{ W}$$

$$\therefore \qquad I_L = \frac{14778}{\sqrt{3} \times 232.4 \times 0.8} = 45.89 \text{ A}$$

(iii)   Excitation voltage $E_f = E_a - I_a (\cos \phi - j\sin \phi)(R_a + jX_s)$

$$= \frac{232.4}{\sqrt{3}} - 45.89(0.8 + j0.6)(0 + j\frac{1.5 \times 840}{1500})$$

$$= 134.17 - 45.89(j0.672 - 0.504)$$

$$= 134.17 + 23.13 - j30.83 = 157.3 - j\,30.83$$

$$= 160.29 \angle -11.09$$

Excitation line voltage $= \sqrt{3} \times 160.29 = 277.63$ V

(iv)   Load angle = **11.09⁰**

(v)   Pull out torque $= \dfrac{3E_a \times E_f}{\omega_s X_s} = \dfrac{3 \times 134.17 \times 160.29}{\dfrac{840 \times 2\pi}{60} \times \dfrac{1.5 \times 840}{1500}} = $ **873.16 N-m**

Example 8.5

A three phase 400 V, 50 Hz, 6 pole star connected round rotor synchronous motor has $Z_s = 0 + j\,2\,\Omega$. Load torque, proportional to speed squared is 340 N-m at rated synchronous speed. The speed of the motor is lowered by keeping V/f constant and maintaining unity power factor by field control of the motor. For the motor operation at 600 rpm, calculate (a) supply voltage (b) the armature current (c) the excitation voltage (d) the load angle and (e) the pull out torque.

Neglect rotational losses.

Solution:

(a)   For 600 rpm, the supply frequency $f = \dfrac{PN_s}{120} = \dfrac{6 \times 600}{120} = 30$ Hz

Line voltage $V_a = \dfrac{400 \times 30}{50} = $ **240 V**

(b)   ***The armature current :*** Power developed $= T_L\,\omega_m$

$$= 340\,(\frac{600}{1000})^2 \times \frac{600 \times 2\pi}{60} = 7690 \text{ W}$$

$3\,V_a I_a \cos \phi = 7690$ W

$$I_a = \frac{7690\sqrt{3}}{3 \times 240 \times 1} = 18.5 \text{ A}$$

(c)   Excitation voltage $E_f = \overline{E_a} - \overline{I_a}\,Z_s$

$$= \frac{240}{\sqrt{3}} - 18.5(1 + j0)(j2)\frac{30}{50}$$

$$= 138.56 - j\,22.2$$

$$= 140.32 \angle -9.10^0$$

Line value of excitation voltage = $\sqrt{3} \times 140.32$ = **243.04 V**

(d) Load angle = **- 9.10⁰**

(e) Pull out torque = $\dfrac{3 \times E_a \times E_f}{\omega_s X_s}$

$$= 3 \times \frac{240}{\sqrt{3}} \times \frac{140.32}{20\pi \times 1.2} = \textbf{773.62 N-m}$$

Example 8.6

A 600 KW, 3 phase, 3.3 KV, 50 Hz, 0.8 lagging pf, 4 pole, star connected synchronous motor has $R_s = 0$ and $X_s = 15\ \Omega$. The rated field current is 12 A. The speed is controlled by variable frequency control. Determine, (i) field current and power developed for a speed of 1800 rpm and unity power factor operations at rated armature current and (ii) Field current and power generated for regenerative braking operation at 2100 rpm and rated armature current unity power factor.

Solution:

(i) Speed =1800

$$E_a = \frac{3.3 \times 10^3}{\sqrt{3}} = 1905.3\ V$$

$$\omega_m = \frac{2\pi \times 1800}{60} = 60\pi\ \text{rad/sec}$$

$$X_s \text{ at 1800 rpm} = \frac{15 \times 1800}{1500} = 18\ \Omega$$

Rated armature current is I_a

$\sqrt{3} V_L I_a \cos\phi = 600\ kW$

$$I_a = \frac{600 \times 10^3}{\sqrt{3} \times 3300 \times 0.8} = 131.21\ A$$

$E_f \text{ rated} = \overline{E_a} - \overline{I_a Z_s}$

$$= 1905\ \angle 0 - 131.21\angle - \cos^{-1} 0.8 \times 15\angle 90^0$$

$$= 1905 - 1968\ \angle 53.13$$

$$= 1905 - 1180 - j1574 = 725 - j1574$$

$$= 1732.9\ \angle - 65.26^{\,0}$$

At speed of 1800 and UPF at rated armature current

$$E_f = \overline{E_a} - \overline{I_a X_s} = 1905 - 131.21\ \angle 0 \times \frac{15 \times 1800}{1500} \angle 90$$

$$= 1905 - j2361 = 3034V \angle - 51.10$$

$$3034 = K1800 \, I_f$$

$$1732.9 = K1500 \times 12$$

$\therefore$ field current $I_f = \dfrac{3034}{1800} \times \dfrac{1500 \times 12}{1732.9} = 17.5 \text{ A}$

Power developed $= \dfrac{3E_a E_f \sin \delta}{X_s}$

$$= \dfrac{3 \times 1905 \times 3034 \times \sin 51.1^0}{18} = \textbf{749.67 KW}$$

(ii)   For regeneration braking at speed = 2100

$$\overline{E_f} = \overline{E_a} + \overline{I_a X_s}$$

$$= 1905 + 131.21 \, \angle 0 \times \dfrac{15 \times 2100}{1500} \angle 90$$

$$= 1905 + j2755$$

$$= 3349.8 \, \angle 55.34^0$$

Field current $I_f = \dfrac{3349.8}{2100} \times \dfrac{1500 \times 12}{1732.9} = 16.56 \text{ A}$

Power developed $= \dfrac{3 \times 1905 \times 3349.8 \times \sin 55.34^0}{21} = \textbf{749.8 KW}$

Example 8.7

A 3 phase, 230 V, 60 Hz, 40 KW, 8 pole star connected salient pole synchronous motor has $X_d = 2.5 \, \Omega$ and $X_q = 0.4 \, \Omega$. The armature resistance is negligible. If the motor operates with an input power of 25 KW at a leading pf of 0.86. Determine,

   (i) The torque angle (ii) The excitation voltage V_f and (iii) The torque T_d

Solution:

Power input = 25KW

$$\sqrt{3} \; V_L I_L \cos \phi = P_m$$

$$\sqrt{3} \times 230 \times I_L \times 0.86 = 25 \times 10^3$$

$$I_L = \dfrac{25 \times 10^3}{\sqrt{3} \times 230 \times 0.86} = 72.97 \text{ A}$$

(i)   Torque angle $\delta = \tan^{-1} \dfrac{I_a X_q \cos\phi}{E_a + I_a X_q \sin \phi}$

$$= \tan^{-1} \frac{72.97 \times 0.4 \times 0.86}{\dfrac{230}{\sqrt{3}} + 72.97 \times 0.4 \times \sin 30.68^0}$$

$$= \tan^{-1} \frac{25.10}{147.59} = 9.65^0$$

(ii)   The phasor diagram of voltages and currents are shown below.

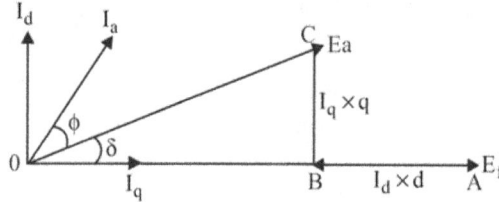

$$\overline{E_f} = \overline{E_a} - j\overline{I_d}X_d - j\overline{I_q}X_q$$

$$OA = OB + BA$$

$$= E_a \cos \delta + I_a \sin(\phi + \delta)X_d \quad \because I_d = I_a \sin(\phi + \delta)$$

$$= 132.79 \cos 9.65 + 72.96 \times \sin(9.65 + 30.68) \times 2.5$$

$$= 130.91 + 118 = 248.9V$$

Line voltage $V_f = \sqrt{3} \times 248.9 = \mathbf{431.1\ V}$

(iii) $N_s = \dfrac{120 \times 60}{8} = 900 \ \text{rpm}$

$$\omega_s = \frac{900}{60} \times 2\pi = 30\pi \ \text{rad/sec}$$

Power input $= T \ \omega_s$

$$T = \frac{25000}{30\pi} = 265.26 \ \text{N-m}$$

Example 8.8

A synchronous motor is controlled by a load-commutated inverter, which in turn is fed from a line commutated converter. Source voltage is 6.6 KV, 50 Hz. Load commutated inverter operates at a constant firing angle α_1 of 140^0 and when rectifying $\alpha_e = 0$. dc link inductor resistance $R_d = 0.1\ \Omega$. Drive operates in self-control mode with a constant V/f ratio. Motor has the details: 8 MW, 3 phase, 6600 V, 6 pole, 50 Hz, unity power factor star connected, $X_s = 2.8\ \Omega$, $R_s = 0$. Determine the source side converter firing angles for the following:

(i) Motor operation at rated current and 500 rpm.

(ii) What will be the power developed by the motor?

(iii) Regenerative braking operation at 500 rpm and rated motor current also

(iv) calculate the power supplied to source.

Solution :

At 50 Hz motor speed = 1000 rpm.

$$V = \frac{6600}{\sqrt{3}} = 3810.5 \text{ V} \quad X_s = 2.8 \ \Omega$$

(i)   Rated power = 8×10^6 w

$$\text{Rated current} = \frac{8 \times 10^6}{3 \times 3810.5 \times 1} = 699.82 \ \text{A}$$

The relationship between the dc link current I_d and ac line current I_s is given by the equation

$$I_d = \frac{\pi}{\sqrt{6}} I_s = \frac{\pi}{\sqrt{6}} \times 699.82 = 897.55 \ \text{A}$$

Load commutated inverter operates at $\alpha_l = 140^0$

The phase angle between $(-I_s)$ and V will be 140^0 and phase angle ϕ between $+ I_s$ and V will be $180 - 140 = 40^0$

For operation at 500 rpm

$$\text{Frequency} = \frac{50 \times 500}{1000} = 25 \ \text{Hz}$$

$$\text{Voltage V} = \frac{25}{50} \times 3810.5 = 1905.25 \ \text{V}$$

Power input $P_m = 3VI_a\cos \phi = 3 \times 1905.25 \times 699.82 \times \cos 40^0$

$$= 3.064 \ \text{MW}$$

Now for three phase load commutated inverter

$$V_{dl} = \frac{3\sqrt{6}}{\pi} V \cos \alpha_e = \frac{3\sqrt{6}}{\pi} \times 1905.25 \cos 140^0 = -3413.9 \text{V}$$

$$V_{ds} = -V_{dl} + I_d R_d = 3413.9 + 897.55 \times 0.1 = 3503.67 \ \text{V}$$

$$V_{ds} = \frac{3\sqrt{6}}{\pi} V_s \cos \alpha_s$$

$$\frac{3\sqrt{6}}{\pi} \times \frac{6600}{\sqrt{3}} \cos \alpha_s = 3503.67$$

$$\cos \alpha_s = 0.393, \qquad \alpha_s = 66.85^0$$

(ii)   Rated current 699.82A

$$I_d = 897.55$$

At 500 rpm,   V=1905.25

During regeneration the load side converter operates at rectifier and $\alpha_e = 0$

The machine operates at $\phi = 0$ or unity power factor

Power developed by motor is P_m

$$P_m = 3V\ I_s\cos\phi = 3 \times 1905.5 \times 699.82 \times 1$$

$$= 3.999996 \text{ MW}$$

(iii)   Assuming negligible loss in both converters power supplied to the source

$$= P_m - I_d{}^2\ R_d$$

$$= 3.999996 - 897.55^2 \times 0.1 = 3.919 \text{ MW}$$

(iv)   For load commutated converter

$$V_{dl} = 1.35\ V_L \cos\alpha = 1.35 \times \sqrt{3} \times 1905.25 \times \cos 0^0 = 4454.9$$

$$V_{ds} = I_d\ R_d - V_{dL} = 897.55 \times 0.1 - 4454.9 = -4363$$

$$1.35\ V_L \cos\alpha_s = V_{ds}$$

where α_s is the source side converter firing angle

$$\cos\alpha_s = \frac{-4363}{1.35 \times 6600} = -0.4896$$

$$\alpha_s = 119.31^0$$

Example 8.9

A three phase 230 V, 60 Hz, 4 pole, star connected reluctance motor has $X_d = 22.5\,\Omega$ and $X_q = 3.5\,\Omega$. The armature resistance is negligible. The load torque is $T_L = 12.5$ N-m. The voltage to frequency ratio is maintained constant at the rated value. If the supply frequency is 60 Hz determine, (a) the torque angle δ (b) the line current I_a and (c) the input power factor.

Solution:

$$E_a = \frac{230}{\sqrt{3}} = 132.79;\ P = 4;\ f = 60 \text{ Hz}$$

Power input = power developed = $T_L\,\omega_s$

$$= 12.5 \times \frac{4\pi \times 60}{4} = 2356.2 \text{ W}$$

(a)   But $p_i = 3 \dfrac{E_a^2}{2} \sin 2\delta (\dfrac{1}{X_q} - \dfrac{1}{X_d})$

where δ is the torque angle.

$$= \dfrac{3}{2} \times 132.79^2 \sin 2\delta (\dfrac{1}{3.5} - \dfrac{1}{22.5}) = 6381 \sin 2\, \delta$$

$\therefore$ $\sin 2\delta = \dfrac{2356.2}{6381} = 0.36922$

$2\, \delta = \sin^{-1} 0.36922 = 21.66$

$\therefore$ $\delta = 10.83^0$

(b) $I_d X_d = E_a \cos \delta$

$I_d = \dfrac{E_a \cos \delta}{X_d} = \dfrac{132.79 \times \cos 10.83}{22.5} = 5.796 \ A$

$I_q X_q = E_a \sin \delta$

$I_q = \dfrac{E_a \sin \delta}{X_q} = \dfrac{132.79 \times \sin 10.83}{3.5} = 7.128 \ A$

$I_a = \sqrt{5.796^2 + 7.128^2} = 9.18 \ A$

(c) $3 E_a I_a \cos \varphi = P_i = 2356.2 \ W$

$\cos \varphi = \dfrac{2356.2}{3 \times 132.79 \times 9.18} = 0.644 \ Lag$

Example 8.10

A 3 Phase, 400 V, 50 Hz, 4 pole star connected reluctance motor with negligible armature resistance has $X_d = 8 \ \Omega$ and $X_q = 2 \ \Omega$. For a load torque of 80 N-m calculate,

(a) The load angle (b) The line current and (c) The input power factor.

Neglect rotational losses.

Solution :

(a)   Synchronous speed $= \dfrac{150 \times 2\pi}{60} = 50\pi \ rad/sec$

Power input $= T \omega_s = 80 \times 50 \ \pi = 12566.4 \ W$

$$P_i = \frac{3}{2}Ea^2\left[\frac{1}{x_q} - \frac{1}{x_d}\right]\sin 2\delta$$

$$= \frac{3}{2}\left(\frac{400}{\sqrt{3}}\right)^2\left[\frac{1}{2} - \frac{1}{8}\right]\sin 2\delta = \frac{400 \times 400}{2} \times \frac{3}{8}\sin 2\delta$$

$$= 30000\sin 2\delta$$

$$\sin 2\delta = \frac{50 \times 50\pi}{30000} = 0.418$$

$$\delta = \frac{1}{2}\sin^{-1}\frac{50 \times 50\pi}{30000} = \mathbf{12.38^0}$$

(b)   Per phase voltage $= \dfrac{400}{\sqrt{3}} = 230.95$ V

$$I_d = \frac{E_a\cos\delta}{X_d} = \frac{230.95 \times \cos 12.38^0}{8} = 28.197 \text{ A}$$

$$I_q = \frac{E_a\sin\delta}{X_q} = \frac{230.95 \times \sin 12.38^0}{2} = 24.761 \text{ A}$$

Armature current $I_a = \sqrt{I_d^2 + I_q^2} = \sqrt{28.197^2 + 24.76^2} = \mathbf{37.53\ A}$

(c) $\sqrt{3}\ V_L\ I_a \cos\phi = T_L\omega_s = 80 \times 50 \times \pi$

$$\cos\phi = \frac{80 \times 50 \times \pi}{\sqrt{3} \times 400 \times 37.53} = \mathbf{0.4833\ Lag}$$

Example 8.11

A 400 kW, three phase, 3.3 KV, 50 Hz, unity power factor, 4 pole, star connected synchronous motor has the following parameters: armature resistance =0, synchronous reactance = 12 ohms, rated field current =10 A. The machine is controlled by variable frequency at constant V/f ratio. Calculate torque and field current for rated armature current, 900 rpm and 0.8 leading power factor.

Solution :

$$\text{Rated armature current} = \frac{400 \times 10^3}{\sqrt{3} \times 3.3 \times 10^3 \times 1} = \mathbf{69.98\ A}$$

$$\text{Rated speed} = \frac{120f}{P} = \frac{120 \times 50}{4} = 1500 \text{ rpm}$$

Rated field current = 10 A

Torque & field current for 900 rpm, rated armature current and 0.8 leading pf is to be found.

$$E_f \text{ rated per phase} = \overline{V} - \overline{I_a X_s}$$

$$= 1905 \ \angle 0 - 69.98\angle 0 \times 12\angle 90^0$$

$$= 1905 - 839.76 \ \angle 90^0$$

$$= 1905 - j\,839.76 = 2081 \ \angle -23.78^0$$

$$\omega_s = \frac{900}{60} \times 2\pi \ = 94.24 \text{ rad/sec}$$

$$E_f = \overline{V} - \overline{I_a X_s} \ \text{V at } 900 = \frac{1905 \times 900}{1500} = 1143 \text{ V};$$

$$\cos \phi = 0.8 \text{ leading}$$

$$X_s = \frac{12 \times 900}{1500} = 7.2\,\Omega$$

$$E_f = 1143 \ \angle 0 - 69.98 \ \angle 36.86 \times 7.2\angle 90^0$$

$$= 1143 - 503.85 \ \angle 126.86$$

$$= 1143 - (-302.24 + j\,403.13) = 1445 - j\,403$$

$$= 1500 \ \angle -15.58^0$$

$$E_f = K\,N\,I_f.$$

We find the value of K at rated values of E_f, N, I_f.

$$2081 = K \times 1500 \times 10$$

$\therefore$
$$K = \frac{2081}{15000} = 0.138$$

I_f at 900 rpm for $E_f = 1500$ is found using the following equation.

$$I_f = \frac{E_f}{KN} = \frac{1500}{0.138 \times 900} = \textbf{12.07 A}$$

At 900 rpm applied voltage $E_a = \dfrac{1905 \times 900}{1500} = 1143V$

$$\text{current} = 69.98 \text{ A}, \qquad \text{Pf} = 0.8 \text{ leading}$$

$\therefore$
$$\text{Power input} = 3 \ E_a \ I_a \cos \phi$$

$$= 3 \times 1143 \times 69.98 \times 0.8 = 191969 \text{ W}$$

$$\text{Speed} = 900 \text{ rpm} = \frac{900}{60} \times 2\pi = 94.27 \text{ rad/sec}$$

$$\text{Torque} = \frac{\text{Power input}}{\text{Speed in rad/sec}}$$

$$= \frac{191969}{94.27} = \textbf{2036.85 N-m}$$

Example 8.12

A Cyclo converter operating at 400 V, 50 Hz supply draws a power of 100 KVA and is driving a synchronous motor at 200 A, 200 V. Calculate firing angle and input pf assuming efficiency of the Cyclo converter as 96% and motor pf = 0.8 lag.

Solution :

Synchronous motor :

$$V = 200V; \ A = 200A; \ Pf = 0.8 \text{ lag}$$

$$\therefore \text{ Power input to synchronous motor} = \sqrt{3} \times V \times A \times \cos \phi$$

$$= \sqrt{3} \times 200 \times 200 \times 0.8$$

$$= 55425.6 \text{ W}$$

Efficiency of Cycloconverter = 96%

$$\therefore \text{ Input to cycloconverter} = \frac{554256}{0.96} = 57735 \text{ W}$$

Cycloconverter input = 100×10^3 VA

$$\therefore \text{ Input P.f} = \frac{57735}{100 \times 10^3} = 0.577 \text{ lag}$$

rms value of output voltage of cycloconverter = 200 V

$$\text{Peak value} = \sqrt{2} \times 200 = 282.8 \text{ V}$$

$$\text{Input line voltage} = 400 \text{ V}$$

$$1.35 \ V_L \cos \alpha = \text{peak value of output voltage}$$

$$= 282.81$$

$$\cos \alpha = \frac{282.8}{1.35 \times 400} = 0.523 \ \alpha = \textbf{58.41}^0$$

Problems

P.1   Describe CSI fed and VSI fed synchronous motor drives in detail with block diagrams and compare them.

P.2   Explain the operation of 3 phase bridge inverter supplying a 3 phase synchronous motor. What are the modes of operation? How is the motor started?

P.3   Describe the converter used for low frequency high power synchronous motor drives with relevant wave forms.

P.4   Explain the operation of a synchronous motor fed from an adjustable frequency current source, with circuit diagram and characteristic curves.

P.5   Discuss various methods of speed control of a synchronous motor in detail.

P.6   Explain the power and control circuits for a three phase synchronous motor under (a) constant terminal voltage and frequency (b) constant line current and frequency. Draw the characteristics of the drive for the two cases and compare.

P.7   Describe the converter and control systems used for (a) constant air gap flux destiny and (b) constant V/f operation of a synchronous motor. Draw the characteristics of the drive for the two cases.

P.8   Describe self-controlled and load-commutated inverter controlled synchronous motor drives in detail and compares them.

P.9   Draw the block diagram of a closed loop synchronous motor drive fed from VSI and explain.

P.10   A 400 kW, three phase, 3.3 kV, 50 Hz, unity power factor, 4 pole, star connected synchronous motor has the following parameters:

armature resistance = 0, synchronous reactance = 12 ohms, rated field current = 10 A. The machine is controlled by variable frequency at constant V/f ratio. Calculate torque and field current for rated armature current, 900 rpm and 0.8 leading power factor. Draw motor characteristics and wave forms under the above method of control.

P.11   Describe the open-loop and closed loop methods of speed control of a synchronous motor using VSI.

P.12   A 6 MW, three phase, 11 kV, 50 Hz, unity power factor, 6 pole, star connected synchronous motor has the following parameters: armature resistance = 0, synchronous reactance = 9 ohms, rated field current =60 A. The machine is controlled by variable frequency at constant V/f ratio up to base speed and at constant V above base speed. Calculate the torque and field current for rated armature current, 750 rpm and 0.8 leading power factor. Draw motor characteristics and wave forms under the above method of control.

P.13   A 500 kW, 3 phase, 6.6 kV, 60 Hz, 6 pole, y-connected wound field synchronous motor has the following parameters: $X_m = 78$, $X_{sf} = 3$, rated pf = 1, n = 5, R_s = negligible. The motor speed is controlled by variable frequency control with a constant V/f ratio up to base speed and rated terminal voltage above base speed. Calculate and plot T, P_m, V, I_m and I_F versus speed for the motor operation at rated armature current and unity pf. What is the range of constant power operation? Neglect friction, windage and core loss. Draw motor characteristics and wave forms under the above method of control.

P.14   How is the output voltage of a VSI improved by PWM techniques? Explain how you will use this converter for speed control of a synchronous motor.

P.15   An 8 MW, three phase, 6.6 kV, 50 Hz, 6 pole, Y-connected wound field synchronous motor has the following parameters: $X_m = 8$ ohm, $X_{sf} = 0.5$ ohms, rated pf = 1, $R_s = 0.01$ ohm, rated field current=180 A. Field winding resistance = 1.2 ohm. Calculate the pf, armature current, efficiency at half the rated torque and at rated field current. Core, friction and windage loss assumed constant at 9 kW. Draw motor characteristics and waveforms under the above method of control.

Objective Type Questions

1. A 6600 V, 50 Hz, 6 pole, 100 kW synchronous motor is running at 500 rpm. Its supply voltage and frequency are

   (a)   4000 v and 25 Hz   (b)   3300 V and 25 Hz

   (c)   3300 V and 50 Hz   (d)   3300 V and 30 Hz   [   ]

2. A 440 V, 3ϕ, 50 Hz, 4-pole synchronous motor is running at 1800 rpm. Its supply voltage and frequency are

   (a)   440 V, 50 Hz   (b)   880 V, 60 Hz

   (c)   440 V, 60 Hz   (d)   528 V, 50 Hz   [   ]

3. When operated with variable frequency a synchronous motor has an advantage over an induction motor in

   (a)   that it is free from torque oscillations

   (b)   that it has very good efficiency

   (c)   that the line power factor can be improved by varying excitation

   (d)   that in certain cases the inverter can be of simpler configuration due to possible load commutation   [   ]

4. By self control of synchronous motor we mean that

   (a)   the elimination of torque ripple

   (b)   the speed of the motor is varied in steps

   (c)   the speed of the motor is function of input frequency

   (d)   the input frequency is controlled from the speed of the motor   [   ]

5. The disadvantage of load commutation is
   (a)   elimination of harmonic torque
   (b)   low efficiency due to losses
   (c)   the speeds from 0 to 10% of base speed are not possible
   (d)   the speed control range is limited to $0 - 10\%$ of base speed   [   ]

6. In salient pole synchronous motor the maximum torque occurs at load angle
   (a)   equal to 90^0 (b)   greater than 90^0
   (c)   less than 90^0 (d)   less than 40^0

7. What are the two types of synchronous motors used for driving high power loads?

8. Name two types of synchronous motors used for driving low power loads.

9. What is the main difference between the salient pole synchronous motor and reluctance motor?

10. Draw per phase equivalent circuit of round rotor synchronous motor.

11. A three phase 4 pole, 50 Hz star connected synchronous motor has $X_S = 6 \ \Omega$, $R_a = 0$, $E_a = 6000$ V, $E_f = 5000$ V all per phase values. What is the value of pullout torque?

12. A salient pole synchronous motor develops maximum torque when the torque angle is 90^0. (True/False)

13. Write the equation for the developed torque in a round rotor synchronous motor having negligible resistance.

14. Write the equation for torque angle δ of a salient pole synchronous motor.

15. It is said that the power input to salient pole synchronous motor has two components. What are they? What are their expressions?

16. A 3 ϕ, 400 V, 50 Hz, 4 pole star-connected reluctance motor with negligible armature resistance has $X_d = 10\Omega$ and $X_q = 3\Omega$. It is operating with torque angle of 10^0. Calculate armature current.

17. What are the different types of control of synchronous motors?

18. In separate control of synchronous motor how is the frequency of supply to motor is changed?

19. In self-control of synchronous motor how is the frequency controlled?

20. A separately controlled synchronous motor is fed from VSI that generates square wave. What type of converter is required on the line side?

21. A separately controlled synchronous motor is fed from a VSI with pulse width modulation. What type of converter is needed on the line side?

22. The rotor of synchronous motor should have damper winding if the motor operates under separate control method. Why?

23. When is a synchronous motor called CLM (Commutator less motor)?

24. When is a synchronous motor called brush less and commutator less motor (BLCLM)?

25. A variable voltage, variable frequency is obtained from square wave VSI for driving synchronous motor. The dc voltage to VSI is obtained from a 3 ϕ, controlled rectifier. What is the major disadvantage of this system if the motor is to be driven at very low speed?

26. If PWM VSI is used for driving synchronous motor the pf on ac mains supply is always unity irrespective of the speed of motor. (True/False)

27. A 600 kW, 3 phase, 3.3 kV, 50 Hz, 0.8 lagging pf, 4 pole, star-connected synchronous motor has Rs. = 0, X_S =15 Ω. The rated field current is 12 A. The speed is controlled by variable frequency control. Determine (i) Field current and power developed for a speed of 1800 rpm and (ii) U.P.F operation at rated armature current.

28. The synchronous motor can work in regenerative braking only at synchronous speed. (True/False)

29. What type of braking is applied for bringing the synchronous motor to rest?

30. Why plugging is not employed in practice for braking synchronous motor?

31. Can a current source inverter be used for driving the synchronous motor in self control?

32. How is the dc link current held constant in CSI drive?

33. How is the dc link current controlled?

34. What type of converters are suitable for driving the synchronous motor in self control mode?

35. What are the applications of cycloconverter driven synchronous motors?

36. Give some advantages of brush less dc motors compared to conventional dc motors?

37. Give some applications of brush less dc motors?

38. From zero to 10% of rated speed of self controlled synchronous motor drive, load commutation is not used. Why?

39. In the diagram given below for self controlled synchronous motor a SCR is shown across the link inductor. What is its purpose?

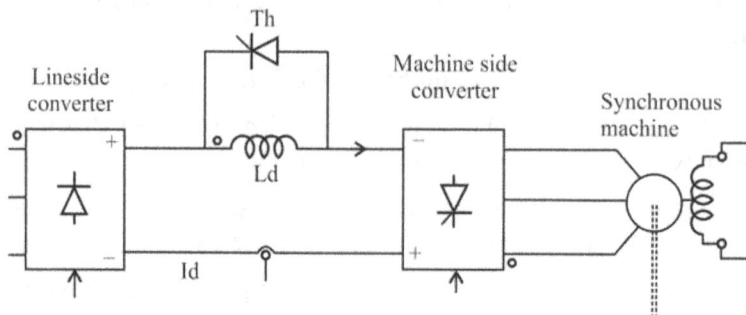

Answers

Chapter – 1

1. (c)

Average voltage $= \dfrac{\sqrt{2}V}{\pi} (1 + \cos \alpha)$

$= \dfrac{\sqrt{2} \times 220}{\pi} (1 + \cos 90^0)$

$= \textbf{99.03 V}$

2. (c)

Average Voltage applied to armature $= \dfrac{\sqrt{2}V}{\pi}(1 + \cos \alpha)$

$= \dfrac{\sqrt{2} \times 220}{\pi} (1 + \cos 60^0) = 148.55V$

Power Drawn $= VI = 148.55 \times 20 = \textbf{2971 Watts}$.

3. (a)

$$V_a = \dfrac{\sqrt{2} \times 220}{\pi} (1 + \cos 45^0) = 169.05V$$

$$E_b = V_a - I_a R_a = 169.05 - 20 \times 0.1 = \textbf{167.05 V}$$

4. (c)

$$T = K_1 I_a \phi = K I_a, \quad \dfrac{T_2}{T_1} = \dfrac{I_2}{I_1}$$

$$T_2 = \dfrac{40 \times 40}{20} = \textbf{80 N-m}$$

5. (b)

$$I \alpha \, T \alpha \, N^2, \quad \dfrac{I_2}{I_1} = \dfrac{N_2^{\,2}}{N_1^2}$$

$$I_2 = 20 \times \left(\dfrac{200}{500}\right)^2 = \textbf{3.2 A}$$

6.   (c)

$$E_a = E_b = \frac{\sqrt{2} \times 220}{\pi} (1 + \cos 60^0) = 148.55V$$

For speed of 600 rpm $E_b = \dfrac{148.55 \times 600}{800} = 111.414$

$$1 + \cos\alpha = \frac{111.414 \times \pi}{\sqrt{2} \times 220} = 1.125$$

$$\cos \alpha = 0.125$$

$$\alpha = \cos^{-1} 0.125 = \mathbf{82.82^0}$$

7.   (d)

E_{b1} at a speed of 1000 rpm is $220 - 10(0.2 + 0.5) = 213$ V

E_{b2} when the current is 20 A is $220 - (20 \times 0.7) = 206$ V

$$\frac{E_{b2}}{E_{b1}} = \frac{N_2\phi_2}{N_1\phi_1} = \frac{N_2 I_2}{N_1 I_1} \quad N_2 = \frac{206 \times 1000 \times 10}{213 \times 20} = \mathbf{483.5}$$

8.   (b)

$$E_{dc} = \frac{2\sqrt{2} \times V}{\pi} \cos \alpha$$

$$= \frac{2\sqrt{2} \times 220}{\pi} \cos 30^0 = 171.53$$

Average current $I_a = \dfrac{E_d - E_b}{(R_a + R_s)} = \dfrac{171.53 - 160}{0.3 + 0.4} = \mathbf{16.47 \ A}$

9.   (c)

$$E_b = 1.9 \times \frac{800}{60} \times 2\pi = 159.19$$

$$R_a = 0, E_a = E_b = 159.19$$

$$\frac{2\sqrt{2}V}{\pi} \cos \alpha = E_a$$

$$\cos \alpha = \frac{159.9 \times \pi}{2\sqrt{2} \times 240} = 0.736$$

$$\alpha = \mathbf{42.55^0}$$

10.   (b)

$$\text{Rated Current} = \frac{3730}{220} = 16.95 \text{ A}$$

∴   RMS Value of supply current $= I_d = \textbf{16.95 A}$

11.   (b)

$$\text{RMS Thyristor current} = \frac{16.95}{\sqrt{2}} = \textbf{11.98 A}$$

12.   (b)

$$\text{The supply power factor} = \frac{\text{Active power}}{\text{Apparent power}}$$

$$= \frac{E_{dc} \times I_{dc}}{V_{rms} \times I_{rms}} = \frac{159.17 \times 16.95}{240 \times 16.95} = \textbf{0.663 lag}$$

13.   (b)

$$E_{dc} = \frac{2\sqrt{2} \times V}{\pi} \cos\alpha = \frac{2\sqrt{2} \times 220}{\pi} \cos 120^0 = -99.03 \text{ V}$$

For Active power $= E_d . I_{dc} = -99.03 \times 20 = \textbf{--1980 W}$

14.   (c)

Torque $\alpha \, I_a$ & $\phi \, \alpha \, I_a^2$, Torque at 80 amps $= \left(\dfrac{80}{100}\right)^2 \times 100 = \textbf{64 N --m}$

15.   False   16.   False   17.   True
18.   True   19.   True   20.   True
21.   0.72
22.   22 to 27 : See notes.

Chapter – 2

1.   [D]

$$V_a = \frac{1.35 \; V_L}{2}(1 + \cos\alpha) = \frac{1.35 \times 400}{2}(1 + \cos 30) = 503.8$$

$$P = V_a I = 503.8 \times 10 = \textbf{5038 W}$$

2.   [C]

$$E_{dc} = 1.35 \; V_L \cos\alpha = 1.35 \times 440 \times \cos 60 = \textbf{297 V}$$

3.   [C]

$$\text{RMS value of linear current} = \sqrt{\frac{2}{3}} I^2_{dc} = I_{dc} \sqrt{\frac{2}{3}} = 20 \sqrt{\frac{2}{3}} = \textbf{16.33 A}$$

4.   [B]

$$E_{dc} = \frac{1.35 \ V_L}{2} \ \text{Cos } \alpha$$

5.   [B]

A 3-ϕ half controlled bridge rectifier is a 3 pulse converter.

6.   [B]

A 3- ϕ fully controlled bridge rectifier is a six pulse converter.

7.   [B]

$$\alpha = 0^0, \ V_a = \frac{1.35 \ V_L}{2} (1 + \cos\alpha) = \frac{1.35 \times 220}{2} (1 + \cos 0^0) = 297.$$

Power delivered = 10000 W

i.e., $E_b \ I_a$ =10000

$$(V_a - I_a \ R_a) \ I_a = 10000, \ (297 - 0.2 I_a) \ I_a = 10000$$

$$0.2 \ I_a^2 - 297 \ I_a + 10000 = 0$$

$$I_a = \frac{297 \pm \sqrt{(297)^2 - 4 \times 0.2 \ 1000}}{2 \times 0.2} = \textbf{34.47 or 1450}$$

8.   [C]

Secondary line voltage V_L should be such that

$$1.35 \ V_L \ \text{Cos } 0^0 = 220$$

$\therefore$ $V_L = 220 \ / \ 1.36 = 162.96$

$$\text{Primary phase voltage} = \frac{440}{\sqrt{3}}$$

$\therefore$ $\text{Turns ratio} = \dfrac{162.96 \times \sqrt{3}}{440} = \textbf{0.641}$

9.   [B]

A Line commutated converter driving a DC motor is a non linear load as the power system contains harmonics.

Chapter – 3

1. (i) Dynamic braking (ii) Plugging (iii) Regenerative braking

2. Otherwise the motor rotates in opposite direction

3. See Fig. 3.16

4. 60^0

5. A dual converter is required to operate in four quadrants. A fully controlled converter is two quadrant converter that can operate either as rectifier or inverter. Regenerative braking can be applied in both directions of motion.

6. Four; (7) Inner loop (8) See Fig.

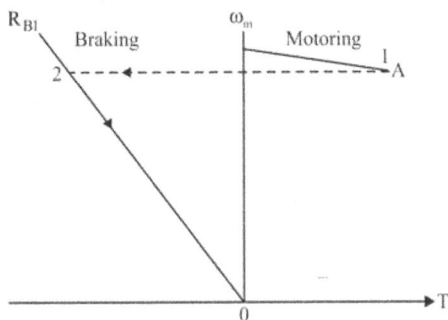

9. See Fig. 3.9 (10) See Fig. 3.11) (11) At zero speed. Otherwise motor rotates in opposite direction (12)No. (13)Zero to 90^0.

14. (a) $1.35V_L \cos\alpha = 1.35 \times 400 \times \cos 60^0 = 270$

    (b) $E_b = 270 - 80 = 190$; . speed = 190/0.5 = 380rpm.

    (c)   zero

    (d)   the speed will be 380 rpm in opposite direction.

15. True   16.   True   17.   True   18.   True   19.   True
20. False   21.   True   22.   False   23.   See Fig. 3.16

Chapter –4

1.(b)   2.(c)   3.(c)   4. (d)

5.   True   6. True

7.   0.5

8.   0.8m.sec

9.   40V

10.   refer 4.1of text

11.   Time ratio control & current limit control

12.   1.225   13.   0.207   14.   48.98V   15.   160A

16.   1.0 to 268.75 rpm. 2.0.0227 to 1   17.   Fig. 4.4   18.   Fig. 4.6

19.   Fig. 4.7   20.   Fig.4.11

21.   IGBT3 is OFF, IGBT4 is continuously ON, ON time of IGBT1 is greater than that of IGBT2 in each cycle.

22.   IGBT3is OFF, IGBT4 is continuously ON, ON time of IGBT2 is greater than that of IGBT1 in each cycle.

23.   IGBT1is OFF, IGBT2 is   continuously ON, ON time of IGBT3 is greater than that of IGBT4 in each cycle.

24.   IGBT1is OFF, IGBT2 is continuously ON, ON time of IGBT4 is greater than that of IGBT3 in each cycle

25.

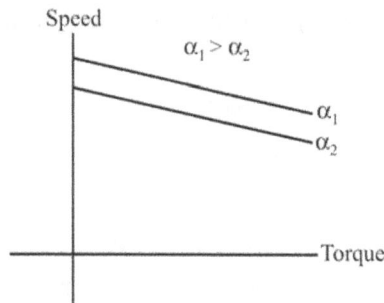

26.

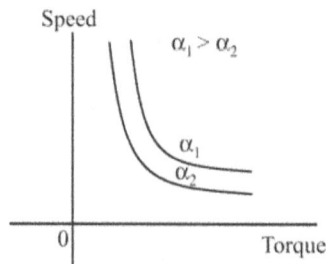

Chapter – 5

1.(c)   2. (d)   3. (d)   4. (a)   5. (b)   6. (d)   7. (c)

8. False   9. True   10. True   11. True   12. Large   13. Plugging

14.   Generator

15.   See Fig. 5.13

16.   See Fig. 5.14

17.   6

18.   10

19.   Stator voltage control; Stator voltage & Frequency control;

20.   See Fig. 5.6 and 5.7

21.   See Fig. 5.5

Chapter - 6

1.   (a)   2.   (a)   3.   (d)   4.   True   5.   True

6.   increase

7.   Reduces

8.   Increased - Held constant

9.   To compensative for the relative higher drop in stator resistance compared to the applied voltage.

10.   1. Voltage source inverter 2. current source inverter 3. cyclo converter

11.   Stator voltage controller ; stator voltage and frequency control

12.   Constant

13.   Ref Fig. 6.6

14.   Ref Fig. 6.7

15.   (a) Series inverter control (b) Pulsewidth modulation control

16.   Ref Fig. 6.10

17.   Ref Fig. 6.9

18.   Absent

19.   Increased

20. $0.45\, E_d = 0.45 * 400 = 180$ V

21. $0.471\ E_d = 0.471*400 = 188.4V$

22. $0.7797*400 = 311.88\ V$

23. $0.8165\ E_d = 0.8165 *400 = 326.6\ V$

24. Ref Fig. 6.3

25. N_s at 40Hz = 1200 rpm therefore $N=N_s\ (1-s) = 1200\ (1-0.04) = 1152$

26. Ref Fig. 6.12

27. SCR s – 6; diodes – 6; capacitors – 6

28. 2

29. $I_{A1} = \dfrac{\sqrt{6}}{\pi} I_d$ 30. Ref Fig. 6.12   31. $I_{R1} = \dfrac{\sqrt{2}}{\pi} I_d$

32. The frequency of CSI is reduced to make synchronous speed less than the actual speed of the motor. Now the 1 M works as induction generator. D.C link voltage polarity reverses. The firing angle of converter is made greater than 90^0 thus making it work as inverter. Thus the d. c link power is transferred to a. c source

33. Ref Fig. 6.15

34. k s f_r (i.e., slip speed) should be held constant

35. Any two from the 5 given in notes.

36. Requires large number of thyristors and hence control is complicated. Harmonics are large at high speeds. Input power factor is poor at low output voltages.

37. False   38. False   39. True   40. True

Chapter – 7

1. Major disadvantage is low efficiency due to additional losses in resistor connected in the rotor circuit.

2. The motor torque capability remains unaltered even at low speeds

3. In driving cranes and other intermittant loads

4. Smooth and step less control, fast response, less maintenance, compact size, simple closed loop control and rotor resistance remains balanced between the three phases for all operating points

5. Refer Fig.7.4

6. $\sqrt{\dfrac{2}{3}}\ I_d;$

7. $i_{R1} = \dfrac{2\sqrt{3}}{\Pi} I_d \sin \omega t.$

8. $I_{1rms} = \dfrac{3}{\Pi} I_{rms}$

9. $R_e = R(1-\alpha)$

10. $V_d = 2.339S\, V_s\, n_m$ where $S = $ slip $V_s = $ Stator phase voltage $n_m = $ Rotor phase turns / stator phase turns.

11. True

12. $I_d = \dfrac{T_L \omega_s}{2.339 V_s n_m}$ Amps.   Where ω_s - synchronous speed in rad/sec

     V_s – Stator phase voltage, n_m – ratio of rotor turns per phase to stator turns per phase

13. $\omega_m = \omega_s \left[1 - \dfrac{T_L \omega_s R(1-\alpha)}{(2.339 V_s n_m)^2} \right]$ rad/sec Where $\omega_s = $ Synchronous speed in rad/sec

     $T_L = $ Load torque in N-m $R = $ Resistance across chopper in ohms $\alpha = $ Duty cycle; $V_s = $ Stator phase voltage $n_m = $ ratio of rotor turns per phase to stator turns per phase.

14. Refer Fig.7.3

15. Refer fig.7.6

16. $3I_r^2 R_r = 600; I_r = \sqrt{\dfrac{600}{3*0.03}} = 81.64$ Amps; $I_d = \sqrt{\dfrac{3}{2}}\, I_r = 81.64\sqrt{\dfrac{3}{2}} = 100$ Amps

17. (d)   18.   (a)   19.   (b)

20. True   21.   (a) 0 and 90^0 , (b) 90^0 and 180^0

22. Converter 1:- 90^0 to 180^0, Converter 2:- 0^0 to 90^0

23. True   24.   True   25.   1000 to 2000 rpm

26. $s = - \dfrac{a_T}{a} \cos\alpha = -\dfrac{0.5}{0.8}\cos 150^0 = 0.541$

27. $3I_r^2 R_r = 300$

     $$I_r = \sqrt{\dfrac{300}{3*0.02}} \quad 70.71 \text{ Amps}$$

     $$I_d = \sqrt{\dfrac{3}{2}}\, I_r = 70.71* \sqrt{1.5} = 86.60 \text{ Amps}$$

28. Refer fig 7.10 (b)

29. $\omega_m = \omega_s \left[1 + \dfrac{a_T}{a} \cos\alpha - \dfrac{T_d . \omega_s R_d}{(2.339 a V_1)^2} \right]$

30.   True

31.   The speed decreases

32.   From synchronous speed to around half of synchronous speed

33.   The speed range will increase

34.   With D.C motor used in basic Kramer drive commutation problems occur. This is eliminated by replacing the d.c motor with synchronous motor and load commutated inverter.

35.   Better power factor and lower harmonics in line current

Chapter – 8

1.   (b)   2. (c)   3. (c)   4. (d)   5. (c)   6. (c)

7.   1. Wound rotor synchronous motor 2. Salient pole synchronous motor

8.   1. Reluctance motor 2)Permanent magnet motor

9.   The rotor of salient pole synchronous motor has winding on the poles and two slip rings on the rotor. The poles are excited by direct current given through slip rings and brushes. The reluctance motor rotor has no field winding. The rotor has projections that act like poles. The reluctance motor works with lagging p.f where as salient pole synchronous motor can work either with lagging or leading p.f currents.

10.   Refer Fig 8.1   11. (b)

12. $T_m = \dfrac{3 \times E_a \times E_f}{X_s \omega_s} = \dfrac{3 \times 6000 \times 5000}{6 \times 50 \times \pi} = 95493 \text{ N –m.}$

13.   False

14. $T_d = \dfrac{3 E_a E_f \sin\delta}{X_s \omega_s}$ N-m   Where E_a – per phase applied voltage E_f – per phase induced emf δ - torque angle X_S – synchronous reactance in ohm ω_s – synchronous speed in rad/sec

15. $\delta = \tan^{-1} \dfrac{I_a X_q Cos\phi}{E_a - I_a X_q Sin\phi}$ where I_a – Armature phase current; E_a- Armature phase

   voltage ; X_q-quadrature axis reactance; $Cos\ \phi$ - Power factor

16. (i) Electromagnetic power $= \dfrac{3 E_a E_f Sin\phi}{X_d}$

   (ii) Reluctance power $= \dfrac{E_a^2}{2}\left(\dfrac{1}{X_q} - \dfrac{1}{X_d}\right) Sin2\delta$

17. $I_a = \sqrt{I^2_d + I_q^2} = \sqrt{22.74^2 + 13.36^2} = 26.37 A$

18. (i) Separate control (ii) self control

19. The frequency is controlled externally from a crystal oscillator or some other electronic circuit.

20. The frequency of the supply to the motor is decided from the information obtained from rotor position or the induced voltages in the stator.

21. A three phase controlled rectifier.

22. A three phase diode rectifier

23. In separate control the frequency is changed independent of rotor speed. Transients occur and hunting takes place. To minimize hunting damper bars are required.

24. When a synchronous motor has its rotor fed from d. c supply and operating in self controlled made it is called CLM

25. The synchronous motor having rotor with permanent magnets or variable reluctance and operating in self control mode is called BLCLM

26. If the motor is to be driven at very low speed the frequency should be low. Since v/f should be held constant supply voltage to motor should be low. This is possible if d. c voltage is low. The firing angle of rectifier should be near 90^0. This results in a very low power factor on the a. c supply mains. A major disadvantage.

27. True

28. Field current $= 17.5$ A Power developed $= 749.2$kw

29. True

30. Dynamic braking.

31. Plugging torque is produced by damper winding. Because of low resistance of damper winding the current drawn is very large but the braking torque is very low. High plugging current can create a severe disturbance in supply links

32. yes

33. By large inductance connected in between line side converter and load side converter.

34. By varying the firing angle of line side converter.

35. VSI, CSI and cyclo converter.

36. They are used for high power applications (up to megawatts range) such as compressors, blowers, fans, conveyors, steel rolling mills and cement plants.

37. Brush less d.c motors require practically no maintenance, have long life, high reliability, low inertia and friction and low radio frequency interference and noise. They have faster acceleration and can be run at much higher speeds upto 100,000 rpm

38. Used for turn table drives in record players, tape drive for video recorders, spindle drives in hard disc drives for computers, low cost and low power drives in computer peripherals, instrument and control systems.

39. From 0 to 10% rated speed, the induced emf in the stator windings of motor is very small and cannot commutate the thyristors of load side converter. Hence load side converter cannot operate as line commutated inverter. Under these conditions forced commutation has to be adopted.

40. When operating at high frequency the commutation time available is less. The link current has to become zero and again build up to required value in this short duration. To make this transition faster this thyristor is used. SCR is trigged when the current has to become zero.

Bibliography

1. Bimal K.Bose-*Modern Power Electronics and AC Drives*- Pearson Education Asia,2003

2. B.K.Bose- *Power electronics control of AC Drives-*

3. Bimbhra. P.S- *Power Electronics*-Khanna Publishers, New Delhi

4. Gopal K Dubey *Power Semiconductor Drives*-PH International Publications

5. G.K.Dubey *Fundamentals of Electric Drives*- Narosa Publications

6. J.M.D.Murphy, F.G.Turnbull- *Power electronics control of AC Motors*-Pergamon Press Oxford,1988

7. Krishnan. R.-*Electric Motor Drives Modeling, Analysis and Control*-Pearson Educaion Asia,2001

8. M D Singh and K.B.Kanchandani-*Power electronics*-Tata –Mcgraw-Hill Publishing company,1998.

9. NK de and PK Sen- *Electric Drives*-Prentice Hall of India Pvt. Ltd.

10. S.K.Pillai-*A first course on Electrical Drives*-New Age International(P) Ltd.

11. SBDewan, G R Selmon and A Straughen- *Power Semiconductor Drives-*

12. Sen P.C-*Thyristorised DC Drives*- John Wiley, New York

13. S.N.Singh-*A Text Book of Power Electronics* Dhanpat Rai & co..

14. Vedam Subramanyam*Thyristor Control of Electric Drives*- Tata –Mcgraw-Hill Publishing company.

15. V.R. Murthi- *Power Electronics devices, circuits, and Industrial Applications*-Oxford University Press.

www.ingramcontent.com/pod-product-compliance
Lightning Source LLC
Chambersburg PA
CBHW061924190326
41458CB00009B/2642

* 9 7 8 9 3 5 2 3 0 0 2 7 3 *